Dialectical Systems Thinking and the Law of Requisite Holism Concerning Innovation

Exploring Unity through Diversity
Volume 3

Exploring Unity through Diversity Volume 3

Dialectical Systems Thinking and the Law of Requisite Holism Concerning Innovation

Matjaž Mulej

with

Stane Božičnik, Vesna Čančer, Anita Hrast, Karin Jurše, Štefan Kajzer, Jožica Knez-Riedl, Tadeja Jere Lazanski, Tatjana Mlakar, Nastja Mulej, Vojko Potočan, Filippina Risopoulos, Bojan Rosi, Gerald Steiner, Tjaša Štrukelj, Duško Uršič, Zdenka Ženko

EMERGENT™
PUBLICATIONS

3810 N 188th Ave
Litchfield Park, AZ 85340

Dialectical Systems Thinking and the Law of Requisite Holism Concerning Innovation
Written by: Matjaž Mulej

Library of Congress Control Number: 2013938074

ISBN: 978-1-938158-09-4

Copyright © 2013 3810 N 188th Ave, Litchfield Park, AZ 85340, USA

All rights reserved. No part of this publication may be reproduced, stored on a retrieval system, or transmitted, in any form or by any means, electronic, mechanical, photocopying, microfilming, recording or otherwise, without written permission from the publisher.

Printed in the United States of America

About the Book Series: Exploring Unity through Diversity

Unity through Diversity is acknowledged to be the leitmotif of Ludwig von Bertalanffy's thinking. It is also the leitmotif of this series; that is, providing space for different perspectives while sharing a common goal in order to promote:

- Systems sciences, cybernetics and sciences of complexity as the most promising approaches towards global challenges humanity is facing in the new millennium;
- Transdisciplinarity and consilience throughout all scientific disciplines;
- The discussion and comparison of different schools of systems thinking;
- Attempts to unify systems thinking and to elaborate a meta-theoretical framework;
- Systems history;
- Critical reflections of the development of systems thinking and the systems movement;
- Revisiting the goals of General System Theory as set by Ludwig von Bertalanffy, Anatol Rapoport, Kenneth Boulding and others;
- Social-scientific, that is, socio-economic, political, cultural and historical applications of systems thinking, including ecological and science-and-technology studies applications;
- Systems philosophy, and;
- Monographs or volumes of collected contributions in systems.

Bertalanffy Center for the Study of Systems Science (Vienna) Book Series
Exploring Unity through Diversity

Series Editor

Wolfgang Hofkirchner (Austria)

Editorial Board

Gabriele Bammer (Australia)
Yaneer Bar-Yam (US)
Gerhard Chroust (Austria)
Arne Collen (USA)†
John Collier (South Africa)
Yagmur Denizhan (Turkey)
Irina Dobronravova (Ukraine)
Klaus Fuchs-Kittowski (Germany)
Ramsés Fuenmayor (Venezuela)
Amanda Gregory (UK)
Ernesto Grün (Argentina)
Jifa Gu (China)
Debora Hammond (US)
Enrique G. Herrscher (Argentina)
Francis Heylighen (Belgium)
Cliff Hooker (Australia)
Magdalena Kalaidjieva (Bulgaria)
Helena Knyazeva (Russia)
George Lasker (Canada)
Allenna Leonard (Canada)
Gary Metcalf (US)
Gerald Midgley (UK)
Gianfranco Minati (Italy)
Edgar Morin (France)
Matjaž Mulej (Slovenia)
Yoshiteru Nakamori (Japan)
Andreas Pickel (Canada)
Michel St. Germain (Canada)
Markus Schwaninger (Switzerland)
Len Troncale (US)
Martha Vahl (UK)
Gertrudis van de Vijver (Belgium)
Jennifer Wilby (UK)
Rainer E. Zimmermann (Germany)

About the Bertalanffy Center for the Study of Systems Science (BCSSS)

Given the global challenges of today, systems science is needed more than ever. Yet system theory is not mainstream. The objective of the BCSSS is to inspire the development of systems science. The BCSSS aims at the advancement of scientific research in the field of systems thinking. In particular, it revisits General System Theory (GST) as founded by Ludwig von Bertalanffy and others in order to reassess it in the light of today's global challenges and to illuminate the course of development systems science has taken since. The BCSSS is open to cooperation with every person or organization supporting the same aim. The BCSSS owns the Ludwig von Bertalanffy archive and possesses a collection of publications of the systems movement.

Website: http://www.bertalanffy.org/

For further information please contact:

Prof. Dr. Wolfgang Hofkirchner,
BCSSS
Paulanergasse 13/5
1040 Vienna
Austria

wolfgang.hofkirchner@tuwien.ac.at

About the Author

Matjaž Mulej, 1941, Slovenian. He retired from University of Maribor as Professor Emeritus in Systems and Innovation Theory. He has +1,600 publications in +40 countries (see: IZUM/Cobiss/Bibliographies, 08082); +60 publications in world-top journals, and close to 400 citations, of which +100 in world-top journals. He was visiting professor abroad for 15 semesters, including Cornell U., Ithaca, NY, and others in Austria, China, Germany, Mexico, USA, about 50 further shorter visits with classes for students around the world. He has served as consultant or speaker in/for enterprises about 500 times in six countries. He authored: *Dialectical Systems Theory* (see: François, 2004, International Encyclopedia); *Innovative Business Paradigm for Countries/Enterprises in Transition* and *Methods of Creative Interdisciplinary Cooperation* USOMID with 6 Thinking Hats. His investigating and publishing runs in teams, mostly. He was nominated member by New York Academy of Sciences (1996), European Academy of Sciences and Arts, Salzburg (2004), European Academy of Sciences, Arts and Humanities, Paris (2004), International Academy for Systems and Cybernetic Sciences, Vienna (2010; president until 2012). In IFSR (International Federation for Systems Research with 46 member associations) he was president in 2006-2010. He has received many rewards for his work on systems approach to innovation in Yugoslavia, Slovenia, Maribor and University of Maribor and many 'Who is Who' entries and invitations, Slovenian and international. His B.A. is in Economic Analysis, M.A. in Development Economics, Doctorates in Economics/Systems Theory, and in Management/Innovation Management. He is married for +50 years and has 2 adult children, 4 grand-children. Addresses: Mail: UM EPF, Razlagova 14, SI-2000 Maribor, Slovenia; E-mail: mulej@uni-mb.si; EPF homepage: epfip.uni-mb.si; IRDO (Institute for development of social responsibility): www.irdo.si.

Co-Authors

Dr. Stane Božičnik, Assistant Professor (Transport Economics), University of Maribor, Faculty of Civil Engineering, Maribor, Slovenia. E-mail: stane.bozicnik@uni-mb.si, Ch. 11.3.7.

Dr. Vesna Čančer, Associate Professor (Quantitative Economic Analyses), University of Maribor, Faculty of Economics and Business, Maribor, Slovenia. E-mail: vesna.cancer@uni-mb.si; Ch. 12.3.6.

Anita Hrast, Manager, IRDO Institute for development of Social Responsibility, Maribor, Slovenia; E-mail: info@irdo.si; Ch. 1.4.

M.A. Karin Jurše, Head of Offices, Municipality of Selnica, Slovenia; E-mail: karin.jurše@selnica.si; Ch. 14.11.

Dr. Tadeja Jere Lazanski, Assistant Professor (Systems, Events Organization). University of Primorska, Faculty of Tourism, Portorož, Slovenia E-mail: tadeja.lazanski@turistica.si; Ch. 11.3.10.

Dr. Štefan Kajzer, Professor Emeritus (Management, Organization, a Cybernetics), University of Maribor, Faculty of Economics and Business. Maribor, Slovenia. E-mail: Stefan.kajzer@uni-mb.si; Ch. 9.3, 11.3.5, 14.12.

Dr. Jožica Knez-Riedl, Professor (Enterprise Economics), University of Maribor, Faculty of Economics and Business, Maribor, Slovenia. E-mail: jozica.knez@uni-mb.si; Ch. 13.8.

Dr. Tatjana Mlakar, Head of Regional Social Insurance Office of Slovenia in Novo mesto Slovenia. E-mail: tatjana.mlakar@zzzs.si. Ch. 11.3.5.

Nastja Mulej, M.S., Entrepreneur. E-mail: nastja@debono.si; Ch. 13.5.

Dr. Vojko Potočan, Associate Professor (Organization and Management), University of Maribor, Faculty of Economics and Business, Maribor, Slovenia. E-mail: vojko.potočan@uni-mb.si; Ch. 1, 9.3, 11.4.5, 13.7, 13.9.

Dr. Filippina Risopoulos, Institute of Geography and Regional Sciences, University of Graz, Graz, Austria; filippina.risopoulos@uni-graz.at; Ch. 13.10.

Dr. Bojan Rosi, Assoc. Professor (Systems Theory and Logistics), University of Maribor, Celje and Krško, Slovenia. E-mail: bojan.rosi@uni-mb.si; Ch. 11.3.4, 14.11.

Dr. Gerald Steiner, Assoc. Prof., Institute of Systems Sciences, Innovation and Sustainability Research, University of Graz, Graz, Austria; gerald.steiner@uni-graz.at; Ch. 13.10.

M.A. Tjaša Štrukelj, Senior lecturer (Management and Strategy), University of Maribor, Faculty of Economics and Business, Maribor, Slovenia. E-mail: tjasa.strukelj@uni-mb.si; Ch. 14.3.

Dr. Duško Uršič, Professor (General Management), University of Maribor, Faculty of Economics and Business, Maribor, Slovenia. E-mail: dusko.ursic@uni-mb.si; Ch. 11.3.1, 11.3.2, 11.3.3.

Dr. Zdenka Ženko, Assistant Professor (Systems Theory and Management), University of Maribor, Faculty of Economics and Business, Maribor, Slovenia. E-mail: zdenka.ženko@uni-mb.si; Ch. 14.7, 14.11.

Contents

Chapter 0—THE SELECTED PROBLEM AND VIEWPOINT OF ITS CONSIDERATION 1

0.1 Circumstances in which the Problem and Viewpoints of its Consideration are selected 1

0.2 Systems Theory as a Theory of Requisitely Holistic Thinking and Worldview 6

Chapter 1—SYSTEMS THINKING: AGAINST CRUCIAL OVERSIGHT CAUSED BY OVER-SPECIALIZATION—CASE OF EVOLUTION TOWARDS SUSTAINABLE AND SOCIALLY RESPONSIBLE ENTERPRISE AND SOCIETY 11

1.1 Requisite Holism in brief 11

1.2 Poor Understanding of Requisite Holism—Background 15

1.3 Sustainability and Social Responsibility— A Way of Requisite Holism 17

1.4 The Essence of Social Responsibility (SR) 21

1.5 Four or Five Phases of Development of the Basis of Competitiveness 24

1.6 Conclusions from Chapter 1 26

Chapter 2—FROM SYSTEMS THINKING TO SYSTEMS THEORY IN THE TURBULENT 20TH CENTURY 29

Chapter 3—SYSTEMS THEORY AND CYBERNETICS— HOLISM AGAINST BIG CRISES CAUSED 33

3.0 The Selected Problem and Viewpoint for Dealing with it in Chapter 3 33

3.1 Holistic Thinking versus Narrow Specialization 33

3.2 Difficulties with Implementation of Holism following Bertalanffy 39

3.3 Economic Reasons Opposing Holistic Thinking in Practice: The Law of Requisite Holism in Decision-Making 43

3.4 Concluding remarks 50

Chapter 4—REDUCTION, REDUCTIONISM, SPECIALIZATION, AND PROBLEMS OF COOPERATION.... 53

Chapter 5—COMPLEXITY, COMPLICATEDNESS, RELATIONS, EMERGENCE, SYNERGY VS. SIMPLICITY AND LOCKED-IN THINKING AND ACTING 57

Chapter 6—KNOWLEDGE, INFORMATION, PROFESSIONS, ORDER 61

Chapter 7—SYSTEMS THEORIES: TOOLS OF HUMAN ACTION AND/OR HUMAN FORMATION 65

Chapter 8—HOLISM VERSUS ONE-SIDEDNESS AND OVERSIGHT: REQUISITE HOLISM... 71

Chapter 9—THE BASIS FOR (REQUISITE) HOLISM TO BE ATTAINED: INTERDEPENDENCE AND ETHICS/VCEN OF INTERDEPENDENCE 77

    9.0 The Selected Problems and Viewpoint of Dealing with it in Chapter 9..77
    9.1 Adam Smith was misread 77
    9.2 General Systems Theory—Short of Ethics of Interdependence (in Practice) 81
    9.3 Ethics and ethics of interdependence..84
    9.4 Lack of Consideration of Interdependence in 20th Century Practice: Causing the need for making of Systems Theory, not only Systems Thinking 87
    9.5 Conclusions from Chapter 9 88

Chapter 10—MODERN (MATERIALISTICALLY) DIALECTICAL THINKING—A FORERUNNER OF SYSTEMS/HOLISTIC THINKING AND CONSIDERATION OF INTERDEPENDENCE 93

Chapter 11—SOME (SOFT) SYSTEMS THEORIES AND THEIR APPLICABILITY TO THE ISSUES OF REQUISITE HOLISM AND ETHICS OF INTERDEPENDENCE— A BRIEF OVERVIEW 99

    11.0 The Selected Problem and Viewpoint of Dealing with it in Chapter 11..99

    11.1 The Chaos Theory and Requisite Holism and Ethics of Interdependence...100

    11.2 The Complexity Theory, Requisite Holism and Ethics of Interdependence...102

    11.3 A Selection of Contemporary Soft Systems Theories— A Comparative View Concerning their Usefulness to the Issues of Requisite Holism and Ethics of Interdependence............. 103

        11.3.0 The Selected Problem and Viewpoint of Dealing with it in Chapter 11.3... 103
        11.3.1. The Viable Systems Theory (VST) 106
        11.3.2 The Soft System Methodology (SSM).. 107
        11.3.3 The Critical Systems Thinking (CST).. 108
        11.3.4 Dialectical Network Thinking 109
        11.3.5 Control System Theory 111
        11.3.6 Cybernetics of the 1st, 2nd, and 3rd Order................................... 115
        11.3.7 A New—4th Order Cybernetics 117
        11.3.8 The Role of Information in Systems Thinking........................... 121
        11.3.9 Some Tools Usable for Simplification of Management/Impact in Complex Processes and Situations... 126
        11.3.10 Historical Parallelism of Systems Thinking: Maya and the Evolution of Consciousness vs. Contemporary Systems Thinking 132
        11.3.10.3. Conclusions from Chapter 11.3.10 140
        11.3.11 Some Comparative Conclusions about the Selected Soft Systems Theories.. 140

    11.4 Some Conclusions about Systems Thinking about the Contemporary Life Problems..141

Chapter 12—DIALECTICAL SYSTEMS THINKING: ABOUT COMPLEXITY, INTERDEPENDENCES, WHOLES AND REQUISITE HOLISM/WHOLENESS INSTEAD OF CRUCIAL OVERSIGHTS 149

12.0 The Selected Problem and Viewpoint of Dealing with it in Chapter 12 149

12.1 Dialectical System (DS) versus the System and the Object of Consideration 150

12.2 The Dialectical Systems Theory (DST) 153

12.3 Dialectical Systems Theory: Grounds of Human Interventions 160

    12.3.0 The Selected Problem and Viewpoint of Dealing with it in Chapter 12.3 160
    12.3.1 The General Ground: Entropy versus Evolution, Human Intervention 160
    12.3.2 Grounds for Human Intervention: Objective and Subjective Starting Points 161
    12.3.3 DST's Ten Guidelines concerning Influence over the Subjective Starting Points 162
    12.3.3.0 Questions to be answered to define the Starting Points 162
    12.3.4 DST's ten Guidelines concerning Implementation of Starting Points 167
    12.3.5 The Law of Requisite Holism 176
    12.3.6 Support to Requisite Holism by Contemporary Operations Research Methods 181
    12.3.7 Typology of Systems and Models 192

12.4 Dialectical Systems Theory and its Application by USOMID 197

Chapter 13—USOMID: AN APPLIED METHODOLOGY OF DIALECTICAL SYSTEMS THINKING 201

13.0 The Selected Problem and Viewpoint of Dealing with it in Chapter 13 201

13.1 The USOMID concept: a Reflection of the Seven Components of the Starting Points 201

13.2 Programoteque 203

13.3 The USOMID-SREDIM as the General Method of Creative Work and Cooperation Procedure 207

13.4 The USOMID Circle—an Organizational Possibility Supportive of Creative Cooperation 210

13.5 Combination of USOMID-SREDIM with the Six Thinking Hats Method 213

13.6 Values and their Influence 219

13.7 Ethics of Interdependence 220

13.8 Social Responsibility—a Way of Requisite Holism and
Ethics of Interdependence 223

13.9 Sustainable Enterprise Ethics (SEE)—Another Way of
Requisite Holism and Ethics of Interdependence 232

13.10 From corporate social responsibility to corporate social
innovation? 235
   13.10.1 Introduction 235
   13.10.2 From CSR to CSI? 237
   13.10.3 Corporate Social Innovation (CSI) 239
   13.10.4 Implications for Corporate Social Innovation 243
   13.10.5 Conclusions and Implications for Research 244

13.11 Some Conclusions from Chapter 13 concerning Ethics 245

13.12 Some Conclusions about USOMID as Applied Dialectical
Systems Thinking 246

Chapter 14—A CASE OF DIALECTICAL SYSTEMS THINKING: REQUISITELY HOLISTIC MANAGEMENT OF THE INVENTION-INNOVATION-DIFFUSION PROCESS 251

14.0 The Selected Topic/Problem and Viewpoint of our
Dealing with it here 251

14.1 The Dialectical System (DS) in Chapter 14 252

14.2 The Invention-Innovation-Diffusion Process (IIDP) 252

14.3 Innovation Typology—an Additional Factor of Complexity
of Requisitely Holistic IIP and KM in it, as well as
Mastering of it 253

14.4 DS of Basic Preconditions for Invention to become
Innovation 257

14.5 Application of USOMID-SREDIM/6 Thinking Hats Method 260

14.6 Application of Co-Laboratories of Democracy 260

14.7 Action rather than Thinking alone Makes Innovation from
Knowledge and Invention—Diffusion of Novelties 260
   14.7.0 The Selected Problem and Viewpoint of Dealing with it in
        Chapter 14.7 260
   14.7.1 Action Based on the Theory of Diffusion of Innovation 261
   14.7.2 Need for Spreading of the Novelty 261
   14.7.3 Diffusion and its Preconditions 261
   14.7.4 Change Agents 266
   14.7.5 Opinion Leaders 266

14.7.6 Concluding Comments on Diffusion of Innovation.................................267

14.8 Innovative Business...271

14.8.0 The Selected Problem and the Viewpoint of its Consideration 271
14.8.1 Definition of the Innovative Business as a Dialectical System 271
14.8.2 Framework Model for Continuous Implementation of
Innovative Business...272
14.8.3 Total/Excellent Quality and Hence Success in Market—Consequences
of Innovation Process Based on Requisite Holism and Ethics of
Interdependence...275
14.8.4 Innovation of Management Style...276

14.9 Innovative Society...278

14.10 Suggestion to Governments of Countries/Regions
Trying to Catch Up 280
14.10.0 The Problem and the Selected Viewpoint for Dealing with it in
Chapter 14.10 280
14.10.1 Creativity, Insight, and Quality Enhancement as Factors in
Innovation Processes in Government and Other Public Offices...................281
14.10.2 Roots of the Lack of Innovation Capacity in Government..........................286
14.10.3 A Dialectical System of Suggestions how to Make Public Officers
Act as Role Models of Innovation...287
14.10.4 Some Concluding Remarks 291
14.11 Four times Ten Guidelines Enabling Innovative Business and
Innovative Society..292
14.12 Some Conclusions about Innovative Business and Innovative Society 296

Chapter 15—SUPPLEMENT: INTRODUCTORY THOUGHTS FROM THE EUROPEAN COMMISSION'S FIRST DOCUMENT AIMED AT ENHANCING INNOVATION AND SYSTEMS THINKING: GREEN PAPER ON INNOVATION (1995)................................299

15.0 Introductory Remark 300

15.1 Innovation, the Firm and the Society.. 300

15.2 Innovation and Public Action... 302

Chapter 16—CAN AND MAY ONE AVOID INNOVATION, ETHICS OF INTERDEPENDENCE AND THE LAW OF REQUISITE HOLISM? NO LONGER!...307

Chapter 17—CONCLUDING REMARKS 313

17.0 The Selected Problem and Viewpoint of Dealing with it in Chapter 17 313

17.1 Growing Dependence of Humankind on High Quality and Hence on Innovation 313

17.2 Growing Complexity of Innovation and Therefore Growing Dependence of Success on Systems Thinking 314

17.3 Need for Careful Choice of Systems Theory 315

17.4 The Seven Principle Attributes of Systems Thinking Applied in this Book 317

17.5 Suggestion concerning Systems Education 320

17.5.1 Introduction 320
17.5.2 Basic Contents to be Covered in Teaching of Systemic Behavior 321
17.5.3 Some Conclusions 325

REFERENCES 329

Chapter 0

THE SELECTED PROBLEM AND VIEWPOINT OF ITS CONSIDERATION

0.1 CIRCUMSTANCES IN WHICH THE PROBLEM AND VIEWPOINTS OF ITS CONSIDERATION ARE SELECTED

Humans, who are living now, are living in the time in which innovation has become as frequent and unavoidable as never before. The most advanced areas of the world—Europe, Northern America, Australia, New Zealand, Japan, and the four Pacific Rim Tigers: Singapore, Hong Kong, Taiwan, and South Korea—comprise the 15-20% of humankind who are living on innovation. The innovative society and economy require humans to master much more entanglement than ever before:

- There are no longer local markets hidden from the global market.
- There are no longer chances for many humans to live with no permanent renewal of their skills.
- There are no longer markets in which supply is not bigger than demand, except for the least advanced areas in which close to a billion people are hungry, while in the other areas about a billion people are too fat to be healthy.
- There are no longer many areas in which humans can live with no innovation and therefore with no demands for requisitely holistic thinking, called systems thinking in systems theory.

- There are no more times in which one can live on the paradigm of industrial values, because they ruin more of the natural resources and other preconditions of survival of the current civilization of humankind than they create. Industrial values became beneficial when the number of humans and their consumption of their natural environment used to be only a small fragment of their current level. Then, the narrow overspecialization that provides for no capacity of interdisciplinary creative cooperation was much less dangerous than today. Narrow specialization is still necessary, but equally so is the said specialists' capacity: cooperation helps humans prevent oversights and resulting failures, because it enables more holistic thinking/behavior.

But there are very few humans around the world, who can/may teach holistic thinking. The role of the narrow specializations which is unavoidable is so strong that people hardly see that holistic thinking—enabled by interdisciplinary creative cooperation—makes specialization of any profession much more beneficial than any operation of the specialization alone. Nobody, whatever their profession, can live well without cooperation with people of other professions.

Some fifty years after the authors of Systems Theory had succeeded in making their theory known, and politicians of the world succeeded in using it (informally) by making the United Nations Organization the most holistic political organization of humankind, the European Union (EU) found it necessary to explicitly link "systemic" views with innovation. The EU, after reminding readers of its previous documents enhancing innovation, states on page 6:

The Action Plan [First Action Plan for Innovation in Europe, 1996, based on Green Paper on Innovation, 1995] was firmly based on the 'systemic' view, in which innovation is seen as arising from complex interactions between many individuals, organizations and environmental factors, rather than as a linear trajectory from new knowledge to new product. Support for this view has deepened in recent years (EU, 2000).

If this has to be stated explicitly in such documents, these questions arise:

- Are we humans capable of the interdisciplinary cooperation that we need almost every moment?

- What is the theoretical basis for those, who are not capable of learning to do this?

The empirical experience- and reference-based answers are:

Very few humans are by their nature and education capable of interdisciplinary cooperation, because specialists teach specialists to be specialists, including being proud of their specialization. This teaching is reasonable, but it is not enough: it may cause one to hide from reality behind the walls of one's specialization and lack respect for other specializations and their need for each other—as well as restricting their capacity to solve real problems by interdisciplinary creative cooperation much better than by separation (see Ackoff, 2001, 2003; Gigch, 2003).

The theoretical basis to learn the skills of the interdisciplinary cooperation stems from the original authors of the Systems Theory and Cybernetics. But many forget that the fathers of the Systems Theory and Cybernetics have created their answers to the burning problems of their and our time through their interdisciplinary approach. This is where our Dialectical Systems Theory (DST) (Mulej, 1974, 1979) originating +3 decades ago allows us to fill the gap.

The well intended and well applied versions of systems theory, which describe a part of reality within the viewpoint of one or another traditional, specialized, scientific discipline, do not match the well stated EU's definition of "systems view". Thus, they help people solve other problems, but not that of the holism of thinking, decision-making, and action, as a precondition of survival of humankind and the planet on which we live, and/or of success in any human action (Geyer *et al.*, 2003).

We will try to demonstrate here that holism of thinking, decision-making, and action is very necessary and has been the point of Ludwig von Bertalanffy (1950, 1968) and his co-establishers of the General Systems Theory and related movement. It is even more needed today, and has been very much forgotten about, but it can be practiced well, if people use what we call DST. This is what this book is about. It tries also to show how much more success one could achieve in the innovation effort especially if one applied more systems thinking. We want to help the catching-up regions of the world most of all because they are facing the "innovation need paradox": those who need innovation most like it the least (Rogers, 1995, 2003). The problem lies firmly in one's mentality: in humans' thinking and worldview as well as within other values and other emotions.

In 2008-2009 the world-wide crisis that is superficially called financial crisis has shown this fact very clearly. We worked on our response to this problem in another book (Božičnik et al., 2008). Here we will only summarize it very briefly.

For several recent centuries the so-called free-market economy has been found more efficient and providing for more economic development and higher living standard of its users than any other socio-economic model of so far. It was in a deep crisis in 1930s (resulting in the WWII, not Keynesian measures only), and has faced it now again under the name of big depression, financial crisis. These labels are too narrow: it is a general social crisis due to a lack of requisitely holistic values/culture/ethic/norms and behavior (made of monitoring, perception, thinking, emotional and spiritual life, decision-making, and action) rather than one-sided and short-term behavior of the influential people and their organizations, including enterprises and governments. This narrowness is based on failure of many to use systemic thinking and behavior due to their over-specialization with a poor capability of interdisciplinary creative cooperation. Consequences include the frequent limitation of the term innovation to the technology innovation alone. Consequences might be dangerous: good technology serving bad/evil/unclear purposes. They can still be avoided—by innovation of culture of one-sided behavior to the one of requisite holism. The current humankind is moving from routine via knowledge to creative society. The latter is based on a new economy and requires new values/culture/ethic/norms—self-interest realized by socially responsible and therefore requisitely holistic behavior. Social responsibility can and must reach far beyond charity towards the end of abuse of power/influence of the influential persons/organizations in their relations with their coworkers, other business and personal partners, broader society, and natural environment as the unavoidable and terribly endangered precondition of human survival, at least in terms of the current civilization. Social responsibility supports innovation also by upgrading criteria of business excellence, by supporting requisitely holistic behavior. Thus, it means also a form of innovation of human values/culture/ethic/norms (VCEN) and knowledge, resulting in a requisitely holistic behavior. In a most optimistic scenario, social responsibility can also provide a way towards peace on Earth. It can lead to covering all these urgent humankind's needs by making coworkers and other people more happy, because it provides to them more feeling of being considered equal and creative rather than abused and/or misused by power-holders. In synergy with ethics of interdependence, because every specialist is complementary to all other specialists as a professional and as a human being, and with the fact than one lives increasingly on creativity, including innovation, social responsibility may innovate society to include social efficiency, social justice and similar values/culture/ethic/norms that, among other references, lie at the core of all social teaching called religions, philosophy of moral and ethical behavior. Technology supports rather than creates future and development into it, and can be used with social responsibility or abused/misused with detrimental consequences. The choice depends on the most influential people and their definition of their self-interest as a background of the new economy and humankind's future. Innovation of VCEN is unavoidable for the current civilization to survive.

So far, the neo-liberal school of economics of Friedman and others in Chicago imposed the VCEN replacing etatistic centralism (rightfully) and Adam Smith's liberalism (wrongly)—by calling free market something that has not been free: the transparency of business, limitation to local markets/businesses, and clear ownership of businesses with no separation of duties and rights by shareholding and limited liability companies used to be presupposed by Adam Smith (Toth, 2008). These preconditions no longer exist. Thus, the current VCEN and resulting circumstances are no Adam Smith's free market, but renewing the pre-capitalist times against which the free market had been estasblished (see also Goerner, 2008). In other words, along with the growth of results and problems of the neo-liberal economics over the recent seven decades or so the biggest-ever impact of one-sidedness has been creating the current crisis; in the same period of time systems theory was created and evolving, but it received a rather poor and neglecting attention and application in economy and society management. This neglecting of holism and wholeness by narrow specialists, who are both unavoidable and insufficient, is the core of the problem of the crisis of 2008-. This neglection must be changed by an innovation of VCEN based on making systemic thinking a usual behavior of all decisive people, organizations and bodies. Otherwise survival of humankind's current civilization is doubtful (Božičnik *et al.*, 2008; Brown, 2009; Harris, 2008; Hilton, 2008; Martin, 2006; Taylor, 2008).

0.2 SYSTEMS THEORY AS A THEORY OF REQUISITELY HOLISTIC THINKING AND WORLDVIEW

We are talking about human worldview and related thinking style, when we are talking about systems thinking. Eduardo de Bono, the world-famous author about creative thinking, said:

"Thinking is the most important human behavior" (De Bono, 2003). We are adding: Holistic and creative thinking rather than partial, one-sided and routine-loving one is what he has had in mind—with full right (see: De Bono, 2005, 2006).

This is what Systems Thinking has surfaced for, millennia ago, and Systems Theory has been created for, half a century ago.

Now there are many systems theories and no unified agreement what is Systems Thinking (for unique overview see: François, 2004). Many of them describe parts of reality from more or less traditional viewpoints and add very interesting new insights. Others provide tools for humans to use for whatever purpose. These limitations are bringing systems theory in trouble (Bailey, 2005; Clusella *et al.*, 2005a, 2005b; Korn, 2003; Hammond, 2003; Loeckenhoff, 2005; Metcalf, 2005; Mulej *et al.*, 2005a, 2005b, 2005c). M. Mulej's DST (Mulej, 1974, 1976, 1977, 1979; Mulej *et al.*, 1992, 2000, 2004, 2008; Mulej & Ženko, 2004a, 2004b) is different and found peculiar (François, 2004, Part I: 169). It is found so rightfully: We do not wish to provide tools or description; we wish to continue where the first author of systems theory as a theory/teaching, Ludwig von Bertalanffy has stopped:

To help people develop their worldview of holism and a methodology supportive of it. We want systems theory to also be usable even informally, i.e. with no formal language of systems theory.[1]

---
1. L. v. Bertalanffy (1950: 142-143) wrote: "General System Theory is a new scientific doctrine of 'wholeness'—a notion which has been hitherto considered vague, muddled and metaphysical. Considered from the viewpoint of philosophy, General Systems Theory is to replace that field which is known as 'theory of categories' by an exact system of logico-matematical laws." Earlier, on p. 134, he indicates at the very beginning of his text: "As we survey the evolution of modern science, we find the remarkable phenomenon that similar general conceptions and viewpoints have evolved independently in the various branches of science, and to begin with these may be indicated as follows: in the past centuries, science tried to explain phenomena by reducing them to an interplay of elementary units which could be investigated independently of each other. In contemporary modern science, we find in all fields conceptions of what is rather vaguely termed 'wholeness.'"—We are thanking Charles François for donating M. Mulej

In this way, we wish to help people to solve current problems caused by the unavoidable specialization becoming over-specialization for several decades, even centuries.[2]

For example, in autumn 2006 the former U.S. vice-president Al Gore, who has been working on a more holistic approach to climate change problem for many years, was warning humankind that an urgent action must be taken for humankind to solve our common problem of sustainable development (SD), with the film "Inconvenient Truth" and many public talks. In 2007 Al Gore and an international institute based in India shared the Nobel Peace Prize. Others were and are warning, too, and requiring action. In 2002 we (Ečimovič et al., 2002) collected many coauthors to suggest a possible solution of this problem by application of systems theory as a basis for interdisciplinary cooperation of various specialists working on the same project, because they feel interdependent and therefore complementary to each other.[3] Ečimovič (2006) made one more step in the same direction. Some of us did the same with him in 2007 and 2008 again (Božičnik et al., 2008; Ečimovič et al., 2007). In 1998 we (Dyck, Mulej and coauthors,) showed how systemic thinking can be well applied in solving economic problems, too.

Al Gore and other authors who are warning of other similar dangers, such as wars, hunger, terror, one-sided misuse and abuse of political and other types of power over the powerless ones, accidents in traffic, factories, mines, epidemic and other health problems, and so on, point to a general common denominator causing all such and similar problems, indirectly:

---

a copy of this article during the 1st ALAS conference in Buenos Aires on 07-09 August, 2006. We find in it, which became later Chapter 3 in Bertalanffy's seminal book (1986, ed. 1979), no definition of wholeness and holism. As we will quote later, Bertalanffy reached beyond "logico-matematical laws" with his General Systems Theory. So do we with DST. The difference reads: Bertalanffy mentions wholeness of insight resulting from a human behavior, Mulej speaks of holism of the human behavior and approach, i.e. observation, perception, thinking, emotional and spiritual life, decision making, and action.—We will come back to Bertalanffy in Ch.3.1.

2. As you may have seen from quotations, this book is another step in +35 years of our efforts toward this aim. In time of our first textbook (Mulej, 1971) the Dialectical Systems Theory has not yet existed (Mulej, 1974).

3. In his Encyclopedia Charles François (2004: 536) said that this book is "an outstanding demonstration of the practical value of systems thinking as applied to vast and complex issues". It reaches beyond limits of wholeness quoted by Bertalanffy (1950) toward holism of thinking/human behavior with no limits inside single sciences. In Bertalanffy's book (1979) there is a basis for it, but no requisitely holistic methodology (like DST) how to attain it.

Oversights of crucial attributes due to one-sided approaches of narrow specialists lacking interdisciplinary creative cooperation, because they are lacking ethics of interdependence and fail to attain the requisite holism, cause most problems.

This is what Ludwig von Bertalanffy has basically addressed about six decades ago. Unfortunately, he remained poorly understood and followed (Hammond, 2003; Davidson, 1983; insight into contents of conferences about systems theory and cybernetics not to speak of other conferences). His teaching was well used within the traditional disciplines, but this is not enough for humankind to solve its burning current and future problems. He seems to have worked alone, but he was a person of several professional backgrounds: philosopher, art historian, theoretical biologist (Drack & Apfalter, 2007); in addition, he cooperated with people from other disciplines and required consideration of "wholeness". Take a precise look at the footnote number 1 there.

Bertalanffy requires interdisciplinary cooperation and finds a basis for it in conceptions, which several disciplines share. He criticizes "reducing to interplay of elementary units which could be investigated independently of each other" as an obsolete manner.

Bertalanffy may have been misread and misinterpreted. Isomorphisms are not the last step in finding bases for bridges between sciences and between professions, but only the first one. They enable various professionals who are different from each other in their specialties, to start working in their interdisciplinary cooperation, rather than only taking notions from each other and keep being closed behind their own discipline's walls.

This is how we understand Bertalanffy and his General Systems Theory, and this is what we are going to try to revive with this book (Mulej, 1974, and later, as quoted above).

Chapter 1

SYSTEMS THINKING: AGAINST CRUCIAL OVERSIGHT CAUSED BY OVER-SPECIALIZATION—CASE OF EVOLUTION TOWARDS SUSTAINABLE AND SOCIALLY RESPONSIBLE ENTERPRISE AND SOCIETY

1.1 REQUISITE HOLISM IN BRIEF

Systems thinking as the practice of holistic rather than one-sided thinking had been many millennia old practice of the successful humans, before systems theory as its theoretical generalization was created. Like most other human capabilities, the practice of systems thinking was informal, first, and then received the form of theory for transfer of good practice through teaching to be easier to make. (Mulej *et al.*, 1998, 2003; Mulej, N. (ed.), 2004;

Potočan et al., 2002). See Table 1 for our definition of holistic thinking (Mulej, in Mulej et al., 1992, reworked here).

Inside an authors' (usually tacitly!) selected viewpoint, one tends to consider the object dealt with on the basis of limitation to one part of the really existing attributes only. When specialists of any profession use the word system to call something a system within their own selected viewpoint—it makes a system fictitiously holistic. It does not include all existing attributes that could be seen from all viewpoints and all their synergies. See Table 2 (Mulej, 2007).

A brief summary of the law of requisite holism may thus read:

The law of requisite holism says that one needs always to try and do, what many, but not all, have the habit to do in their behavior—do one's best towards avoiding the exaggeration of both types: 1) the fictitious holism, which observers cause by limiting themselves to one single viewpoint in consideration of complex features and processes; 2) the total holism, which observers cause by trying to include totally all attributes with no limitation to any selection of a system of viewpoints in consideration of complex features and processes. Instead, the middle ground between both exaggerations should be covered, which can be achieved by using a "dialectical system", made by the author/s as a system as (i.e., network) an entity or network of all essential and only essential viewpoints.

For the requisite holism to be achieved three preconditions, at least, matter:

1. Both specialists and generalists are needed, working in teams that feel ethics of interdependence and cooperate.
2. They include professionals from all and only essential professions / disciplines.
3. Their values are expressed in their ethics of interdependence and practiced in a creative teamwork, task force, session(s) based on an equal-footed cooperation rather than top-down one-way commanding.

Requisitely holistic thinking cannot include the global attributes only, because they make a part of the really existing attributes only, although they matter very much and tend to be subject to oversight by specialists. Neither can requisitely holistic thinking include the parts' attributes only, although they

Interdependent actual general groups of real features' attributes	Interdependent attributes of the requisitely holistic consideration of real features	Considered attributes of thinking about real features	Attributes of participants of consideration at stake	Surfacing of all these attributes in a given case
Complexity	Systemic	Consideration of the whole's attributes that no part of it has alone	Interdisciplinary team	The final shared model resulting from research as a dialectical system of partial models
Complicatedness	Systematic	Consideration of the single parts' attributes that the whole does not have	One-discipline team / group or individual	Partial models resulting from one-viewpoint based investigation
Relations—basis for complexity	Dialectical	Consideration of interdependences of parts that make parts unite into the new whole—emerging (in process) into synergy (in its outcome)	Ethics and practice of interdependence—path from one-discipline approach to the interdisciplinary teamwork	Shared attributes and complementary different attributes, which interact to make new synergetic attributes, i.e. from systemic to systemic ones
Essence—basis for requisite realism and holism of consideration	All essential	Consideration that selection of the systems of viewpoints must consider reality in line with the law of requisite holism for results of consideration to be applicable – by reduced reductionism	Capability of researchers to deviate from reality as little as possible in order to understand reality, including systemic, systematic and dialectical attributes of it	Findings applicable in practice, due to/ although resulting from theoretical considerations

Table 1 *Dialectical system* of basic attributes of requisite holism/realism of thinking, decision-making, and action*

* A dialectical system comprises in a network/system all crucial viewpoints in order to help the observer attain a requisite holism, once a total, i.e. real holism with all viewpoints, synergies and attributes reaches beyond the human capacity. See Table 2 for definition of requisite holism. We will come back to some details in Ch. 10 and 12.

←--→		
Fictitious holism/realism (inside a single viewpoint)	Requisite holism/realism (a dialectical system of all essential viewpoints)	Total = real holism/realism (a system of all viewpoints)

Table 2 *The selected level of holism and realism of consideration of the selected topic between the fictitious, requisite, and total holism and realism*

matter very much and tend to be focused by specialists of single disciplines and professions. Oversight of relations, especially interdependences causing influences of parts over each other, may not be forgotten about in (requisitely) holistic thinking; especially specialists, who have not developed the habit to consider specialists different from themselves, tend to make crucial oversights in this respect. This experience means that they are not realistic enough. See Tables 1 and 17.

Take a look at experience around you and discover (again): success has always resulted from absence of oversights with crucial impact. And failure has always resulted from crucial oversights, be it in business, scientific experiments, education, medical care, environmental care, invention-to-innovation-to-diffusion processes, or wars, all way to World Wars of the 20th century, or the world-wide economic crises.

Holism of thinking is aimed at avoiding crucial oversights. Systems thinking should better be called holistic thinking and be the worldview and methodology of holism, or better and more realistic: requisite holism.

Systemic, i.e., (requisitely) holistic, thinking matters due to scientific reasons, for individual success in whatever activity, and for economic reasons, too. See Tables 3 and 4 for a quick look at the historic and recent changes requiring (requisitely) holistic thinking more and more today e.g., in relation to humans' natural environment, on which humankind's survival depends, but humankind threatens it by one-sided behavior, which causes its destruction. (See: www.climatecrisis.net; Ečimovič *et al.*, 2002; Stuhler *et al.* (eds.), 1995; Ečimovič *et al.*, 2007; Božičnik *et al.*, 2008; Brown, 2009; Taylor, 2008).

1.2 POOR UNDERSTANDING OF REQUISITE HOLISM—BACKGROUND

Why are facts in Tables 1-4 so alien to so many contemporary people?

For most of the time of the recent 100,000 years of its history, humankind has lived in self-sustained economy with a random market, e.g., in the form of fairs. Innovation did not matter; requisite holism was reduced to local and family relations, mostly, so was ethics of interdependence. In producers' market innovation and holism and/or sustainable development did not matter either, because competition was negligible; cases may include medieval guilds, strong trade unions, or market monopolists of other types. Once their power had been broken, after 1870s (Rosenberg & Birdzell, 1986) innovation and hence requisite holism and VCEN of interdependence gradually became crucial—in the emerging customers' and state supported customers' market. Hence, in a very short period of time people have become supposed to change millennia old habits—add innovation to routine, and requisite holism to growing narrow specialization, as well as interdisciplinary cooperation to self-sufficiency of specialists. Narrow specialization that is unavoidable today, must add to it VCEN ethics of interdependence rather than self-sufficiency; the latter makes specialization dangerous, not only beneficial.

Prescribed standards, such as ISO 9000 (quality), ISO 14000 (environment), are cases of this change in the customers' market situation. In addition, in recent decades market changes became much quicker (Table 4). People of today are overwhelmed by market demands for change, which they must match with innovation and hence requisite holism and hence VCEN of interdependence, like never before. The change has happened in one-generation time, rather than as slowly as people were used to changing earlier, and is keeping this speed.

Over the decades after World War II, market requirements have been changing more quickly than the human capacity to unlearn the old and accept the new VCEN. In every next decade, rather than a two-generation cycle of about 70 years[1] new attributes preconditioned success in addition to

---

1. Historical data about how much time has passed between critical historic events, such as from the liberation of USA from United Kingdom and USA becoming a country of its own until USA civil war, and then to the great depression, and to the current role of USA as the only superpower in the world, including the world-wide crisis beginning to show in USA in 2008, demonstrate that the critical modernization of the prevailing VCEN has tended to last for two

Viewpoints Type of Market	Basic Relation/s between Production and Consumption	Impact of Humans on Natural Environment	Humankind's Interdependence with Natural Environment
RANDOM MARKET	Producers' own consumption and occasional exchange of random surpluses	Minimal impact, growing as humankind grows in number and needs/requirements	Intuitive human consideration of nature based on experience in agriculture, gathering, hunting, wood cutting, fishing and mining
SELLERS' / PRODUCERS' PREVAILING POWER = PRODUCERS' MARKET	Growing production for poorly considered, known or unknown, customers, who lack impact on suppliers (supply smaller than demand)	Specialization and narrow thinking grow and so does the humans' detrimental impact on nature (especially by industrialized production)	Nature is subordinated to profit, jobs depend less on nature, more on growing urbanization and manufacturing as well as industrialized agriculture
BUYERS' / CUSTOMERS' PREVAILING POWER = BUYERS' MARKET	Growing impact of customers requiring satisfaction/total quality of products and services, and conditions of life (supply bigger than demand)	Specialization and its bad one-sided impact on nature keep growing, so does biased application of science, causing need for inter-disciplinary cooperation	Nature is still subordinated to profit, but nature is thought of more due to cost, caused by backlash of oversights caused for profit; inter-disciplinary insight grows
STATE / GOVERNMENT SUPPORTED BUYERS' MARKET	Increasingly organized/legalized impact of customers demanding total quality of products, services and conditions of life (supply much bigger than demand)	Growing awareness about the terrible impact of humankind's one-sided impact on nature & its dramatic consequences for humans' survival	Same as before, but world-wide official documents and actions urge governments and businesses as well as humans to be more holistic; so does a part of market (e.g. by requiring social responsibility)
GOVERNANCE AND MARKET USING SOCIAL RESPONSIBILITY / REQUISITE HOLISM AND WHOLENESS	Further increase in customers' impact introduces more and more honesty and requisite holism because monopolistic abuse becomes too expensive	Application of awareness of bad consequences of one-sidedness for economic action and investment to innovate the natural preconditions for humans to survive	Humans' poor care for the natural preconditions of their survival is old history: replaced by requisite holism in both businesses' and government's behavior, based on VCEN of interdependence

Table 3 *Development of market relations and environmental care quality: A case of growing awareness of the requisite holism as a precondition of humankind's survival*

the previous ones. Every phase after 1960, the West (and Japan, Taiwan, South Korea, Hong Kong, Singapore, Australia and New Zealand) with 20% of the world population, lives in the customers' and state supported buyers' market (in Table 3). Competition keeps causing lower cost, including a lack of care for natural environment, if short-term and one-sided views prevail. A need results for costly eco-remediation, health care, organizational, managerial, business and technological innovation concerning e.g., emissions in air and water and their prevention under ISO 14000 standards family.

Concretely, we can find: too one-sided considerations in past times caused oversight that the technological progress causes along with beneficial also detrimental consequences. One-sided estimations often oversee side-effects, which are essential in their long-term consequences. Data say, among others, that the growth of richness of the West over the recent good half a century, at least, has been much bigger in one-sided book-keeping than in long-term economics, since the West has been only postponing rather than covering the cost of saving humankind's natural environment, which makes these cost accumulate to sums showing the growth of richness is fictitious (Božičnik, 2007). Economic consequences of such short-term abuse of the law of external economics are calculated as enormous (Stern, interview: Stein, 2007: 14–15): if humankind does not tackle the climate change very quickly and radically, they will cause humankind's cost as high as 5.500 (five thousand five hundred) billion Euros, which is more than the cost of both World Wars combined. Without measures to reduce greenhouse gases the world-wide GDP will fall for 5%, possibly for 20%. Sustainable enterprises are needed and must develop to socially responsible ones for the current civilization of humankind to survive; the industrial VCEN must be replaced (Table 4).

1.3 SUSTAINABILITY AND SOCIAL RESPONSIBILITY— A WAY OF REQUISITE HOLISM

Consequently, with full right, humankind needs the development level of sustainable and socially responsible enterprises (in Table 4: Decades of 2000- and 2010-). It requires requisitely holistic understanding of the current reality and of the role and importance of all humans in that reality, especially of the critical entities such as enterprises. This means that humans must use

---

generations in the transition from the pre-industrial to the modern society. This can be seen in other areas too (M. Mulej, in Mulej et al., 2000: 108-116). We call this the law of two-generation cycles (Mulej, 1994; we first used the term in 1989).

Decade	Market & Social Requirements	Enterprise's Ways To Meet Requirements	Type of Enterprise
1945–	Covering of post-war conditions of scarcity, rebuilding, etc.	Supply of anything; supply does not yet exceed demand	Supplying enterprise
1960–	Suitable price (as judged by customers)	Internal efficiency, i.e. cost management	Efficient enterprise
1970–	Suitable price X quality (as judged by customers)	Efficiency X technical & commercial quality management	Quality enterprise
1980–	Suitable price X quality X range (as judged by customers)	Efficiency X technical & commercial quality X flexibility management	Flexible enterprise
1990–	Suitable price X quality X range X uniqueness (as judged by customers)	Efficiency X technical & commercial quality X flexibility X innovativeness management	Innovative enterprise
2000	Suitable price X quality X range X uniqueness X contribution to SD (as judged by customers)	Efficiency X technical & commercial quality X flexibility X innovativeness X SD	Sustainable enterprise
2010–	Suitable price X quality X range X uniqueness X contribution to SD (as judged by customers) X social responsibility	Efficiency X technical & commercial quality X flexibility X innovativeness X SD X honesty reaching requisite holism and wholeness beyond legal demands	Socially responsible requisitely holistic enterprise

Table 4 *From a supplying to a socially responsible requisitely holistic enterprise, and a new definition of the concrete contents of requisite holism*

requisitely holistic thinking (Tables 1, 2 and 8) in their behavior for humankind to survive; they hardly can use it without ethics of interdependence. (For details see: Knez-Riedl, 2000a; Knez-Riedl et al., 2001; Mulej, 1979; Mulej et al., 2000; Mulej, Kajzer, 1998,1998a; Potočan, 2000; Potočan et al., 2005; Potočan, Mulej, 2007a, b)

How can enterprises and other organizations of so far become sustainable and then socially responsible requisitely holistic enterprises? According to data in Tables 1-4, especially Table 4 (esp. decades 2000- and 2010-), humans, as consumers, buyers, citizens, and competitors, need and require enterprises to take a new, more/requisitely holistic and future-anticipatory, criterion of their own long-term viability. Consequences of one-sidedness in enterprises' decisions are clear: the economic and other crises of recent decades and 2008–, which include high cost of sustainable development that has become unavoidable. It is much easier to make decisions than to think requisitely holistically (Table 12). More attention must be paid to a requisitely holistic preparation, definition and realization of goals including long-term SD in order for humankind to overcome its permanent and costly economic crises and to survive.

Bosses and other members of modern enterprises are, hence, facing a basic question: How should they define their new development and future business? By sustainable development principles (Potočan & Mulej, 2006) and by social responsibility principles (EU, 2001; and later; Hrast et al., 2006, 2007, 2008): the most probable alternative of requisite holism is one-sidedness including crucial oversights and hence new crises due to which very few new firms live more than a few years (Gerber, 2004). Enterprises exist and develop best if their actions are requisitely holistic. However, in both, theory and practice, we detected no holistic model of business that provides a requisitely holistic, harmonized, and goal-oriented development. The sustainable development concept offers a (possible) solution, at least, to achieve a sustainable orientation of human activities (Potočan et al., 2005). Even more holistic approach is enabled by social responsibility principles.

On the basis of theoretical cognitions and our own experiences in business practice, one can define a sustainable enterprise, in the most general sense, as an enterprise attaining a synergetic whole of economic, ecological, social, and ethical dimensions (e.g., goals) of its business, along with the requirements listed in Tables 4 (esp. decade 2000-), 5 and 6 (Ackoff & Rovin, 2003; Brandon & Lombardi, 2005; Breu & Hemingway, 2005; Drucker, 1985; EU, 2005; Goerner, 2004; Lunati, 1997; Potočan, 2002; Schermerhorn & Chappell, 2000; WBCSD, 2004; WCED, 1987). Socially responsible enterprises attain these goals beyond legal requirements—Table 4, decade of 2010-.

Table 5 shows the basic aspects and resulting criteria of what are sustainable enterprises, and possible means of implementing market and social requirements as imperatives in and beyond the decade of 2000-. A sustainable enterprise tries to conceive and run its working and behavior in a way that meets both, human and environmental needs and requirements. (For details concerning each aspect and its criteria, see also: Ackoff & Rovin, 2003; Brandon & Lombardi, 2005; Cooper & Vargas, 2004; Daft, 2000; Dees & Emerson, 2002; Drucker, 1985; Ečimovič et al., 2002; Edwards & Orr, 2005; EU, 2005; Florida, 2002; Goerner, 2004; Koch, 1998; Lunati, 1997; McIntryre, 2005; Mulej et al., 2002; Potočan & Mulej, 2003, 2005; Rhimesmith, 1999; SIC, 2001;

Schermerhorn & Chappell, 2000; UNESCO, 2000; WBCSD, 2004, 2005; WCED, 1987.) Humans namely live on four basic levels to be considered in sustainable development, therefore by sustainable ethics: individual level; enterprise (e.g., corporate) level; closer environment (e.g., natural, social, and ethical) level; and broader (i.e., global) environmental level. On all four of them four main criteria make the dialectical system to be considered as in Table 5.

Aspect	General Criteria
Economic imperative	Competitiveness
Ecological imperative	Habitability
Social imperative	Community
Ethical imperative	Legitimacy
All aspects	Combined criteria

Table 5 *Sustainable enterprise's basic aspects and main criteria of its quality level*

These needs require sustainable enterprises to conceive, formulate, and use requisitely holistic criteria, and to evaluate their business critically. Table 6 summarizes some basic criteria to evaluate sustainable enterprises' business from some critical viewpoints.

Criteria Aspects	Individual Performance Criterion	Corporate Performance Criterion	Societal Performance Criterion	Global Performance Criterion
Economic Imperative	Individual prosperity	Corporate profitability	Societal wealth	Global wealth
Ecological Imperative	Individual eco-efficiency	Corporate eco-efficiency	Societal eco-efficiency	Global eco-efficiency
Social Imperative	Individual quality of life	Corporate reputation	Societal quality of life	Global quality of life
Ethical Imperative	Individual values	Corporate values	Societal values	Humankind values
All aspects in synergy	Individual sustainable life index	Corporate sustainable behavior index	Societal sustainable development index	Global sustainable development index

Table 6 *Basic criteria for the evaluation of a sustainable enterprise: A suggestion*

Hence, a sustainable enterprise attains the highest level of requisite holism and destroys the human condition for survival the least of all enterprises. A sustainable enterprise does not only command with the most modern and comprehensive knowledge, but uses VCEN that allow sustainable enterprises to do no/to do the least harm, such as sustainable VCEN resulting from sustainable development principles. We will come back to VCEN in Ch. 9.

Social responsibility adds the VCEN of interest of enterprises to do more than the law requires officially because it helps them outcompete the others by more requisite holism of their approach and wholeness of their outcomes.

1.4 THE ESSENCE OF SOCIAL RESPONSIBILITY (SR)

We are viewing SR here in perspective of systems theory as a science on attainment of requisitely holistic (RH) behavior aimed at requisite wholeness of insights and outcomes. We use the latter also to deal with innovation and we see a practical connecting point between them and SR in the daily experience—VCEN need innovation towards more holism meaning less selfishness for selfish reasons. Namely: a narrow selfishness does not protect us from envy and protests all way to terrorism on part of those who feel that the decision-makers do not decide with SR, but with a narrow and short-term, if any responsibility except a fictitious one.

SR does not ask whether or not there are e.g., entrepreneurs and more or less high and even questionable awards for managers, but it asks about criteria that should be felt among people as, at the same time:

- Requisitely honest and based on real achievements, hence acceptable without envy, i.e., as ethically correct;
- Achievements enabling economic and social advancement including a RH quality of a requisitely big majority, and;
- Attained by methods/products that do not ruin natural conditions for life of humans and other living beings without which humans cannot live, such as, e.g., bees.

People, times and conditions define differently what is a socially acceptable, i.e., SR behavior. Criteria have always depended on VCEN of the most influential ones, the power holders. Their values became culture, ethic, and norms, when attracting people as followers by appeal or force (Potočan & Mulej, 2007). Their VCEN were expressed in ideologies, e.g., religions and similar tools of power providing ownership and joy to the most influential ones. These VCEN, according to official definition of SR tackle manners of the influential ones in treatment of (EU, 2001, 2006 a, b):

- Their coworkers;
- Their other business partners;
- Their government, non-governmental organizations, i.e., broader social environments, and;
- Their natural environment as the natural precondition of survival.

In all four aspects the influential ones must attain more RH behavior than earlier, i.e., innovate their practice.

Thus, SR is a process of social innovation and its objective for humankind is to find its way out of the current blind alley. Success of this process depends on humans, of course, especially on the influential ones.

Influential people can use their influence to define criteria for what is wrong or right, sometimes with a too narrow and short-sighted egoism. Then, they do not prove their SR, and they lose their power, ownership and joy, gradually at least. During the latter process, SR and legal responsibility tend to mix up, but they can differ: the power-holders are influential enough to be able to adapt legal rules to their interests, including narrow, one-sided, biased, and short-term interests. They often do so more easily than accept VCEN with SR based on broader defined and perceived RH. This may bring them in trouble. Thus, the famous Friedman's definition that SR is unacceptable is wrong: companies must care for their profit and benefit of their owners, but not with narrow and short-sighted criteria only (Goerner et al., 2008; Toth, 2008). Friedman won his Nobel prize for economy in 1970 for his theory of conservative neo-liberalism, which now proves to be out-dated and detrimental for enterprises and society at large. It does not match the old proverb that "The first profit does not go in the pocket"—a short-term benefit based on narrow and short-sighted criteria often costs much in a longer term.

For millennia, people also used many religions to foster SR, and they do so today. There has always been a mixing, networking, and fighting of the concepts of more narrow and short-term interests on one hand (read: interests concerning now and here) and of the more long-term and broader interests, on the other hand, reaching beyond now and here (Rudel, 2008; Wu, 2004).

Slave-owning and feudal societies clearly enforced narrow and short-term interests, as their opponents said. These short-term interests led both, societies to a life that in criteria of quality of life of today has experienced a poor economic efficiency and quality of life of a big majority of people, and in extreme differences between the rulers and subordinates, around the world. Before the Western Industrial Revolution China and India supplied 80% of all global production, but today they are coming close to 10% (Bošković, 2006). The industrial and post-industrial/entrepreneurial society differs from the previous ones by its principle of equal chance of everybody to expose their skills and interests and to contribute to the quality of life of them-selves and others. Practice shows that in terms of book-keeping data the entrepreneurial society seems successful in raising the standard of living, but the differences in quality of life are again very similar to those in feudal times: if only two hundred of the richest individuals donated less than five percent of their properties, four million children a year would not die for hunger and illness (Crowther *et al.*, 2004b). Similar are other data (e.g., Nixon, 2004; Toth, 2008). Private owners enforce their interests, so do governmental ones, although formally legally there are no owners. Ownership is no problem, but the short-sighted and narrow definition of interests of the influential ones, who forget about SR's longer-term and broader effects, or failure of using SR concepts.

Thus the crucial issue of SR reads: the influential ones abuse/misuse rather than use with RH behavior their chances hidden behind legal responsibility and protection; abuse/misuse fails to lead to SR, but to its opposite. Hence, in our perception, the essence of SR in practice is the prevention of misuses/abuses of legal, economic, and natural laws, and enforcement of replacement of the narrow and short-sighted criteria of right and wrong for broader or even RH criteria. Actually, this is what A. Smith has been speaking for, although today they ascribe him the opposite opinion.

Rare authors (such as Walker, 1978) say that Adam Smith and Karl Marx have aimed in their research at a way to preserve the village-solidarity of earlier times after transition from the village to the entrepreneurial society. They did not succeed. Nobody did. Therefore the effort called SR is showing up today to help influential people think in longer-terms and broader criteria. No wonder, SR has hard times to become a general VCEN. The short-term and narrow views of decision-makers make obstacles all the time, and there is neither a theory to replace the current economics, although this leads humankind to a blind alley.

People who abuse the label of liberalism to cover the huge modern differences in richness, health, famine, etc. and destruction of humans' natural environment—preconditions of humankind's survival—fail to see that A. Smith does not favour narrow and short-term interests. The invisible hand expresses the logic of economic interdependence: you must delight your customer to have him return and make you happy as a supplier. The fact that people enforce under the label of A. Smith economic thoughts and interests opposing his ideas, is visible in conditions concerning the human care for natural preconditions of life and survival of the current civilization: this care is worrying even in global official data.

These data express abuse of the law of external economics. This law can often be beneficial, but has been applied to nature with expensive consequences. They will obviously damage generations to come soon—our children and grandchildren already. The influential ones act like if they hated their off-springs, when they act on a narrow basis and with no SR.

Thus, SR enforces own benefits/interests of people, but not merely the narrow and short-term ones, but also or even first of all the long-term and broad ones. People need to reinforce them in the form of national and international legislation and VCEN of their enterprises and other organizations for the human civilization of today to survive. Market—as an institution aimed at reinforcing the invisible hand—needs support. Not all private or governmental owners should be off, but the ones without SR. They make to much damage to the coming and their own generations

Let us hence be less selfish for selfish reasons. We are not independent, but an interdependent part of nature on the planet Earth. The development of the basis of competitiveness tends to go the same direction.

1.5 FOUR OR FIVE PHASES OF DEVELOPMENT OF THE BASIS OF COMPETITIVENESS

There is an interesting view of economic development phases, in terms of the changing basis of competitiveness, that stresses the notions that are summarized in Tables 1-6. It sees four phases: (1) the factors phase means that a nation or region lives on natural resources and cheap labor, providing for a rather poor life for millennia; (2) the investment phase means that a nation or region lives on foreign investment into its economic development and can hardly compete; (3) the innovation phase means that a nation or region lives on its own progress and attains a better and better standard of living; (4) the affluence phase means that people have finally become rich, which makes them happy, but also lose ambition. Thus, phase 4 is not the highest development phase only, but also the phase of growing problems of employment, supporting everybody (Porter, quoted in Mulej, 2006). Conclusion: one must attain and keep capacity of requisitely holistic approach in order to enter the innovation phase quickly and remain in it as long as possible, or may return to it from phase 4, probably via phases 1 and 2, like history has already shown e.g., in the case of the Roman Empire as well as other societies that have attained affluence and complacency. What offers a solution?

We can talk about companies (Collins, 2001; Collins & Porras, 1997; Gerber, 2004; etc), individuals, countries, or regions. Florida (2005) in his field research about the reasons of differences in economic prosperity between regions of United States found two basic causes of them:

1. In the U.S., the creative class is rising from 5 (five) percent a century ago to +30% in 1999, with 12% in its super creative core, while the working class is dropping from 40% at its peak several decades ago to 25% now. The largest social group is the service class, but it does not earn much, because it only provides preconditions for the creative class to create most of all (Florida, 2005: 90-99).[2]
2. In the U.S., the most prosperous regions have the highest 3T indicator: tolerance for difference between neighbors all way from traditional families to gays etc; talents that are attracted by tolerance and chances to be creative; technology invested (Florida, 2005: 257–273).[3]

Tolerance is a relation making room for differences between humans to complement each other, thus to help them to avoid oversights and to attain more holism. Talents make the basis for creativity, including innovation, which in turn can best result from cooperation of specialists different from each

---

2. In addition, the creativity of the rather poorly paid people is overseen in this definition. But they must be very creative, although with another contents of creativity, to survive.
3. Tolerance to failures in business risk-taking is much bigger in USA than for example in Europe. This makes USA much more innovative. USA is a product of the most entrepreneurial Europeans, who left Europe to take their risk more freely. The routine-lovers remained in Europe and their culture keeps prevailing in it. (See Mulej, 2006a, and some notes later in this book.)

other (as this book will show later briefly). Investment in technology supports them, and receives support from them: if various and different talents work hand in hand, results of their creativity have more chance to attain requisite wholeness and therefore to succeed.

1.6 CONCLUSIONS FROM CHAPTER 1

Tables 4-6 may lead us to an additional finding: the decade of 2010 is defined by new attributes. It may well be marked by new efforts for informal systems thinking aimed at requisite holism in order to solve the current problems of humankind such as ISO 26000. These efforts may be seen in the concepts of (corporate) social responsibility (Hrast *et al.*, (eds.), 2006, 2007, 2008; Mulej & Prosenak, 2007; Prosenak & Mulej, 2008; Prosenak *et al.*, 2008) and in total responsibility management (Waddock & Bodwell, 2007; Gorenak, 2008).

In other words: (informal) systems thinking is the background of the creative class and creative society/regions. But it causes difference, obviously, because not all people are equally capable of holistic thinking.

What makes people incapable of requisitely holistic thinking?

Chapter 2

FROM SYSTEMS THINKING TO SYSTEMS THEORY IN THE TURBULENT 20TH CENTURY

For millennia, the division of labor has been growing for people to become more productive and rational in their effort to meet their needs. After 1820s (Bošković, 2006) the Industrial Revolution became a new tool in this effort. After 1870s (Rosenberg & Birdzell, 1986) abolishment of monopolies of guilds over economy and of church over thinking was the next one. It made room for entrepreneurship and hence innovation and hence narrow specialization both, in practice and science. Efficiency was growing, but results of narrow, unholistic, thinking included terrible effects such as World Wars and the world-wide economic crisis in 1914-1945.

In such circumstances, Ludwig von Bertalanffy is the father of the General Systems Theory, the oldest one among systems theories, which are many now. He lived from 1901 to 1972, which made him live through both World Wars and the world-wide economic crisis between them in 1914-1945. He was a philosopher, art historian and theoretical biologist. This multidisciplinary capacity led him to a big interdisciplinary result—the General Systems Theory.

He namely experienced, that the human way of fighting our problems is also the cause of our problems. Humankind has developed, millennia ago, the one-sided attitude that the human being is the master of the rest of nature, rather than a part of nature and adapting to its natural environment. This is how agriculture and handicrafts started replacing hunting, gathering, and nomadic economy and life. Later on manufacturing industries followed, now services do. All these evolutionary results of human creativity are called progress. It was able to feed more people, but it required more and more knowledge, which made room for more and more one-sided specialization, including oversights along with deep insights. Since then, and especially in the 20th century, we have—as humankind—developed huge lots of insights into the laws of nature, including society, and the methods/technologies and techniques of using them, due to specialization and resulting concentration of specialists in small parts of reality.

We benefit from them; we have never lived a better live, in our own criteria[1]. But we can no longer really understand and master our lives, because we—as humankind—know so much, that we—as individuals—must be narrowly specialized. What is the consequence?

This narrow specialization to single professions, life in single areas and in poor tolerance and VCEN of independence rather than interdependence, consideration of reality from a man's or woman's single perspective or from children's or adults' viewpoints alone, only, causes oversights. Nature does not exist along with physical, chemical, biological laws in separation, but in their interaction resulting from interdependence of parts and attributes of nature, including humans. And we do not live as humankind, only, but as individuals and groups, first of all.

The whole world is not fragmented into parts, which may no longer be able to become a whole, but we humans see it in parts rather than as one whole. The resulting oversights cause crucial problems, because we humans make our decisions on the basis of our perceptions.

What can systems theory and cybernetics do against the resulting problems?

---

1. Other parts of nature have different criteria, and this difference between humans and the rest of nature is causing climate change problems, other environmental problems, diseases, etc.

Chapter 3

SYSTEMS THEORY AND CYBERNETICS—HOLISM AGAINST BIG CRISES CAUSED

3.0 THE SELECTED PROBLEM AND VIEWPOINT FOR DEALING WITH IT IN CHAPTER 3

Why is there so little practice of systems thinking, such that there are wars all the time, less than 1-2 percent of inventions become innovations, very few companies survive more than 3-5 years, there is such a huge and dramatic climate change that many glaciers no longer exist, Arctic and Antarctic areas are melting, there are epidemics of new types, the list goes on?

3.1 HOLISTIC THINKING VERSUS NARROW SPECIALIZATION

A good half a century ago, right after the end of the "World War I & World Economic Crisis & World War II (1914-1945)" period, scientists such as L. von Bertalanffy, N. Wiener and their colleagues (from several disciplines and in cooperation) found a new response to the terrible consequences of one-sidedness visible in events of this period, again: holistic rather than fragmented thinking, decision-making and action. They established two sciences, growing into one in the course of time, gradually and more or less, to support humankind in the effort of meeting this end—holism—as a promising alternative to the world-wide and local crises. These were (General)

Systems Theory and Cybernetics, of course. With full right they are called the science of synthesis (Hammond, 2003). System was and is the word entitled to represent the whole.[1] One fights one-sidedness in order to survive. Bertalanffy wrote very clearly (1986, (ed.) 1979: VII):

Systems science ... is predominantly a development in engineering sciences in the broad sense, necessitated by the complexity of 'systems' in modern technology .... Systems theory, in this sense, is pre-eminently a mathematical field, offering partly novel and highly sophisticated techniques ... and essentially determined by the requirement to cope with a new sort of problem that has been appearing.

What may be obscured in these developments—important as they are—is the fact that systems theory is a broad view which far transcends technological problems and demands, a reorientation that has become necessary in science in general and in the gamut of disciplines ... It ... heralds a new world view of considerable impact. The student in 'systems science' receives a technical training which makes systems theory—**originally intended to overcome current over-specialization** *(exposure of these words by bolding is ours, M. Mulej) into another of the hundreds of academic specialties...* (Bertalanffy, 1979: VII)

It presents a novel 'paradigm' in scientific thinking ... the concept of system can be defined and developed in different ways as required by the objective of research, and as reflecting different aspects of the central notion. (Ibid.: XVII)... General systems theory, then, is scientific explorations of 'wholes' and 'wholeness' which, not so long ago, were considered to be metaphysical notions transcending the boundaries of science... 'Systems' problems are problems of interrelations of a great number of 'variables'... models, conceptualizations and principles—as, for example, the concept of information, feedback, control, stability, circuit theory, etc. by far transcend specialist boundaries, were of an interdisciplinary nature ... (Ibid.: XX)

This fact about the Systems Theory itself speaks of the "uncommon sense" Bertalanffy has been speaking for (Davidson, 1983): he was fighting the common current practices of one-sidedness, because they were dangerous

---

1. We may never forget this definition of the system. There are many contents of the word system in colloquial usage, unfortunately. There are 15 groups of definition of its contents in vocabulary (Webster, 1978). See Ch. 12.3.7.1.

and still are so with a growing trend. The great author on creativity, E. De Bono might say that Bertalanffy has been using the lateral thinking rather the vertical one (De Bono, 2006). Systems thinking was and is fighting the vertical thinking that only follows rules like e.g., in solving crosswords; instead, it requires creative thinking along an unknown path, i.e., lateral thinking to become a normal human habit along and in combination with vertical thinking. Let us return to Bertalanffy!

What is to be defined and described as a system is not a question with an obvious or trivial answer. It will be readily agreed that a galaxy, a dog, a cell and an atom are real systems; that is, entities perceived in or inferred from observation, and existing independently of an observer. On the other hand, there are conceptual systems such as logic, mathematics (but e.g., also including music) which essentially are symbolic constructs; with abstracted systems (science) as a subclass of the latter, i.e., conceptual systems corresponding with reality. However, the distinction is by no means as sharp and clear as it would appear. … The distinction between 'real' objects and systems as given in observation and 'conceptual' constructs and systems cannot be drawn in any commonsense way. (Bertalanffy, 1979: XXI–XXII)

This supports our understanding of the term system (Mulej, 1979: 10): **Systems are mental pictures of real or abstract entities as objects of human thinking; they are concepts that represent something existing from a selected perspective/viewpoint/aspect.** Thus:

In mathematical formal terms, a system is a round-off entity consisting of elements and relations, which makes it holistic. In terms of contents, a system depends on its authors' selected viewpoint[2]; hence, it does not comprise all attributes of the object under consideration, but only the selected part of them. This fact makes a system both, holistic (formally, with no contents, or inside the selected viewpoint only) and one-sided (due to the unavoidable selection of a viewpoint).[3]

---

2. A town can be considered from a city-planning viewpoint, from inhabitants' one, from traffic one, from air-pollution one, from employment chances one, from educational one, from medical services one, from commercial centers one, from factories one, from water supply one, and from many others as well. Each and every one of them matters, but none is holistic, because it is unavoidably limited to the boundaries inside the selected viewpoint. This is why we suggest the "dialectical system" to put all essential viewpoints in synergy and come closer to holism. We will add to comments in Chapter 1 along with Table 2 about the "law of requisite holism" (Mulej & Kajzer, 1998; Rebernik & Mulej, 2000; Mulej, 2006; 2007) in Ch. 12.

3. Therefore, in terms of contents no system is holistic, but limited to one part of the really

See Table 7 for a brief presentation of these relations. We will come back to them in Ch. 12.

Objects exist, and humans watch and manipulate them with different levels of holism. Total holism makes the object and the system as someone's mental picture of the object totally equal, but it reaches beyond human natural capacity. This is why humans are specialized and limited to single viewpoints causing humans' limitation of consideration of any object to a one-viewpoint system. By cooperation, normally by an inter-disciplinary one that includes several essential professions in a synergetic effort, a team can attain more holism—by a dialectical system. Both a system and a dialectical system exist inside the human mental world, in human thinking and feeling; they can be expressed in models for other humans and other living beings to receive information about humans' thinking and feeling. We will add details to Table 7 in Ch. 12.

The essence of the concept of the dialectical system and related law of requisite holism/realism and wholeness is well expressed by Wilby (2005: 388), although she leaves open the question of viewpoints selected and thereby determining the boundaries of study: "The goal of holistic study is not to look at 'everything'. Instead it is to make a decision about what is relevant to the study and what is not and to know and understand why those choices were made. The biases and interests affect the choice of what is likely to be included and excluded (i.e., what is in the system as opposed to what is relegated in the environment of the system)." What Wilby calls holistic, we call requisitely holistic, to be precise and clear.

Why is requisite holism important? There are scientists attempting to say that their discipline offers the only unique and unifying basis for dealing with systems. They do not speak of worldview, like Bertalanffy does, but of professional/scientific disciplines. Can they be right? Yes, in their own perspective they can. Can they be sufficient for holism? They can be so rarely, exceptionally. Nobody can be really, i.e., totally, holistic: teams can perhaps be requisitely holistic with interdisciplinary creative cooperation. (See cases in

---

existing attributes of the object or topic under consideration. A system can anyway be composed of two kinds of smaller systems: (a) *subsystems* cover attributes, due to which they differ from each other (such as countries of a continent, or production units of a factory, or bonds from blood vessels, etc.); (b) *partial systems* cover attributes which the different parts share (such as a number of uniting organizations of a continent, human resources issues of an office or factory, etc.).

Level of realism of consideration of the selected topic	Level of simplification of consideration	Viewpoints of consideration taken in account	Components taken in account in consideration	Relations taken in account in consideration
Existing object to be dealt with	None	All existing	All existing	All existing
Dialectical system	Small - requisite	All essential	All essential	All essential
One-viewpoint system	Big due to specialization	Single – selected by specialization	Selected within the boundaries set by the selected viewpoint	Selected within the boundaries set by the selected viewpoint
Model of the one-viewpoint system	Big due to specialization and modeling aimed at clear presentation	Single – selected by specialization and simplified to be clear	Selected within the boundaries set by the selected viewpoint and shown in a simplified-modeled way	

Table 7 *Relation between reality and holism/realism of human consideration of it*

e.g.: Gu & Chroust (eds.), 2005; Gu & Tang (eds.), 2006; Rebernik & Mulej (eds.), 2006; Trappl (ed.), 2006; Wilby & Allen (eds.), 2005; Wilby (ed.), 2006.)

Elohim (1999) quotes Bertalanffy requiring people to behave as citizens of the entire world rather than of single countries and consider the entire biosphere rather than its local parts only; this is a precondition for humankind to survive. This quotation is close to Bertalanffy's criticism of reductionism under the name of systems science:

Physics itself tells us that there are no ultimate entities like corpuscles or waves, existing independent of the observer. This leads us to a 'perspective' philosophy for which physics, fully acknowledging its achievements in its own and related fields, is not a monopolistic way of knowledge. Against reductionism and theories declaring that reality is 'nothing but' (a heap of physical particles, genes, reflexes, drives, or whatever the case may be), we see science as one of the 'perspectives' man with his biological, cultural and linguistic endowment and bondage, has created to deal with the universe he is 'thrown in', or rather to which he is adapted owing to evolution and history. (Bertalanffy, 1979: XXII).

So Bertalanffy believed that the overall fate of the world depends on the adoption by humanity of a new set of values, based on general systems Weltanschauung (= worldview). He wrote:

We are seeking another basic outlook: the world as organization. This [outlook] would profoundly change the categories of our thinking and influence our practical attitudes. We must envision the biosphere as a whole ... with mutually reinforcing or mutually destructive interdependencies. [We need] a global system of mutually symbiotic societies, mapping new conditions into a flexible institutional structure and dealing with change through constructive reorganization.

Bertalanffy advocated that we dare to broaden our loyalty from nation to globe..., that we become patriots of the planet, endeavoring to think and act primarily as members of humanity..., that we must begin protecting the individual and cultural identity of others. He advocated a new global morality: "an ethos which does not center on individual goods and individual value alone, but on the adaptation of Humankind, as a global system, to its new environment." The need for this new morality, he said, was imperative:

We are dealing with emergent realities; no longer with isolated groups of men, but with a systematically interdependent global community: it is this level of [reality] which we must keep before our eyes if we are able to inspire larger-scale action, designed to assure our collective and hence our individual survival. (Davidson, 1983, quoted from: Elohim, 1999)[4]

---

4. The crisis of 2008- suggests that Bertalanffy has been right: requisite holism might better

Bertalanffy—as you see—stresses the whole, wholeness, and interdependencies, rather then parts and independencies or dependencies. This necessary worldview fights reductionism, which has been very helpful over the recent centuries, but causing oversights as well, with consequences causing World Wars, climate change, and economic and social crises.

3.2 DIFFICULTIES WITH IMPLEMENTATION OF HOLISM FOLLOWING BERTALANFFY

We could summarize our own local and international experience concerning backgrounds of making the Bertalanffian holism a practice as follows:

Bertalanffy explicitly speaks about holism as a worldview, which is important, but not enough. In this way, he may be read as limiting his approach to an impact over values,[5] and leaving knowledge of advocates and users of his approach to themselves, as individuals and groups who are unavoidably specialized in a discipline of their own choice. He also leaves the possibilities aside, which might have to be created for his concept to have more influence.[6]

Notions, which come along with Bertalanffy's GST as a worldview, are rather few and not elaborated enough in serious details to become a prescribing operational vocabulary and method(ology). The worldview of holism has hence a too poor cognitive support to easily make it widely accepted and work along with narrow specialization/s.

Specialization is not a matter of bad will or bad behavior of contemporary individuals, be they scientists or practitioners, but an unavoidable consequence of the immense and growing stock of humankind's knowledge. Inside their own horizons those scientists and practitioners tend to cover all they find important—to them and to their topic under consideration from their selected viewpoint/s.

---

prevent the current crisis from happening than the economic and political one-sidedness does.
5. We will be back to values/VCEN later on. See Chs. 12 and 13. Values subjectively reflect the objective needs as perceived by the given person, and combine the cognitive and emotional aspects of her personality traits in a synergetic way.
6. We will come back to the notions of values, knowledge and possibilities later on. (See: Mulej *et al.*, 2000 and earlier, since 1974, and Mulej, 2000) See Tables 11, 16, and Chs. 12 and 13.

Holistic thinking is a vague concept rather than a precisely defined one. Does it cover wholes without holes, or holes without wholes? In other words, a person is supposed to behave as a citizen of the world, not of a single country only and to consider the entire biosphere (like Bertalanffy requires[7]). It is easy to say Bertalanffy is right, in the time after two world wars and the time of a permanent danger of a nuclear war. But (total, real) holism is still beyond the human capacity to practice in the daily life and work of every human being alone.[8]

Generality on the basis of isomorphisms (= similarities), actually, leaves the specialization alive and unquestioned, although Bertalanffy quotes the danger of over-specialization in the very Foreword of his seminal book. Isomorphisms namely limit themselves to carrying some concepts from one science to the area of another science, but Bertalanffy does not seem to explicitly ask the question of interdependence of sciences or practices[9].

Viewpoints from which e.g., the same biosphere, including humankind, is looked at, select the attributes to be found crucial among many. They do not erase other attributes, but forget about them, find them (fictitiously and wrongly, sometimes) unimportant or even non-existent, at least "belonging to another discipline" rather than to the same nature, biosphere, organization (in Bertalanffy's terms). Holism becomes rather fictitious, if a single viewpoint is selected, in terms of the requirement that e.g., the entire biosphere should be considered.

Interdependence of viewpoints is in this way forgotten about, so are synergies that result/emerge from the overseen impacts over each other resulting from interactions based on interdependencies of the fictitiously separated attributes of reality.

Complexity of the real life tends to be forgotten about, too. Generality is emphasized; this seems to be a version of understanding of the so-called

---

7. Let us quote from Bertalanffy again (1979: 187-188): "Considered in the light of history, our technology and even our society are based on the physical world picture which found an early synthesis in Kant's work. Physics is still the paragon of science, the basis of our idea of society and our image of man."
8. Teamwork may do a better job concerning holism, but Bertalanffy does not elaborate on it. We will come back to this issue later on. See Ch. 13.
9. Bertalanffy does not seem to think a lot about an interdisciplinary approach as a way toward holism, although he speaks of interdependence and organization. We will come back to this issue later on. See Ch. 12 and 13.

transdisciplinary approach by several later authors. But generality, especially the one linked with isomorphisms of several disciplines/viewpoints concerned with parts of attributes of nature, is unavoidably limited to the general part/subsystem of the entire system of attributes, thus leaving aside the group-specific and individual subsystems of attributes (Table 23). This is a serious simplification, based on admitting (realistically!) the definition that science simplifies and is based on reductionism. Besides, on another occasion Bertalanffy (1979: 259) defines GST as a new field in science as a logical mathematical field; this is not very much in line with the definition that GST is supposed to be a worldview and fight over-specialization (as quoted here earlier in Ch. 3.1.).

His concept of organization involves interdependence and interaction of components of the same entity different from each other[10]. It should imply the same interdependence concerning the viewpoints (which have evolved, among other effects, to several ten thousand specialized scientific disciplines and professions). Maybe, it does so implicitly rather than explicitly when he disagrees with GST becoming one of many academic specialties and maintains that GST transcends technological problems and demands and presents a new paradigm in scientific thinking. If GST should help solving real problems of holism versus over-specialization, which it was created for, it should not go for simplification by isomorphisms only, but add consideration of the real complexity, too. This can hardly be done without a lot of (the requisite) interdisciplinary work, which enables specialists to be what they are and to attain the requisite level of holism, too. Which level is the requisite one, depends on the decisive persons; they unavoidably take the risk of success versus failure.

Bertalanffy speaks of goal-directedness (Bertalanffy, 1979: 258) with the usual weak point of many: there is no discussion about what is the basis for goals to be defined. Mulej's DST (Mulej, 2000; Mulej et al., 2000; and earlier, since 1974; Ch. 12 here, in brief) showed that lots of attention must be paid to the definition of starting points, on which the definition of the selection of the (dialectical system of the requisite) viewpoint/s depends, and so in turn does

---

10. The perhaps best case of a real-life application of Bertalanffy's concept is the worldwide control over the *nuclear weapons* since the end of World War II. Global superpowers possess many of them, but keep them out of use. The United Nations declaration on *sustainable development* has received much more pressure of narrow interests. So did the *climate change*. Social responsibility is left to enterprises' voluntary decisions. Etc.

the definition of goals (Table 19). Many misunderstandings also arise from the Bertalanffy's putting equal the object under consideration and the system as its mental representation[11], although he warns about it (as Ch. 3.1); in DST this is corrected by the explicit care for the selection of viewpoints and author's responsibility for it and its consequences.

One could continue, but let us stop here; we hope it is clear enough.

Being a wonderful concept, GST has not come to be elaborated enough for many later authors and users to really understand Bertalanffy. What results is far from Bertalanffy's basic intentions:

Oversights are replacing insights, the broader ones, at least. And they do so under the name of system and holism, which is very misleading.

A seminal idea of the holistic thinking does not become a really influential concept, even less so a reality in which the exaggerated specialization would receive a way out of its (and humankind's) blind alley. The crises, including the one of 2008- have been unavoidable.

Something must be added. We have done so by developing Mulej's DST and related findings of other members of our group. We will come to DST in Chapter 12.

Let us now try to understand the practical lack of implementation of Bertalanffy's concepts from the viewpoint of economic reasons against them.

---

11. Bertalanffy, and many other authors, implicitly or explicitly put the system *equal to object* under consideration by saying no word about the *viewpoint* from which they watch the object when they call it a system. See T. 7.

3.3 ECONOMIC[12] REASONS OPPOSING HOLISTIC THINKING IN PRACTICE: THE LAW OF REQUISITE HOLISM IN DECISION-MAKING

Bertalanffy spoke for a very broad and deep kind of systemic thinking, maybe even for a total system. He mentions planet, humanity, biosphere as a whole, global system of mutually symbiotic societies, global community, inspiring large scale action at this level of reality, our collective and hence individual survival, etc. (Davidson, 1983). Davidson rightly speaks of the Bertalanffy's "uncommon sense" in the title of his book about him. The common sense is the opposite, one-sided, narrow one. Which one is right? It depends on the viewpoints. To narrow specialists the one-sided view looks like enough, systems thinkers should be broader. There are reasons backing both practices, of course.

The available room allows only for a summary of economic reasons as backgrounds of rationality, efficiency and effectiveness of decisions and action, which humans of today take as business persons. Under the label of economic reasons long-term views are rarely preferred to short-term ones; broad views are rarely preferred to local/narrow ones, in the main-stream criteria of what is the basis of rationality, efficiency and effectiveness. In longer-term practical terms, this orientation is a critical oversight: the invisible hand praises only suppliers who please their customers well enough (Cornell, 2005; Petzinger, 2000). And this means that suppliers must seriously consider their interdependence with their customers and other co-opetitors in order to attain their own benefit. They do not watch their own benefit only, literally. If they do, others see it and punish them, making them have to be less selfish for selfish reasons, in a real market (Quinn, 2006). But the multinational companies cause a fictitious market with a poor insight in their working and public and socially responsible impact on it. Crises are unavoidable consequences because interdependence is subject to oversight other than essential of VCEN.

---

12. Economic reasons are meant to be bases and tools of judgment, which effort matches the foreseen and/or attained outcomes best. It makes sense, if the outcomes/outputs outweigh the efforts/inputs, in terms of criteria chosen by the decision-makers. They may be entrepreneurs, inventors, ecologists, quality consultants, researchers, developers, family decision-makers etc, by their concrete roles in the concrete cases and conditions. Climate e.g., may be included into criteria or left aside, as the less important attributes of inputs, processes, structures, and outputs, etc. are. So may be humankind's sustainable future (Ečimovič et al., 2007)

Consideration of interdependence and VCEN of interdependence when narrow and short-term specialists were reading Adam Smith's sentence on the invisible hand has not prevailed over a more biased one for several reasons. These reasons include the following experiences.

It is decision-making which leads life. And it is based on interests and other values and emotions, and on talents, knowledge to answer the questions "what?" and "how?" making an interdependent/ordered set, i.e., system (called "subjective starting points" in DST—see Chapter 12). With them one perceives the outer circumstances, conditions and preconditions. This is the basis of selection of viewpoints to be considered, and of the resulting decision-making, acting, and benefiting. From economic viewpoints, which prevail in the humankind's reality of today, marginal benefit and marginal cost have a high priority (Rebernik & Repovz, 2000). "In the long term, we will all be dead", humankind's economists have been taught by the most influential economist of the 20th century, Lord J. M. Keynes. Thus: What is, "sustainable development and future" compared to the current benefit?—If such a question is replied on business terms of the free-market economy, the Bertalanffy's concept of the GST as a/the worldview can hardly enter the scene.

Life is supported by technology of any work process in production and all other activities. Technology is no longer based on the real-life experience (of trial and error) only, but more and more on science, for humankind to avoid disasters. Scientists have the obvious tendency to attain objective scientific insights, fortunately. But they tend to be naturally limited to a selected viewpoint for the amount of problems to be manageable, unfortunately. Interdependencies under consideration are limited to the selected object under consideration and the selected viewpoint or set of viewpoints or, in a best case scenario, to a dialectical system of viewpoints (see: Ch. 12 and Tables 2, 7, 8 and 19). The selections are quite far from the worldwide scope and scale, which Bertalanffy required, normally, because Bertalanffy requires a (nearly?) total system (as we conclude from the above quotations).

One reason for the above selection is the (unavoidable!) specialization of science, knowledge, and skills to single disciplines and professions, for the amount of data, messages, and information of today to be manageable. It is well known that in the last few decades more scientific and other findings have been published, more books and articles, than in all millennia before the 1950s. About 200,000 articles are published a year, some of them are classified as the ones in best scientific journals, many further journals and languages are neglected in this process; so are books. It has become impossible for individual specialists, both, scientists (in labs) and practitioners (in the real world), to match the Bertalanffian scope and scale. Flood (1999) puts it nicely: humans are somewhere between mystery and mastery of life, knowing of the unknowable, managing with the unmanageable.

The unavoidable specialization has been the reason for one more fact: reduction of the GST ideas to an understanding of its notion of transdisciplinarity, which tends to support formal isomorphisms (see, e.g., Jackson, 1991). They are rather transferable from one single science to another single science than making them cooperate. Thus, the concepts claimed by GST come to be limited to application inside single viewpoints and objects selected for consideration as small fractions of the entire biosphere. The so-called generality of such isomorphisms mostly becomes a formality, in terms of the above understanding/quotation of Bertalanffy. Specialists are unavoidable, but they should better add to their role the one of interdependent mono-disciplinary specializations needing and practicing inter-disciplinary creative cooperation as a basic way towards a trans-disciplinarity, which would not be too reductionist or too generalistic. Thus, the notion of GST would be better applied.

But, over the last three centuries, since the times when enlightenment and rationalism had replaced the medieval concluding with no proving, reductionism has been a very helpful style of both, scientific thinking and practical work. Appreciation of the many good experiences of the reductionist thinking resulted, because it certainly did provide and still does provide many important insights and capacities to control the selected parts of reality (see de Zeeuw, 1999). But they though tend to be only partially realistic and hence only partially scientific and usable and useful due to their reductionist (i.e., narrow, limited) basis of analysis and hence of outcomes. But the Bertalanffian concept, if misunderstood due to e.g., narrow specialization, causes another danger: no focus, which most people might be able to cope with (Tables 2 and 8).

Though, there seems to hardly be another way out. Specialization has caused specialists to educate specialists. The dilemma has been clear: you are either a rather narrow specialist, who knows a lot/enough/something about a small fraction of reality, or you know nothing deeply enough. The problem could be overcome by team work. But, only individuals who are able and willing to work with specialists different from each other know that they are interdependent with the ones different from them. The others live with a negation of dialectics[13] as a practice and a philosophical science of interdependence, which does not suit the closed-in/"tunnel-vision" thinking of specialists in their own ivory towers[14]. Besides, dialectics would best be applicable in the formulation offered by F. Engels (1953), which has been even more frequently misread and misused than the ancient Greek and Hegelian formulations.[15] The ancient Chinese "yin and yang" is a version of the same concept, which is also very much forgotten about or misunderstood by the West (Delgado & Banathy, 1993). The West prefers systematic work style to the systemic one (Mueller-Merbach, 1992), i.e., thinking in parts with no interdependencies between them rather than in wholes. That's why Davidson (1983) rightly comments that Bertalanffian thinking is found uncommon sense.

Bertalanffy might not have used dialectics, formally and directly, but he quotes Latin words expressing dialectics as "interdependence" in one of our quotes from his work. Anyway, in order to introduce the notion of interdependence, Bertalanffy introduced his own understanding of the notion "organization". He stresses interdependence, not the subordination or the legal aspect of an organization. Over millennia, there has been a lot of reduction

---

13. Dialectics is the ancient Greek word for interdependence, which is the ancient Latin word for humans' need for each other, which is due to difference from each other due to which A is complementary to B and B is so to A. (See: Bai & Lindberg, 1997; Britovšek et al., 1960; Delgado & Banathy (eds.), 1993; Mulej, 1974 and later). We will come back to it in Chs. 12 and 13.

14. See, e.g., Creech (1994) etc. for—dangerous—practical consequences of one-sided thinking in organizations. The crisis of 2008- is a clear global economic and political example of them.

15. Engels wrote his "Dialectic of Nature" in the 1870s, but the book was published only in 1925 in Stalin's Moscow (Ziherl, in Foreword to the Slovenian edition, 1953). Engels expressed a big respect to Hegel and his creation of dialectic as the only author after Aristotle, but he wondered why had Hegel represented the view, that the spirit, idea, thought, is primary, and the real world is only a footprint of the idea, while using very many real-life cases from nature and history, which demonstrate that ideas depend on life conditions and reflect them as facts rather than vice versa (ibid.: esp. 46-67); awareness is the highest form of natural, i.e., material facts/processes. Engels did not defend, but attacked the "vulgar materialism" transferring laws that science has discovered in nature mechanically on society—as "social Darwinism" all way to "racist theories of contemporary imperialism" (ibid.: 7-9). This means that the notion of thesis, antithesis and synthesis does not cover thinking only, but exists in nature and technical artifacts and society, too. On p. 70 Engels ascribes to Hegel even—if we use the language of systems theory—the awareness of the notion of synergy as a natural fact/process. Thus, Engels did not attack, but completed Hegel.

of the notion of organization to the issues of power and bosses' control over subordinates' motivation and work processes (Schmidt, 1993). It had the form of organizational schemes stressing hierarchy of the commanding ("higher") position rather than hierarchy of processes (sequences, following order of process steps, succession) and interdependence inside them and between them. Subordinates were visibly dependent on their bosses, who could hire and fire them. But bosses hired them, when they needed their help, work. In order to give them good orders, bosses also needed data, messages and information from their subordinates. Interdependence has been around, but bosses preferred to see it as a one-way dependence, which has been only the legal viewpoint, actually. It is necessary, but not enough, to match the Bertalanffian concept of organization.

Subordination has also been practiced for millennia since humankind's transition from the nomadic economy to the settled agricultural one. This innovation enabled many more humans to survive, but it also made humans (wrongly) feel as bosses over the nature rather than as an equal and interdependent part of nature. Neglecting, as well as a lack of, data about interdependencies in nature, including humans, has been practiced for millennia. This may have been less of a problem as long as humans were much less numerous, production and wastes were much smaller, and nature was still able to digest it all. The industrial period on the basis of free entrepreneurship made people in the West much richer by innovation (Rosenberg & Birdzell, 1986), and nature much more unable to digest all the waste. One-sided innovation caused soil to become less fruitful, air and water etc. to become sources of illnesses such as pest, cancer, etc., although humans aimed at the opposite outcome. But relations between causes and consequences, inputs and outputs, have been hard to see, especially all the so-called side-effects. Only later on, recently, such data were offered by Gaia, even much after the main bulk of the work had been done by Bertalanffy; Gaia has been reaching some understanding and broader publication only recently (Myers, 1991). The United Nations conference in Rio de Janeiro, 1992, claimed the concept of "sustainable development" and thus required interdependence of humans and other parts of nature to be considered better, i.e., more holistically. All later conferences, though, gave poor outcomes so far. The concept of division of humans and "natural environment" prevails over the concept of "co-world"[16].

---

16. Pless (1998) speaks of "Mitwelt" rather than "Umwelt". She is much closer to Bertalanffy than other authors. Božičnik et al. (2008), Ečimovič et al. (2007) speak of sustainable future, not of sustainable development only, while the latter should be only a means to sustainable

Similar is the prevailing of the contemporary disregard for interdependencies in relations of humans to humans. What happens is a reduction of, e.g., the notion of democracy from an organizational form enabling (interdisciplinary) cooperation, creativity, and coming closer to (requisite) holism, to a tool of one-sided control over the governmental relations on the level of political entities, only/mostly. Why democracy as a concept of equal rights and duties should be reduced to the general part of concerns of a society and forget about relations in all other entities? Democracy provides for a possibility for all members of an organization to make an impact, indirectly and synergistically, at least. From the Bertalanffian viewpoint, this would mean an important chance for everybody to be a responsible citizen rather than a passive member of the unit under consideration. Networks of friends function much better than hierarchical subordinations, experience has demonstrated. Hierarchical subordination surfaced millennia ago, when they were building pyramids as graves of the pharaohs in Egypt (Schmidt, 1993). It was an innovation, then, giving the broader professionals the upper hand over the narrower ones. It also provided for a type of motivation. Both reasons have survived until the very contemporary times. Scientific management (Taylor, 1967) resulted, but in times when there were many less highly educated and trained persons around than today, in the advanced part of the world, at least. In the market, businesses, other organizations, families, schools, local communities, democracy makes much more sense than hierarchy as somebody's right to order others to work for his or her own benefit[17]. But it goes against the grain of those persons who forget about their own specialization and resulting interdependencies. They cannot be expected to care for a Bertalanffian role as citizens of the entire world, unless awarded for this role and punished for the opposite behavior.

Our recent analysis of what is offered under the concept of "system" in the "systems thinking/science/theory and cybernetics community" lets us see quite clearly (again) that very few authors tend:

---

future.

17. Private ownership is found the most efficient, today. Rightly, in economic data. But which one? There are at least two concepts. The ancient Roman definition says that ownership is the owners' right to use and abuse. There is no responsibility, which has surfaced later on in one part of the world (see: Uršič, 1996). The crisis of 2008– shows a lot of abuse is around both in the private and public ownership practices—due to a lack of requisitely holistic behavior, expressed e.g., as social responsibility

- To define explicitly what they mean by a "system"[18], despite or because of its long list of meanings;
- To work in interdisciplinary teams rather than individually and inside their own specialization (see contributions in Trappl (ed.), 2008 and earlier, since 1972).

What results, is the oversight of the crucial role of the selected (system/s or even dialectical system/s of) viewpoint/s in the definition,

- Which parts of reality are considered, and what about their interdependencies with other parts of reality;
- Which parts of the really given systems of attributes of the feature/object/thing/event/process under consideration are found worth consideration, and which ones are less/not so.

This is actually the basis of reductionism. This would be less of a problem, if it did not happen under the name of systemic thinking, frequently even explicitly under the name of the GST (see, e.g., ISSS conferences). It happens for practical reasons, summarized as the conferences prevalence of the real complexity and complicatedness over the natural human capacities to see and think holistically, especially in terms of Bertalanffy. See Table 12.

The consequences are oversights, disappointing surprises, several kinds of blindness (Oshry, 1996). The crisis of 2008- is an example of them, being only the top of the iceberg called one-sidedness rather than requisite holism of behavior of decisive persons/organizations/bodies.

What might be worth consideration here, last but not least, might be the issue: "What comes first, analysis or synthesis?" We have put this question very many times over e.g., Mulej's nearly forty years of dealing with systems theory and applying it to innovation management, including consulting. The usual response is: "Analysis comes first; synthesis follows it to summarize its findings." This is not hard to prove. But there is another question: "Why do

---

18. One can easily be confused by the very different meanings of the word *system*. It may mean an existing object, a mental and/or emotional representation of such an object from a selected viewpoint, a method, a usual way of working, a defined order, a mathematical ordered set, a complex feature, an emerging feature, a synergetic feature, a round-off entity of any contents, a tool or aide, etc. Therefore, a clear, exact and explicit definition of the selected content is necessary, whenever the word system is used. See Table 7 and Ch. 12.3.7.1.

different individuals / groups reach different conclusions when analyzing the same situation / event / process / thing / individual / group / object / feature...?" There is the crucial role of the fact that every analysis does not only lead to a synthesis (of conclusions, results), but also depends on a synthesis (of the starting points of the analysis), which causes the orientation of analysis to be either more holistic or more reductionist or less so. Thus, there are two types of synthesis, in a schematic approach. The one before analysis is, of course, impacted by the synthesis of results from a foregoing analysis, which has also been based on synthesis of the starting points for the analysis under consideration. This means, that Bertalanffy was very right, when he wanted GST to be a new worldview and a new methodology, both at the same time[19].

3.4 CONCLUDING REMARKS

We better stop here and make a conclusion from the collected insights. It reads:

Bertalanffy's concept is very necessary to humankind, but both, the understanding and the implementation of it depend on humans, especially on their/our capacity to think, decide, and act requisitely holistically.

But people learned to live in another world, the one of reductionism.

---

19. See Mulej, on DST, subjective and objective starting points, 1974 and later, only a few years after Bertalanffy's book, which was not known or used in Yugoslavia then. Bertalanffy's revised edition of 1979 was published at the same time as Mulej's book on DST (Mulej, 1979) which sold in 3,000 copies in a nation of two million. It was a bestseller among professional books of the time. See Ch. 12.

Chapter 4

REDUCTION, REDUCTIONISM, SPECIALIZATION, AND PROBLEMS OF COOPERATION

The whole and interdependencies, what does this mean? This is what many ask. Why? For millennia, humans have been trying to assure their survival, like all living creatures do. The difference is, that humans happened to become creative, to start and continue to change the parts of nature in which they were living. Sometimes they were successful and sometimes they were not, on their own criteria. Thus, they have learned from their own and others' experience. They learned more and more. Eventually and gradually, they came to be overwhelmed with the bulky quantity of insights available.

The humans' way of solving this problem was and still is obvious: every person has unavoidably to specialize in a selected part of the entire bulk of humankind's knowledge. This specialization, in the next step, unavoidably causes reduction of the entire quantity of insights into a rather manageable one. Manageability of one's life has been very appealing; it helped and helps person/s and group/s produce insights and outcomes, which they deemed necessary.

Thus, the next step was to make the reduction the right way of thinking and acting, in the human attitudes/VCEN. Reductionism became the dominating school about the way of thinking, especially in research, science, but also in the real life practice. It was very helpful, and still is so, when one tries to discover details, which are obviously very important for humans to understand and manage abilities and life. By it, mystery has come more and more to be mastery (see Flood, 1999), which was found great, of course, and still is so. But limitation to a single viewpoint, such as a single profession, place of living, or job causes the mental picture of the existing reality to overlook many attributes, which one could see, if one selected another viewpoint; e.g. in hospital a medical doctor focuses on other attributes than an economist, lawyer, engineer, cleaner, or nurse. Nobody sees the really existing whole, therefore mystery becomes mastery inside the selected limits only, and the whole is mastered fictitiously rather than holistically, or requisitely holistically, at least.

But: What happens to the whole, if everybody considers a selected part of details only? The whole still exists. And it has attributes, which make it different from every individual detail, part, e.g., a watch is different from each and every of its parts, even from their sum. It has—as one case of a whole—its own attributes, which are not produced by its parts alone, but by their interdependencies, essential relations: every part of the watch needs other parts of the watch and is needed by them for the watch to be a watch. Another case: the edible salt is made of sodium and chlorine, which are two poisons, but no longer so in synergy called edible salt. Besides this: a watch or edible salt have different attributes focused on, if an engineer, an economist, or a lawyer deals with them, and nobody covers all of them and can hence never be holistic in his or her behavior. In addition, specialists tend to fail to see their interdependencies with other, different, specialists. They may use the word system, but as a mental picture of a selected part of all real attributes this system cannot be equal to the whole object to be dealt with (see Tables 1, 2 and 8 again).

But can we individuals, being various and different specialists, consider interdependencies and the outcomes of their mutual influences, if we do not cooperate? We cannot, because there are several thousand professions and we tend to master somehow one of them, normally.

How much have you been learning about cooperation, especially about cooperation with persons, who differ from you in their knowledge and values, so far? Not much, if you have been in a usual school. What is the consequence?

Chapter 5

COMPLEXITY, COMPLICATEDNESS, RELATIONS, EMERGENCE, SYNERGY VS. SIMPLICITY AND LOCKED-IN THINKING AND ACTING

The consequence of the lack of cooperation, especially of an interdisciplinary one, is oversights, several kinds of blindness: humans do not see the real reality[1] (Oshry, 1996). Specialization without cooperation beyond the borders of that specialization locks humans in their own arena. If we are e.g., economists, we are—for obvious natural reasons, limits of time and capability—not able to think and act in the role, and from viewpoints, of mechanical engineers, medical doctors, cooks, cleaners, unless we learn another skill. Nobody can learn all professions and skills, which are around these days. So, we tend to go on specializing and getting more and more locked in our own cage in our behavior. In this we achieve, that the reality around us seems to be simple enough to be manageable with our own skill.

---

1. The real/objective reality exists, but humans construct—in their thinking and feeling rather than in 'reality out there' as it is—their own imagined 'realities' by setting boundaries and excluding attributes of reality beyond these boundaries; thus, they tend to live in a fictitious, one-sided 'reality', and hence experience (often unpleasant or detrimental) surprises, because the overlooked attributes of reality keep existing and influencing, although they are not considered.

But: Is this simplicity true or false? It is locked in a single viewpoint (see Tables 1, 2, 7, 8). In the real world, we soon discover that other people think and act differently, quite frequently. This neither means that we are wrong, nor does it mean that they are wrong; but we (and they) are just over-simplifying: we tend to call objects that exist, systems (i.e., wholes), while our mental processes cannot provide for holism, but focus on single selected parts of attributes of objects rather than on whole objects, i.e., on all their attributes. So are other specialists, but from other viewpoints. Nobody is totally right. The real life is much more complex and complicated. It is complicated in terms of the huge amount of attributes, which exist inside single parts of reality, and it is complex in terms of attributes, which are caused by relations between parts of reality.[2] (Schiemenz, in Mulej et al., 2000). Reality is not as simple as it seems to be, if we consider it from a single viewpoint alone and forget about relations, emerging attitudes and synergies. Everything and everybody exists in relations, not alone.

This means: within a single viewpoint (e.g., economic) and a single viewpoint-system (e.g., focusing only on economic attributes of a watch or a hospital) one tends to over-simplify and hence see complicatedness only, rather than complexity as well (see Tables 1, 2, 7, 8).

This unavoidable and natural limitation has made humans forget about over-simplification too much. It is herein, that the cause for most contemporary human problems lies: we humans tend to over-simplify instead of thinking in systems as wholes (rather than fictitious wholes, limited to selected viewpoints), interconnectedness based on interdependencies between several specialized disciplines; we tend to forget about real complications and complexities. And then we imagine, that we are mastering our mysteries and managing the unmanageable (rather than with it), but we hardly know of the unknowable (Flood, 1999). And persons, who try to teach us to consider the reality without simplification, are difficult to understand and accept (Oshry, 1996). As a result, system thinking has been created, but it remained poorly used, since it did not help humans to simplify, but rather forced them to face complex reality (Molander & Sisavic, 1994).

---

2. E.g.: parts of a watch are parts of reality. Their relations of a specific type, such as technology, economy, law, physics, chemistry, sociology, advertising, etc., may make the watch *emerge as a new whole and a part of a bigger whole*, which has new attributes, i.e., a case of *synergy*. The same insight is true of what ever.

So far, knowledge has tended to be developed and used for humans to master their problems within specializations. What knowledge can help humans overcome this blind alley?

Chapter 6

KNOWLEDGE, INFORMATION, PROFESSIONS, ORDER

Humankind has been developing for about one hundred thousand years (Ečimovič *et al.*, 2007). Life has been more and more complex rather than simpler, be it in terms of biology, in terms of economy, in terms of sociology, in terms of technology, in terms of communication and languages. All these "terms"—viewpoints have led to specialized parts of knowledge, which humankind has developed over all those millennia, and especially in the recent decades.

This development is a response to the fact that life has been increasingly interesting and difficult to understand and master. New and new data, messages, and resulting information have been added; this at least has been the intention. What actually has happened and still is happening is production of new data rather than messages and information.

Data are made when signs are put in an order, e.g., letters in a word, single sounds in music. They exist and wait for somebody to understand them. Once this happens, the person receives a message. Seeing a word, hearing a piece of music, deciphering attributes of a stone, or a plant, may make a message from data, if data's meaning, content becomes clear. This is still no information, as long as it still fails to make impact, causing an action, such as memorizing of the message, linking its meaning with some other messages and their meanings into a newly emerging synergy of messages, resulting in new understandings, new behaviors. Influential data and messages only may be called information (see Rosicky, in Mulej *et al.*, 2000).

Data, messages, and information—all of them contain some order, thus making the reality easier to manage, simpler. As long as the usual order reoccurs, and so do its circumstances, conditions, and preconditions, knowledge, data, messages, information are all helpful—unless they are too one-sided, partial rather than requisitely holistic. Contemporary professions tend to be one-sided rather than holistic, unless professionals develop the habit of systems thinking, which reaches beyond the limits of their own professions, in the interdisciplinary context and related interdisciplinary creative cooperation.

This tendency means: knowledge alone is not enough; VCEN all way to worldviews matter, too. Requisite holism depends on both, knowledge and values/VCEN. So do all sciences, but they mostly tend to limit themselves to knowledge, which causes one more type of their one-sidedness.

Systems theory as the science of synthesis aimed at holism becomes necessary to help humans fight their one-sidedness (Hammond, 2003). Synthesis results from building bridges between specialists of different kinds and with different contents due to different specializations.

We will come back to the topic of information later. Now we will try to answer a dilemma arising from findings acquired so far: What is systems thinking as a capacity and systems theory as a type of knowledge aimed at?

Chapter 7

SYSTEMS THEORIES—TOOLS OF HUMAN ACTION AND/OR HUMAN FORMATION

Forgetting about life as the context outside one's own specialization and selected viewpoint is very easy to do. Humans are all specialists in small parts of reality; all other parts of this reality make all humans strangers almost everywhere. But we saw in reality and in the cases offered as examples here, that parts of reality matter, and that they are interdependent with other parts of reality. This means that context matters even more than parts alone[1].

The growth of, and closeness in, humans' specialization caused humans to often forget about contexts of their own lives, actions, specialties, views, opinions, and experiences. Is it not interesting, that systems theory, as a theory of considering the wholes, has surfaced briefly after a few decades in the 20th century, in which:

- Humankind's knowledge has been growing tremendously, and has been causing an increasingly narrow specialization into single parts of knowledge, with very rare and poorly developed habits of interdisciplinary cooperation;
- Humankind suffered from the biggest crises ever, having the form of two World Wars and a world wide economic crisis between them?

---

1. If you see a lion or another wild animal, does it matter, whether it is on the other side of a strong fence or on your side? It certainly does. You better know the context, too, not only an isolated fact.

The combination of a lot of knowledge and a lot of one-sidedness has proved to tend to be very dangerous.

For example, in World War II, Hitler was using tremendous amounts of knowledge, tremendous amounts of order, tremendous amounts of products, innovation, science and its application as technology, but he obviously was not using systems thinking sufficiently. Systems theory surfaced later on. It was a response to mistakes of humankind, such as letting such a one-sided person come to a position with such general and crucial influences. His values were refused, although too late for many millions of humans.

Hitler's ideas were obviously very different from the ones of Bertalanffy, which we have quoted earlier. So have been ideas of many others, over the history of humankind. Even today persons of Hitler's kind are rather easy to detect among politicians, entrepreneurs, managers, educators, everywhere where people can enjoy power and resulting benefits. And their success is often poor (Ackoff & Rovin, 2003). But: today systems theory is already around, and it has been so for beyond half a century. What happened? Why is it not used more?

Bertalanffy felt the need to find a way out of the blind alley of the combination of a lot of knowledge and a lot of one-sidedness. For this reason, we feel, he wanted to innovate the humans' worldview: we should all see the entire world and its entire biosphere as our only home, a very complex organization, i.e. a whole with many interdependencies. This worldview is what he called the general systems teaching[2]. What many other authors call the general systems is not so much a worldview, but rather a methodology of its own. It transfers knowledge, data, messages and information from one discipline to another discipline (see, e.g., Frank, 1962), but leaves disciplines locked in behind their own bars. It does not create interdisciplinary cooperation, but rather stays with formal bridges between disciplines. They are called isomorphisms, which mean attributes and tools of consideration, which one can apply in several different disciplines[3].

---

[2]. We see an important difference between notions of systems theory and systems teaching: teaching includes impact, while theory only offers generalized insight into a selected part of reality. Bertalanffy wanted to impact humans!

[3]. The mathematical language is a very good case, and a very useful one, as long as the mathematical formulation does not become more important than the real complexity and real differences in viewpoints. E.g., inside economics and business, some relations are suitable for a useful application of mathematics, others are not so. Finances are much more so than

From the development of systems theories, which we cannot consider in any detail here[4], we can see that von Bertalanffy has tried to create systems teaching (translated into English as theory) as a tool of human formation. He tried to help the world to survive by establishing a broad (rather than narrow-minded) worldview, a holistic one, by this theory (which he called teaching, originally, with a clear reason). Other authors and practitioners mostly did not follow him, and they changed systems theories into tools of human action for anyone to use as they like.

They achieved many important results within single disciplines, which became more holistic, as well as facing and clarifying more complexity and complicatedness, in their own field/s and related viewpoint/s. Less has so far been achieved in interdisciplinary cooperation and hence in consideration of broader contexts, although this would bring humankind closer to von Bertalanffy's and Elohim's as well as our own:

Warning: "we are ruining more than creating a solution for humankind and the planet Earth". The cause is failure to use systems/holistic thinking in broad contexts, with ethics of interdependence in mind, and attaining the quite high requisite holism. Limitation of single specialized disciplines behind their selected boundaries tends to leave uncovered areas beyond these boundaries.

Environmentalists, meteorologists and many others share the same warning, even the highest political bodies of humankind do—the United Nations Organization. But they keep staying short when facing the rather narrow and concrete interests (see, e.g., our contributions in English in: Ečimovič *et al.*, 2002, 2007, 2008 and feel free to download them.) Hence:

Human formation is prerequisite for human's humane use of systems theories as tools of human action and a more holistic and therefore beneficial use of traditional disciplines, professions, skills, and experience-based capabilities.

Humankind keeps struggling between two kinds of its own interests: the broader and longer-term ones versus the narrower and shorter-term ones. We might call this struggle the dilemma of holism versus one-sidedness, perhaps. Let us see. It may help us understand the essence of the contents of the

---

marketing relations or labor relations. Etc. Engineers and medical doctors were able to create many important devices to help disabled persons on the basis of isomorphisms, but in interdisciplinary creative cooperation rather than in separation. (Likar *et al.*, 2006).

4. See (Hammond, 2003) for it.

Tables 1, 2 and 8 better and to direct systems theory towards the formation of humans' humane attributes.

Chapter 8

HOLISM VERSUS ONE-SIDEDNESS AND OVERSIGHT: REQUISITE HOLISM

Holism is an easy word to use, as long as we do not try to exactly define it (see Tables 1 and 2 again; add Table 8). Different authors tend to have different definitions (see Checkland, 1981; Delgado & Banathy, 1993; Dyck & Mulej, 1998; Flood, 1999; François, 1992; Frank, 1962; Jackson, 1991; Miege & Mahieux (eds.), 1989; Mulej, 1979, 2007; Mulej et al., 1992, 2000; Schiemenz, 1994; Wilby, 2005), implicitly or explicitly.

In a strict sense, a whole contains everything, all parts, all their relations, all viewpoints and all their relations, and resulting synergies. But: which—everything?

Experience demonstrates, as we already have noticed, that humans do not seem to be either able or willing to think in the breadth, which von Bertalanffy goes for. To specialists of other disciplines the entire biosphere does not mean as much as it does to von Bertalanffy, who was a biologist, art historian, and philosopher.

It turns out that everybody feels entitled to define what is a whole to them on their own criteria.

What can we do? Let us repeat first. The mathematical basis, which was introduced for some help, said that a whole (= system) is an ordered set, which means that it is made of two sets, the set of elements (parts) and the set of relations (interdependencies included). This is a generally valid isomorphism. It serves very well, as long as contents are less important than the mathematical basis for description of the "system". Then we see, that the same piece of reality can be described with many different "systems", i.e., as many different "wholes".[1] If wholes can be so many and so different in concrete contents, when is the behavior holistic, and when is it one-sided?

If the Bertalanffy's concept of GST has not taken roots in the real practice of the contemporary humankind with all its many and diverse specialists, which one has?

None did with a general validity, as the conferences about systems thinking and theory as well as their application let us find out. Specialization is too important to be forgotten about and sacrificed to a kind of holism, which might tend to leave specialization aside. But, it is neither necessary to sacrifice specialization nor enough to be specialized; specialization is unavoidable and it causes also many oversights, because each and every specialization can cover only a single viewpoint or a few of them, at best, while other attributes remain outside the scene and the screen. (But they do no stay outside reality and influence!)

Consequences include the dilemma:

1. Shall we have a complex approach, which causes quite a lot of work for insights to attain requisite wholeness and provide for a realistic basis for behavior? This is difficult, but this work may predominantly result in outcomes, which meet requisite wholism and do not cause unpleasant side-effects and surprises; in other words: a complex approach and work have simple consequences.
2. Shall we have a simplified and simplifying approach, which does not cause lots of work for behavior to seem to be rather holistic, but to be most probably not requisitely holistic? This is easy, but may predominantly

---

1. Let us return to the simple case of the watch. One viewpoint can be the technology to produce each of its parts. Another viewpoint can be assembling the watch from parts in terms of technology, another in terms of economics, another in terms of supply, another in terms of organizing, etc. Several more viewpoints can have to do with legal affairs, marketing, human resources, accounting, safeguarding, patents and similar rights, etc. Last, but not least: there can also be the issue of interlinking all those aspects for the whole business and use of the watch to emerge as their synergy. Another case: the mathematical formula $a=b+c.x$ means what ever in mathematical theory; it expresses the general model of cost in economics, and the one of lever in civil engineering. Hence: the *selected viewpoint must always be explicitly stated*.

result in serious oversights, because its outcomes cannot meet requisite wholeness; in other words: a (too) simple approach can have complex consequences.[2]

Obviously, in each and every case a decision has to be made on: Which level of holism is good enough to solve the above dilemma well enough? One should avoid both exaggerations (Tables 2 and 8):

- The total holism, trying to include all attributes from all viewpoints, and interlink all viewpoints into the system of all viewpoints with no selection; the result may be a lack of focus and hence a lack of response, which outcomes of behavior are essential; besides, this reaches beyond the human capacity;
- The fictitious holism, trying to limit the concentration to one single viewpoint, which might be too much of a selection of a narrow kind; the result may be a good focus, leading to a lot of knowledge about nearly nothing.

A middle way is close enough to reality in terms of both, need and possibilities. This is what we called (Mulej & Kajzer, 1998b) the requisite holism. It turned out to be a law: successful persons and organizations (tacitly) live with this law (see Ženko, 1999). See Table 8.

What is the point of Tables 2, 7 and 8? Reality has many attributes, and they are many more than thinking about reality, be it tangible or intangible, can cover (Table 12). We are trying to condense reality and thinking in four interdependent groups of attributes (Table 1). Every entity under consideration has them, and every researcher, manager, or customer should better consider them in order to avoid oversights, the crucial ones at least, by attainment of requisite holism of approach and requisite wholeness of outcomes.

What is the way to meet the law of requisite holism (Tables 1, 2, 7 and 8)? Rarely, a single person can know, and have time, enough to meet it, perhaps in the case of a very simple activity, such as a job on the assembly line, or a partial book keeping; such jobs are less and less around, as quoted earlier (Florida, 2005). More normally, several specialists will need to cooperate as representatives of several viewpoints, e.g., professions different from each other. But: What is the basis for them to agree to enter into such a cooperation,

---

2. A clear and dangerous case of over-simplification today is the climate change; references were mentioned earlier. So is the 2008– crisis. Etc.

which is much less simple to do than working inside a single discipline and alone? See Table 7 and go to Chs. 9, 10 and 11.

APPROACH TO DEALING WITH AN OBJECT AS A TOPIC OF BEHAVIOR	One-sidedness by a single viewpoint	Requisite holism by cooperation of all essential professionals and only them	Total holism by consideration of totally all viewpoints, insights from all of them and synergies of all of them
TYPE OF APPROACH	(Too) simple	Requisitely simple	Very entangled
TYPE OF SYSTEM AS A MENTAL PICTURE OF THE OBJECT DEALT WITH	Single-viewpoint based system	Dialectical system	Total system
ATTRIBUTES OF OBJECT INCLUDED IN SYSTEM	(Very) few	All essential	All
RESULT OF APPROACH	Fictitious holism (in most cases)	Requisite holism (good in most cases)	Total holism (ideal)
FOCUS MADE POSSIBLE	(Too) Narrow focus (in most cases)	Requisitely holistic focus	Lack of focus
NUMBER OF PROFESSIONS	One single	Requisitely many	Literally all
TYPE OF WORK	Individual	Mixed team of requisite and different experts	All humankind in cooperation/synergy
CONSEQUENCES	Complex due to crucial oversights, dangerous	No problem due to no crucial oversights	Simple due to no oversights
AVAILABILITY	(Too) Frequent in real life	Possible in real life	Not possible in real life

Table 8 *Law of requisite holism (Table 2) in some details*

Chapter 9

THE BASIS FOR (REQUISITE) HOLISM TO BE ATTAINED: INTERDEPENDENCE AND ETHICS/VCEN OF INTERDEPENDENCE

9.0 THE SELECTED PROBLEMS AND VIEWPOINT OF DEALING WITH IT IN CHAPTER 9

We have mentioned requisite holism/wholeness and ethics of interdependence frequently so far. It is time to devote them some more attention.

9.1 ADAM SMITH WAS MISREAD

Humankind has been taught for millennia to act as the proverbs say:

- Help your-self on your own, and God will help you.
- Depend on yourself and your own horse.

It means: you neither care for anybody nor depend on anybody! Be as selfish as everybody else is! Be independent rather than dependent! Interdependence is not mentioned much.

This notion entered the economic theory with the sentence of the theorist of the political economy of the (early!) market economy Adam Smith that is still repeated all the time, today:

Everybody is supposed to care for his or her self-interests only. The invisible hand (= the market forces) alone takes care of the common good of the entire society, since it praises the ones, who please their customers/buyers better than the others do, and punishes the ones, who do not do so.

But: if you read carefully, you see that the Smith's sentence says something else than the two proverbs; it seems to have not been well understood. It says that the supplier takes good care of his own interest only, when he or she takes into account that he or she is not alone neither independent, but interdependent: he or she needs his or her customers (to cover cost and make profit and to go on working), they need him or her for goods he or she offers. The same relation of need for each other exists between the managers, entrepreneurs, owners, and employees, between professionals of different professions and offices inside the same firms or other organizations or inside different ones. Why? That's why:

Everybody is specialized into a small part of the entire stock of knowledge and skills; this capacity makes him or her needed by others; the same limitations make him or her need others who are capable of something else.

In a summary, what is natural is interdependence rather than independence or dependence. Power holders tend to teach us another story. They consider the life from another viewpoint: Who may and can impose something on the others, and who may and can take something away from the others, because they depend more than the power holder does (or feels so)? The same situation is expressed in economics with the theory of bargaining power in the supply and demand process setting prices, i.e., in market.

What is interdependence and its good consequences, is expressed by other proverbs, such as (see Ilich, 2004, for more proverbs)?

1. Hands wash each other, and they wash the face together.
2. Union is strength.
3. Single straws can be broken easily, a bundle cannot.

Once interdependence exists and works, we cooperate and attain an effect reaching beyond the sum of single effects of single components—synergy results by emergence. So is it in nature, in machines, in human beings and groups and organizations of all kinds and sizes.

The problem arises with the natural difference between the existence of synergies resulting from interdependencies, on one hand, and the narrowly educated specialists tending to make oversights of their own interdependencies with other specialists. The traditional education of specialists by specialists did and does not provide for ethics of interdependence (Kajzer & Mulej, 1997, published in 1999; Mulej & Kajzer, 1998a, b), partly because there has until half a century ago been a lack of systems thinking and no systems theory. Now, systems theories exist, but many influential people still do not care for them or for requisite holism. This oversight is a crucial cause of all crises, including the one of 2008-. How important this oversight is in the simple daily life too, can be well seen from the following example by Balle (1994):

If we wanted to train a horse to run better and we used the traditional methods instead of the [requisitely] holistic thinking and theory, we would try to train each leg extra. We might teach them something, but cooperation of the legs among themselves as well as with other parts of the body would be forgotten about. The outcome would, most probably, not meet expectations.

The same would apply e.g., to innovation: all twelve groups of preconditions briefed in Table 9 and discussed in this book in Ch. 14 must be considered interdependent and resulting in synergies (this is why their relation is denoted with X (= multiplication): none may be zero for the entire outcome to be more than zero).

Innovation = (invention × entrepreneurship and entrepreneurial spirit × requisite holism × management × co-workers × innovation-friendly culture × customers × competitors × suppliers × natural environment × socio-economic environment and other outer, i.e. objective conditions × random factors, such as luck)

Table 9 *Equation of preconditions of innovation: A case of a dialectical system*

But people used to live in self-sufficient farms and small communities with little innovation and a rather small degree of specialization and division of labor for 100,000 years except the very recent short time of 1-3 centuries. This fact did not teach them to consider specifically the great importance of interdependence except within their own family and village, perhaps. It was natural. Life was more difficult than today in terms of the modern standard of living, but simpler in other aspects.[1] There was much more self-sufficiency.[2] There was much less need for knowledge and products of strangers.

A superficial reading of A. Smith's sentence quoted above as well as of the proverbs mentioned above made one part of the story. Its second part was the experience that strangers might be dangerous. Hence, if we enter some links or cooperations, we better choose people/s similar or equal to ourselves, rather than the ones who are different[3].

Of course, it is true that we cannot live and work together with those from whom we are different in all our attributes. But it is also true, that we do not need each other, if we are equal in all our attributes. In the political part of human lives this natural fact is expressed by the experience that all unanimities (religious, political tyrannies) have failed, but democracies (accepting and growing the intertwining of both, equal and different attributes) have survived: they have permitted for more creativity and holism, because they admitted differences and interdependencies and interactions. This blocks one-sidedness and resulting oversights.[4]

---

1. We hardly feel that in old times there has been research finding that, in the world, one billion people are too fat and close to one billion are starving; this was a big message in daily journals in November 2006 and later.

2. Today one speaks of a global economy. Data in the foregoing footnote clarify differences in it to some extent. They are consequences of the Industrial Revolution and its transition into the innovative society and economy, which has tackled various parts of the world to very different extents. Agnus Magnuson, an economic historian, is quoted to provide data that for long eighteen centuries India and China, until around 1820, used to present four fifths of the world economy. Then the Industrial Revolution left them behind. Now they are catching up. (Bošković, 2006). Not all countries are. The difference between the richest and the poorest country of the world, measured in national income per inhabitant, has grown from 3:1 around 1870 to 150:1 in 1970, and to +500:1 today. The most innovative countries are at the top, the most traditional and self-sufficient ones at the bottom (World Bank data, in Mulej, 2006a). On December 10, 2006, Radio Slovenia reported that 1% of people own 50%, and 10% own 85% of the world's property, while the poor 50% own one single percent. The current "feudal capitalism" needs to be innovated for market to be less fictitious and in trouble.

3. How many interdisciplinary professional associations do you know, e.g.? With whom do you find a debate easier to conduct: with persons of your own profession, life style, experience, age, interest, views, values, ethics, etc., or with different ones? But, on the other hand: from whom can you learn more, whom do you need more in a creativity-and-innovation oriented process—the equal or the different ones?

4. This is why we quoted R. Florida's (2005) finding that the 3T lead to success: tolerance, talents, technology. But let us add: democracy in which the number of votes counts more than requisitely holistic proofs/arguments do, is fictitious, not requisitely holistic.

The (General) Systems Theory was produced against over-specialization, hence to support the blocking of such one-sidedness and to help people prevent such dangerous oversights. But: Is systems theory applied for this aim today?

9.2 GENERAL SYSTEMS THEORY—SHORT OF ETHICS OF INTERDEPENDENCE (IN PRACTICE)

The General Systems Theory (GST), the about six decades old original version of systems theory, has tried to introduce a common basis for all disciplines of science by introduction of the concept of isomorphisms (Davidson, 1983; Jackson, 1991). This effort produces a formal bridge between different professions, which has had the form of a unified vocabulary (Delgado & Banathy (eds.), 1993; François, 1992 and later, including 2004; Frank, 1962; Kukoleča, 1969). It is useful as a partial basis of system thinking and its theoretical generalization in systems theory, but less so than a seriously and generally accepted methodology of interdisciplinary creative cooperation would be.[5]

The most known conferences on systems theory, such as EMCSR (Trappl (ed.), 2008, and earlier, since 1972), tend to rather keep systems theory inside single traditional fields of inquiry.[6] Only less than ten percent of papers to conferences on systems theory and cybernetics are written by interdisciplinary coauthors (Mulej et al., 2006). The same seems to be true of other conferences:

---

5. We know of only few international conferences of systems scientists with an explicit interdisciplinary orientation:

1. G. de Zeeuw's et al.'s "Problems of ..." which has been held since 1979 biannually in Amsterdam, NL, and it is there no more;
2. M. Rebernik's and M. Mulej's "STIQE" which has been held since 1992 biannually in Maribor, SI;
3. G. Chroust's, P. Doucek's et al.'s "IDIMT" which has been held since 1993 yearly in Czech Republic, earlier in Zadov, later in Budweis, now in Jindřichův Hradec, CZ.
4. GESI's rather traditional work chaired by Charles François in Buenos Aires and the newer FundArIngenio systemic group's in Santiago del Estero, titled Permanent Seminar on Transdisciplinarity, in Argentina.

6. One of the most prominent authors of systems theory as a mathematical discipline, our good friend Prof. Dr. Franz Pichler from Linz, Austria, even analyses an author who compares different traditional disciplines in terms of how much of systems thinking is required inside each of them. They both say no single word about interdisciplinary approach as an essential precondition of trans-disciplinary work as an essential precondition of holism and aim/topic of GST (see Pichler, 1997).

most of them cover a single discipline anyway. Why should there be a difference between the traditional and systemic contributions?

The background of systems thinking (such as Delgado in the 1950s, Churchman in the 1960s, Mulej in the 1970s—see Delgado & Banathy, (eds.), 1993) is dialectics (Table 10). Dialectics is the ancient Greek word for interdependence (Britovšek *et al.*, 1960). It is also the philosophical science about it and resulting processes of becoming rather than being (Engels, 1953: 17, 49-52, 66-72). This attribute shows up in unity and impact of opposites on each other called today attractors causing emergence of synergetic properties or transcendence, which makes systems complex entities made as synergies of their components. The process can happen because they are interdependent in their relations, i.e., in need of each other[7].

Some new authors are now joining us (such as Bai & Lindberg, 1998; Wu, 1996). But it is still unclear whether they are limiting what they call a contradiction, conflict, complementarities, or opposites and we call interdependence due to differences between components, inside a single profession, or put it in an interdisciplinary perspective, too.

If a narrow view is used, holism is not attained on the level of the aspiration of the fathers of the GST. They practiced requisite holism based on interdependence, because they found it necessary for the science to overcome the fragmentation into rather independent and unrelated disciplines of inquiry per parts only (Hammond, 2003). They did not attack specialization, neither do we attack it. It is unavoidable in contemporary conditions, but not enough.[8]

What they and we attack is a tendency of specialists to forget about their specialization and resulting limitation leaving aside many links among parts, which actually do exist in both, nature and culture, i.e., man-made products and society. Even more, those links are rightfully found essentially influential.[9] Working with no consideration of those links is dangerous, as one sees in the

---

7. Like man and woman, parts of body, parts of machine, business factors, business partners, parts of un-tackled nature existing in the same area, or even far away (see: Meyrs, 1991)—see Table 10.

8. De Bono (2005) says the same story this way: The left front tyre of a car is important, but not enough for the car to work.

9. It is exactly forgetting about them, which, to a high degree, causes the emergence of *strange attractors* instead of the expected ones. See literature about the Chaos Theory.

Interdependence (= background of any process: we need each other for complementary differences)	$\longrightarrow$	Unity and mutual impact of opposites / differences in attributes (= surfacing of interdependence); attractors*
$\uparrow$		$\downarrow$
Dialectical negation (= existence of a new quality after a phase of change); synergy	$\longleftarrow$	Process of changing from the given quality (= essential attributes) towards another one via quantitative (= small, less essential) changes; emergence

Table 10 *The permanent dialectical process in the language of both, dialectics and systems theory.*

* Bertalanffy (1950: 154) is very close to this statement: "Every whole is based upon the competition of its elements, and presupposes the 'struggle between parts' (Roux). The latter is a general principle of organization in simple physico-chemical systems as well as in organisms and social units, and it is, in the last resort, an expression of the coincidencia oppositorum that reality presents."

current situation of climate change, massive health problems, massive hunger problems, massive employment problems, massive discrimination problems, massive feeling of exploitation and response to this type of terror with another type of terror, the resulting crisis of 2008-. Ethics/VCEN of interdependence could provide for requisite holism of human behavior, if it was added to the unavoidable narrow specialization as an equally unavoidable attribute of humans.

The mainstream GST, as practiced nowadays, we may conclude, has proven useful in the traditional disciplines, but has not demonstrated much interest or capacity to really support the efforts aimed at the requisite holism (Tables 1, 2, 7, and 8) reaching beyond the rather partial or even fictitious holism limited to a single-discipline viewpoint. Thus, it is not too much of a wonder that a number of new systems theories have surfaced in the 1970s and later on (Jackson, 1991). Even more: many scientific disciplines and professions with a big prominence, such as automation, management, have forgotten that their roots have grown from Systems Theory and Cybernetics (Umpleby, 2005).

Here they are less interesting to us; we need to explain what we actually mean by ethics/VCEN of interdependence. We find it crucial for the real interdependence to be detected and felt normal as well as crucial and worth pursuing in practice, by experience.

9.3 ETHICS AND ETHICS OF INTERDEPENDENCE

Ethics is a feeling rather than a component of the left-brain rationality, knowledge or skill. It enables us to distinguish right from wrong (Enciklopedija, 1959; Koletnik, 1998; Mulej et al., 2006; Potočan & Mulej, 2003; Sruk, 1986; Schnaber, 1998). Empirical researchers consider ethics a synergy of behaviors, which tend to be preferred in a society or community, as a social group, for long enough periods of time to come to be kind of codified (Wilson, 1998). Moral rules result, as a formal next step (Sruk, 1999). They co-create a culture, be it the one of social sub-groups, organizational units, organizations as wholes, or the one region, nations, social classes, professions (Mesner Andolšek, 1995).

Thus, something, which is originally an individual characteristic, comes to be objectified as a component of the objective conditions (i.e., external to single individuals). It becomes a part of broader requirements imposed on the individuals, and tends to return, in this way, back to individuals as a part of their values, i.e., their emotional perception of the objective needs or requirements they face (see Trstenjak, 1981). Thus it enters (or re-enters) the individual's starting points, which influence perception, definition of preferences, their realization in the form of goals, later on tasks, procedures of realizing the tasks (Table 11).

Individual values (interdependent with knowledge)	→	Culture = values shared by many, habits making them a rounded-off social group
↑	×	↓
Norms = prescribed ethics on right and wrong in a social group	←	Ethics = prevailing culture about right and wrong in a social group

Table 11 *Interdependence of values, culture, ethics, and norms, i.e. VCEN*

It means that ethics is equally essential as professional knowledge and skills, for any human activity. We even found all three of them (this means: knowledge to the answer "What?" knowledge to the answer "How?" and values and other emotions embracing the non-rational components of the human personality) interdependent (see Table 19), and so did our colleagues (Mulej, 1974 and later).

Our experience, and quite probably your experience as well, tends to practically confirm the observation made by Wilson (1998) saying that researchers find that the ethical norms have been changing in an evolutionary process based on the interplay of biological and cultural factors. This finding may hold and explain many things, some of which may be quite relevant in the context of this contribution, such as:

Ethical principles of pre-industrial societies are based on their experience that the solidarity of the extended family (and community) helps them survive, and does so better than the ethics of the individualistic competitiveness of the industrial and post-industrial societies/communities. This is so because the (pre-industrial, non-western) ethics of solidarity tends not to forget about interdependence (see Fromm, 1994). It may be a source of the current Japanese troubles that "in Japan an individual finds no acceptance if he or she does not belong to a group", even in the very competitive times of nowadays (Ženko, 1999; Mulej & Ženko 1998).

It is well known that Adam Smith wrote his book "Theory of Moral Sentiments" (1759) before his "An Inquiry into the Nature and Causes of the Wealth of Nations" (1776), from which the sentence in Paragraph 9.1 above is taken. As a professor of ethics and moral he presupposed ethics of altruism would help people overcome their natural selfishness, which was and is making them forget about solidarity and interdependence, once they feel that a narrow individualism might help them better than solidarity. Today, altruism is no more appealing than it used to be to most people in A. Smith's times as well as in industrial and postindustrial capitalist times. But it can well be replaced, even in the hard, very competitive business world, by ethics/VCEN of interdependence which surfaces as creditworthiness, trustworthiness, credibility, reliability, and so on (Kajzer & Mulej, 1997; Knez-Riedl, 1998; Koletnik, 1998; Rozman & Kovač (eds.), 2005; Thomen, 1996). Partners, who are worth trust, cause much less cost of checking, control, and the like. (Palčič & Mulej, 1991). The contemporary efforts for more social responsibility are aimed to work in the same direction (Hrast et al. (eds.), 2006, 2007, 2008; Rozman & Kovač (eds.), 2006; Waddock & Bodwell, 2007); we will come back to it in Ch. 13.8.

Democracy expresses ethics of equal legal rights of all men and women, if the latter are covered by its definition of the entitled members/participants of the democratic processes. In antique Greece it covered adult healthy men slave-owning family-heads only. After the "Magna Charta Libertatum" feudalists and clergy came to be included. The French Revolution and other so called bourgeois revolutions brought the entrepreneurs, later on other professionals, and later on broader masses of both genders, in political democracy. All subordinates came to be included, once the previous power-holders started feeling too much interdependence between power-holders and subordinates to keep forgetting about them and letting them cause troubles. Nowadays, in the advanced parts of the world, at least, political democracy is completed up with economic, organizational, family, local community and similar kinds of democracy, in order to channel the human creativity away from causing too much trouble, etc. If ethics/VCEN of interdependence is overseen, and the one-sidedness enters the scene again, then a new kind of trouble surfaces. Then, democracy stops fighting for holism and becomes one-sided imposing of partial interests and insights, which it has been established against (Christakis & Bausch, 2006).[10]

Systems thinking may, on these terms, be said to be an expression of the ethics and habit of seeing the wholes rather then parts only (Oshry, 1996) and thus of their interdependence (Mulej, 1974, and later). Like democracy, systems thinking suits better the persons with broader horizons and bigger flexibility, as well as persons, organizations, nations and other communities living in rather complex entities and circumstances, rather than the ones with narrower horizons, poorer flexibility, and living in simpler conditions with less division of labor and resulting practical interdependence. It is so because interdependence is the background of complexity, open mindedness and flexibility, not only as a practice, but also as ethics or VCEN and applied perception (subconscious, frequently).

Once interdependence enters ethics, it enters sub-consciousness, which means that it becomes, like any other ethic, rather a master of humans' personality than our tool only (Mesner Andolšek, 1995). Then, it comes to be practiced automatically, as something very normal, unquestionable. This is why ethics/VCEN is so crucial (Table 11). And this is why it must include interdependence. This is better achieved, if there is a theory to respond to the need, not merely an experience, which is hard to transfer, or teach.

How bad was the 20th century in this perspective to give birth to the need for systems theory in 1950s?

---

10. In the current practice the number of voices of a party or coalition matters more then the power of arguments and their requisite wholeness, unfortunately.

9.4 LACK OF CONSIDERATION OF INTERDEPENDENCE IN 20TH CENTURY PRACTICE: CAUSING THE NEED FOR MAKING OF SYSTEMS THEORY, NOT ONLY SYSTEMS THINKING

Systems thinking has been (as a rare practice) around forever[11] (not so systems theory, of course), since the human capacity to influence had outgrown the human capacity to holistically understand the events, objects and/or processes under influencing (Table 12). Then (informal) systems thinking became necessary for survival, but could not be always attained. Sometimes humans happened to be requisitely holistic in their behavior to make a requisitely holistic impact rather than a mistake, but mostly they/we have not been so lucky and caused oversights causing new difficulties while solving some earlier ones (Ečimovič et al., 2002; Knez-Riedl, 1998a; Mlakar, 1998; Mulej et al., 1998; Mulej, N., 1998; Potočan & Mulej, 2003; Potočan & Rebernik, 1998).

In the 20th century, two world wars and the world wide economic crisis (in 1914-1945) made a series of events lasting far beyond one generation, i.e., beyond 35 years of permanent big trouble for very many humans, including a premature death of several ten million persons.[12] These times demonstrated how terrible the consequences may be, if influential humans lack systems thinking, be it in societal affairs, business, science, education, engineering, or other fields of life. So does the crisis of 2008-.

Human capacity of influencing > Human capacity of a holistic understanding

Table 12 *The difference in two crucial human capacities: The root of failures and of the need for systems theory supportive of systems thinking*

It can hardly be a coincidence only that it was exactly at the end of the 1914-1945 +30 years of culminating troubles when the efforts to create systems theory and cybernetics enjoyed support, which were aimed at overcoming the prevalence of the tunnel vision (i.e., one-sidedness, one-handedness, instead of holism) of human behavior. Conditions were ripe for (General) Systems Theory (GST) and Cybernetics to be born and established. Their intention

---

11. See chapter 11.3.10 about the Maya culture. One could also write a lot about the ancient Chinese culture expressing the systems thinking style by yin/yang (Delgado & Banathy, 1993) or the ancient Greek culture expressing it by philosophy of interdependence (Britovšek, 1960).

12. A documentary film on how the Nazis had treated Jews showed that only in Treblinka concentration camp Nazis had murdered close to one million people; only one percent spent in this concentration camp more than two hours.

and mission was to replace the tunnel vision with holism, i.e., to establish the science and ethics of the requisite holism. These words, though, do not seem to have been used (see Bertalanffy, 1950, 1979).

9.5 CONCLUSIONS FROM CHAPTER 9

It turned out that the practice of many GST users went too far into the concept of isomorphism based on natural sciences to be able to fulfill GST's mission fully. GST's definition of the notion of system was that the system is an ordered set[13] and represents a whole entity. The notion was fine and simple enough, while having no content, to suit all sciences and practices, which means the danger that the tunnel vision returns to power (because specialization is unavoidable)[14]. What the very creative and important authors, fathers of the GST and Cybernetics have left aside too much, was the self-evident fact that a system, as an ordered set, could let the specialists with a tunnel-vision kind of behavior keep forgetting about their own interdependence with specialists different from them. Namely, the same object under consideration and/or control can be looked upon from several viewpoints. Each and every system (= mental picture) uncovers a different part of the really existing characteristics of the object under consideration and/or control (Mueller-Merbach, 1992; Mulej, 1979: 83-84). Only an absolute/total system, which can normally not be attained, could make the system really equal to the object represented by the system, which is what the GST has actually and unsuccessfully tried to achieve (see: Checkland, 1981 and earlier; Jackson 1991; Mulej, 1974, 1976, 1979 and later, including Mulej et al., 1992, 1998). An absolute/total system would be really holistic, but it, actually, presupposes a system of all systems/viewpoints (not only a single one, neither a system of all essential, nor of all possible systems/viewpoints). No individual can do it alone[15] neither can

---

13. Let us repeat: An ordered set is made of a set of components and a set of relations, thus making a system—in the mathematical sense of the word. A *mathematical equation, a human body, a sentence, a book, a country, the entire universe*, etc. all meet this mathematical criterion. Even more: they all meet it *from every single viewpoint*: a partial insight replaces the foreseen holism and makes room for fictitious holism (See Tables 1, 2, 7, and 8 again).

14. The return of a narrow specialization instead of holism, which was expected, is possible: oversight was made that the natural human capacity of knowledge and skills and interests dictates the humans to limit themselves to *their selected viewpoints* when considering a whole entity. Hence, every profession perceives a single fragment of the actual attributes of the whole entity (e.g., students, professors, clerks, ministries, neighbors of a school, etc. all find different fragments of the actual attributes of the school worth their own consideration). See Table 12 again.

15. There are several thousand professions around. Numbers considering different interests

normal cooperating teams. This was overseen (Tables 1, 2, 7, and 8).

Four unfortunate consequences resulted from this oversight, at least:

1. Systems Theory has not become really general or useful for linking different disciplines, or both at the same time, i.e., to be a bridge from the single-disciplinary and multi-disciplinary towards trans-disciplinary by the inter-disciplinary approach to monitoring, perception, thinking, emotional and spiritual life, decision-making, and action, i.e., entire human behavior.
2. Systems Theory has not become enough of a mass movement (Molander & Sisavic, 1994) to come to be applied generally and to establish ethics/VCEN of non-fictitious holism.
3. Systems thinking is in crisis (Hofkirchner & Elohim (eds.), 2001; Midgley, 2004; Mulej et al., 2003, 2004; Troncale, 2002; Warfield, 2003).
4. Humankind is in crisis, therefore, because one-sidedness is not prevented.

These developments replaced the intention, which used to make (G)ST surface and become important in last good six decades. Instead, practice limited (but not destroyed) its usability and usefulness and application mostly to the arena of inside-discipline-holism. As long as scientists as well as practitioners of any discipline do not feel interdependent, but rather independent, the problem of a rather real holism can hardly be solved.[16]

In other words: times and conditions have come for ethics of interdependence to be acknowledged as the way out of the current crisis of both, systems theory and humankind, and to start prevailing in the best interest of humankind and of each and every individual. Thus, one may have a chance to come closer to the requisite holism and wholeness.

The requisite holism is, in practice, obviously defined by individuals and/or groups defining the scope of their behavior. Thus, it is these persons who take responsibility for consequences, if the requisite holism is not defined

---

are not available. The number of all viewpoints to be included in an absolute/total system can hence neither be known nor attained.

16. The same findings hold of e.g., business functions in organizations, of business partners in markets, individuals in families, organizations and communities, regional units such as counties, states, countries, European and other Unions and Associations, etc. The longest step toward ethics of interdependence seems to be made by the organizational paradigm *network of autonomous units replacing the commanding hierarchy* (Schmidt, 1993). Schmidt compares this network with the network of reliable friends, on the basis of his own life and consultancy experience: it works better.

broadly, on an interdisciplinary basis, enough. This unfortunate consequence can easily happen if these persons practice systems thinking, which is more oriented to:

- Formalities and internal attributes of the event, process, person, group, organization, society, part of nature etc. dealt with while no environment is considered, and
- Dealing with "systems" within one single viewpoint rather than more broadly, including the ethics/VCEN and practice of interdependence.

This can happen quite easily due to the normal human need for specialization. It is helpful and unavoidable, but not enough.

The background of both, surfacing of this problem, and of its solution, needs some more of our attention. We may understand the situation more profoundly and therefore find a way out of the blind alley more easily.

Chapter 10

MODERN (MATERIALISTICALLY) DIALECTICAL THINKING—A FORERUNNER OF SYSTEMS/HOLISTIC THINKING AND CONSIDERATION OF INTERDEPENDENCE

Once upon the time there were many less persons on this planet Earth, and there were also many less contacts and mutual impacts among persons. Thus, it is not too difficult to understand that in the pre-industrial times the common denominator of the style of thinking used to be quite different from the one which was summarized by the philosophy of the 19th century's industrial times. In a summary the difference in four attributes reads; see Table 13.

The above philosophical difference reflects different life realities and causes essentially different consequences, when applied today. Both types of consequences are summarized in Table 14. Link Tables 13 and 14 with Table 10!

MEDIEVAL METAPHYSICS	CONTEMPORARY DIALECTICS
Isolation	1. Interdependence
Absence of opposites	2. Unity and fight of opposites
No changeability	3. Continuous changing
Total negation	4. Dialectical negation

Table 13 *A summary of the pre-industrial and modern basis of thinking in four interdependent attributes*

One cannot master the reality, if one perceives and impacts only some parts of reality rather than all of it. If one oversees interdependencies, one cannot see reality, mistakes are unavoidable. New solutions are needed, therefore, again and again.[1]

One cannot master reality, if one expects no combination of unity (in one part of attributes) and opposites, i.e., crucial differences (in the other parts of attributes) as well as their impact on each other (called "fight" by philosophers).[2]

One cannot master the reality, if one expects the given solutions to last and be useful forever, because no new conditions are supposed to surface. Mistakes are again unavoidable, if oversights happen.[3]

One cannot master the reality, if one expects completely all attributes to disappear if any change comes ("total negation").[4]

In other words, one cannot master reality, if one is not realistic (the right column in Tables 13 and 14).

---

1. Interdependencies also change, in reality. As a child, one is interdependent with one's parents, later on with one's partner, later on with those helping the old and disabled persons out, etc. In the case of a product, the first phase sees interdependencies of the designers and investors, then of the supply offices and suppliers, then of the stock and production units, later on of the sales office and production units, and the sales office and buyers, etc.
2. Parts making an engine must be different from each other and complementary to make it. The engine is their synergy, the system made of those parts (in the above mathematical perception of the system). Members of a sport team do the same. And so do two teams competing in a game: both team unities are their shared kind of sport, but they are opponents to each other as competitors for the victory. The same philosophy holds of parts of a body, of an organization, United Nations, etc., everything.
3. Would you like to receive the same kind of a birthday gift at the age of 2, 20, and 50? There is also a life cycle of needs and wishes.
4. If this expectation was true, one could make a good wooden house from rotten iron waste, e.g., Etc.

MEDIEVAL METAPHYSICS	CONTEMPORARY DIALECTICS
Isolation = reflects the old times practice of a lack of contacts, impossible today = provides a basis for a lack of cooperation badly needed by specialists today	Interdependence = reality of specialists as well as of all parts of both, nature and society, applies also to technical products; independence of parts is quite unusual, we are all parts of larger entities
Full harmony = monolithism and monopolism, unanimity, i.e. the boss cannot be wrong, subordinates must not think = provides a basis for the individualities to be overseen and only the general attributes to be admitted, and a basis for mutual complementarities to be disregarded	Unity and fight of adverse/opposites = one part of the attributes is common to all parts of the entity, other parts are different from each other and complementary by their differences, because of which they influence each other; both parts of attributes are natural, so is their interplay and its permanence, no final harmony, only processes
No changeability = reflects the old times routine-loving life, favoring tradition to novelty (which is "illogical" because it opposes tradition) = provides for a lack of room for creativity and innovation	Continuous changing* = due to influences and interplays changing is normal, its speed and direction may differ; new needs cause creation of new possibilities, which along with new troubles create new needs, creativity and innovation
Total negation = reflects the fear that everything changes, if any part changes, leading the entire old, accustomed practice to disappearance = provides for opposition to innovation, even incremental	Dialectical negation = in the process of change the obsolete parts of attributes disappear, the still valid, usable and useful ones survive; the change may be evolutionary (step by step) or revolutionary (leap-froging)

Table 14 *Practical grounds and consequences of the medieval metaphysics and contemporary dialectics*

* *In the language of philosophy: the given "quality", existing essence, changes by small steps, "quantitatively", to a new "quality". The seed of a tree, e.g. changes by growing to a tree, which later on changes to fruits and wood, etc., that later on change to something else again, but they never fully lose their original attributes. A problem changes to a felt problem and initiative to produce an invention, which may later on be developed to a prototype or a similar potential innovation, that may later on in customers' acceptance change to innovation, bringing satisfaction, profit and new problems and/or initiatives, as well as diffusion. See Table 10.*

Let us recall the process of working of the four laws of modern dialectics[5] in Table 10.

But narrow specialists as well as routine-lovers prefer unchangeable tradition and isolation to have an easier life[6], as long as there is a market for

---

5. There are also *categories of dialectics,* which also help us understand interdependencies better. But we do not have room and time to deal with them here.

6. There has never in the known history been so much of changing as today, so much science, so much data and chances of information (i.e., data making sense and impact). The generation

their products and services, which demands very little cooperation with other specialists, and interdependence of humans or products' parts is simple. It is rare now.

Innovative business, no innovative society, no global economy, information network and life are all here now; they all denote the same situation: interdependence is a growing daily practice. Thinking in terms of interdependencies, i.e., modern dialectical thinking is the realistic choice. But the dialectical thinking is also more demanding:

- It shows the reality in a less simple light, and demonstrates its real complexity.
- It allows for less room for one-sidedness and leisure and a lazy, routine-loving behavior.
- It does not allow e.g., specialists to run away from cooperation with other, different specialists, but rather requires them to seek mutual impacts because they differ from each other.
- It allows no boss to be sure that he or she is the only person entitled (and obliged) to think and capable of doing and fixing things. The boss should rather find him- or herself capable of one part of the entire process to be mastered, while his or her subordinates are actually his or her team members and partners who complete each other up in a shared process.

Thus, dialectics suits the complex modern life much better than the medieval metaphysics, but not in totally all cases[7]. By letting us see reality in more complex terms, dialectics does not make our life more complex, but the opposite: it solves the problem of an exaggerated over-simplification, which

---

has not yet fully retired which was trained to serve the conveyor belt in mass production rather than to be creative and even innovative on their jobs. Even in USA they are not satisfied with the adaptation of the education to the current needs of the manufacturing and other industries. The same applies to Europe (sSee: Green Paper on Innovation, 1995; Green Paper on Information Age, 1996; EU, 2000; etc.) and Japan (Zenko, 1999). The remaining 80% of humankind are much more traditional in their practice. Only a few countries are changing their habits quite rapidly, such as China, India, and Brazil, as specialists of contemporary economics let us know (Boškovićč, 2006).

7. Cases in which behavior in terms of isolation, full harmony, no changing and total negation is the appropriate behavior are rare, but they do exist. E.g., in a medical blood lab, blood of different patients may never be mixed up. A worker on the conveyor belt is interdependent with other workers, but he never intervenes with their jobs, unless something very exceptional happens. Etc. A very reliable and trustful boss allows for more of such behavior.

may end in very unpleasant surprises[8]. Let us repeat: the consequence may look as a paradox: (over-)simplification causes complexities and complications, consideration of complexities and complications, in time, simplifies (Table 8).

Interdependence is a more complex, but hence more realistic, picture of reality and a more realistic basis of behavior. It does not exist among ideas only (as the "idealistic dialectics" used to teach, being limited to the world of ideas and disregarding the real life's causes and roots of ideas), but in the entire nature and society, including the world of ideas. This is expressed by the notion of "materialistic dialectics" in philosophy (Britovšek et al., 1960; Bahm, 1988; Engels, 1953; Waldrop, 1992; Delgado & Banathy (eds.), 1993). Thus, we no longer separate the world of thinking from the world of practice, but see their interdependencies.

Systems thinking came into the scene, when dialectics was forgotten about, and the result of this forgetting was the terrible period of 1914-1945.

Now, there are many systems theories. Some of them, at least, have their (informal) background in dialectics. Let us move to them.

---

8. Oversights, e.g., when crossing a road, when making an investment, concluding a contract, mixing up chemicals, using several medicines together, or disregarding side-effects of any kind, etc. are clear examples. On the Internet you can find cases of inventions patented, but falling rather in this category than among innovations. Only one percent of patented inventions become innovations, which is due to some crucial oversights, in general. See Table 9. We will come back to it in Ch. 14.

Chapter 11

SOME (SOFT) SYSTEMS THEORIES AND THEIR APPLICABILITY TO THE ISSUES OF REQUISITE HOLISM AND ETHICS OF INTERDEPENDENCE— A BRIEF OVERVIEW

11.0 THE SELECTED PROBLEM AND VIEWPOINT OF DEALING WITH IT IN CHAPTER 11

Systems theory has developed as a methodology, too, not only as a worldview. Experience says, that the extremely broad thinking, which Bertalanffy has required from humans, reaches beyond limits of capabilities, VCEN/interest/s of nearly all men and women on the planet Earth today. Humans need what they find a requisitely realistic approach (Table 23), less limited to the general part of attributes only (and hence less over-simplifying, too reductionist), except in the case of the aim to find out the most frequently repeated attributes (which is good enough in a phase of the scientific work, but less so in other phases of it and in the "real world"). This fact might belong to reasons that the GST is not much supported, in practice. New theories of a very similar orientation, which is towards general isomorphism and common policies and/or rules have shown up over the last years. The most recent include the chaos and complexity theories.

11.1 THE CHAOS THEORY AND REQUISITE HOLISM AND ETHICS OF INTERDEPENDENCE

The Chaos Theory first emerged in physics and was considered a mathematical discipline, but it tends to be widely used in other areas (= i.e., from other viewpoints), too. In this book users of Chaos Theory may find that this is so, because its authors tend to (informally and independently!) apply the (dialectical) systems thinking, although such purpose is not explicit. Of course, newer attempts bring new capabilities of insight, and one can say that in some respects, Chaos Theory is renewing systems thinking (see: Huang, in Vezjak et al. (eds.), 1997; Mulej et al., 2000; Mulej & Kajzer, 1994; Shirhall, in Dyck et al., 1998, 1999). This is clearly visible from the quotation (Kiel & Elliot (eds.), 1997). Chaos Theory is dealing with,

> *real units or entities, which cannot be foreseen and are not linear, and were excluded by the Newtonian science.*

In other words, the processes are chaotic, if their behavior is partly deterministic and partly in-deterministic, hence partly foreseeable. In the language of dialectics, one would say that both types of behavior are a consequence of interdependence which is always visible as the unity and fight of opposites, (see Tables 10, 13 and 14). On this basis, unforeseeable processes may have foreseeable consequences, and foreseeable processes may have unforeseeable ones. This is why complex processes cannot be watched and managed totally realistically in a linear manner, especially not, if the observers/managers limit themselves to a single viewpoint. Consequences of complex processes are normally caused by interactions and come to be emerging as synergies from interactions of life components which are partly interdependent, partly independent, and partly dependent, at least in a given moment/period of time, rarely for ever. Besides, these units tend to include in the process different parts of their own attributes on different occasions.

The same finding can be found about human attempts to master a process or a situation (and even to do it once for ever). There is usually a certain level of self-organization, especially in nature, but also in enterprises and other man-made organizations. Research, even a lab-based research can provide insights, which are usable on a framework/model basis rather than directly and/or linearly. This experience reflects the fact that life is nonlinear, and so is everything else, which is worth human consideration today (Kiel & Elliot (eds.),

1997). Nonlinear processes are "sensitive to the starting conditions" and to the mutual interactions between the processes under consideration and their environment/s, which may show up later on.

In other words, foreseeability/predictability and its lack exist at the same time and in a networked style. On the level of individual attributes, the processes are unforeseeable, on a more general level statistics can show averages and deviation from them, thus more attributes of the entity as a whole can be foreseen more easily (if conditions do not change too much). In the language of the Newtonian science the entity as a whole would be called a disorder. But statistics show the level of possibility and probability that an order shows up anyway. Actually, as we can see, both, order and disorder exist at the same time, are interdependent and interactive, which causes and enables, at least, the complicatedness and complexity as well as the existence of the real life and its parts. Thus, although the Chaos Theory is realistic, in terms of the Newtonian, hard-systemic, thinking the Chaos Theory is an unusual theory (Richards in Kiel & Elliot, 1997). It requires its users to think systemically rather than traditionally, and thus to give up research per parts and the summing up of partial findings with no emerging synergies. In reality, the entire object cannot be partioned into parts which no longer influence each other, and are still able to be what it/they used to be.

If we had time for details, we could see that the Chaos Theory presents an interesting further step of those kinds of the systems theory/ies, which have not evolved into tools of single rather narrowly specialized disciplines, but have kept a more general methodological validity. Let us hope that its development is not over yet; humankind of today lacks methodologies, which would be able to combine exactness and usefulness in one, which most topics could use quite well (e.g., Loeckenhoff, 2005).

If compared with DST, the Chaos Theory does not explicitly deal with the formation of human personality—the subjective starting points—, but with the natural phenomena and their general attributes. This makes it useful for the understanding of some aspects of life, but less for providing to humans of today a clear orientation what to do about it and how to approach it.

Another modern version of systems thinking, which also deals with complex problems and fights the traditional thinking style, is called the Complexity Theory.

11.2 THE COMPLEXITY THEORY, REQUISITE HOLISM AND ETHICS OF INTERDEPENDENCE

Dent (1999) published a good summary of the basic attributes of the Complexity Science. The purpose of his article is said to be "to offer a simple definition for Complexity Science and to demonstrate the shift in worldview necessary for Complexity Science to become second nature to people as traditional science is now—simply put,

Complexity Science is an approach to research, study, and perspective that makes the philosophical assumptions of the emerging worldview (EWV)— these include holism, participant observation, mutual causation, relationship as unit of analysis, and others; see Table 1.

Here we stress that systems theory/thinking is receiving another endeavor/ support, after many had forgotten about it. "Classical science, as practiced in the 20th century, for the most part makes the philosophical assumptions that will be labeled here the traditional worldview (TWV)—which include underlying assumptions of **reductionism, objective observation, linear causation, entity as unit of analysis, and others**" (bold print ours again, as above). "This TWV, which has allowed people to make significant achievements in many fields, is no longer serving as a reliable guide." (Dent, 1999: 5-6)—His entire Table 1 demonstrates the contrasting differences.

An additional quote may make sense (Dent, 1999: 7):

It is important to note that theorists are not suggesting that the traditional underlying assumptions are wrong. In fact, many of them seem to be useful in localized settings. For example, Prigogine and Stengers (1984: xxiii) see **determinism and indeterminism not as irreconcilable opposites but "each playing its role as a partner in destiny"** *(stressing ours, as above). Between bifurcation points, determinism is operative. At a bifurcation point, however, indeterminism takes over. Consequently, indeterminism (which does not dismiss localized determinism) and other emerging assumptions seem to be more useful abstract concepts. They reflect reality more accurately in a large number of instances.*

Experiences, which we have made with two conferences applying Complexity Theory to Management, in 1999 and 2000, and a long discussion over the internet ever since, demonstrate that complexity is reviving the good principles of systems thinking very well and with good promises (NESCI, 1999, 2000). But, in terms of applied method(ologie)s supportive of complexity as a worldview, at the moment we still better turn to the DST and some other contemporary Systems Theories. Alternatives have yet to show up.

In terms of the ethics/VCEN of interdependence and requisite holism, the quoted attributes are coming quite close to it, as we have briefly but very indirectly shown in the footnote above. A more recent experience with a conference (ANZSYS, 2005) in Christchurch, New Zealand, showed an equal lack of interdisciplinary contributions and an equal limitation to one-viewpoint systems (i.e., selected mental picture of objects/phenomena under consideration) as other conferences do. Thus, the co/operation of many agents in nature incl. humankind of today, such as different enterprises of the same area or different industries from the simple old-fashioned farming and trading to the most modern high-tech ones, local and other authorities of several administrative levels, non-governmental organizations, population (as individuals, families, interest groups, and the like) is only made possible. Complexity Science, as a EWV, does apply—but it may need a methodological support to be useful and beneficial, from these viewpoints, too.

Hence, let us turn to some other contemporary soft systems theories.

11.3 A SELECTION OF CONTEMPORARY SOFT SYSTEMS THEORIES—A COMPARATIVE VIEW CONCERNING THEIR USEFULNESS TO THE ISSUES OF REQUISITE HOLISM AND ETHICS OF INTERDEPENDENCE

11.3.0 The Selected Problem and Viewpoint of Dealing with it in Chapter 11.3

In its first decades, until the 1970s, the traditional (i.e., linear, descriptive) systems theories prevailed or even were the only ones around (Jackson, 1991). Later on, several authors found that this was leading to an over-generalization, which tends to be poorly realistic. One interesting distinction, among many, showing differences between the existing systems theories is the delimitation between soft and hard systems theories.

The soft systems theories, according to Checkland (1981), include the ones, which presuppose that the output depends on the input probabilistically rather than deterministically.

In machines and similar technical artifacts, hard-systems behavior is needed, expected, and aimed at, which is why mechanics, technical cybernetics, including automation and computerization, are supposed to produce very reliable outcomes as responses to the foreseen impacts (such as switching on the electrical lightening at home, or starting a car). The notion of soft systems resulted from the experience that a manager cannot expect his team members to respond to his or her initiatives in a hard-systems style. Checkland expressed the difference also by denoting the topic of the soft-system approach as action and action research (rather than discovering natural facts or designing tools on their basis). Human action differs from processes not running on conscious self-organization or organization, but on laws, which reflect/are natural facts and are studied from the viewpoints in which single natural and engineering sciences are specialized. The previous concept of the General Systems Theory and Cybernetics, that there are generally valid isomorphisms, resulted into its limitation to smaller groups of features, events and processes. Humans, as individuals and as groups, are the part of nature, which is not designed, but active. They act along the lines of their own ethics, interests, knowledge, skills, and perceptions of real circumstances, of laws of physical and other reality, their own selection/s of preferences in terms of both, needs and possibilities.

Recent decades saw more differences within the systems movement, not only the limitation of the General Systems Theory to a small portion of the attributes considered as general ones, and not only the differentiation into hard and soft systems theories. We, the authors of this summary are not feeling qualified to discuss any details of the hard systems theories; they are mostly topics of engineers aiming to design and produce reliable and helpful tools such as machines. Our education and experiences limit us to the soft systems theories, because they help us deal with social aspects of the real life and the human action in it. Within this set of systems theories there are a number of them, which we want to brief in this contribution, in order to help the readers to make their choices of the (mental) tools, when they will be dealing with the problems and other topics of the modern life of the 21st century. Another contribution should support a similar selection among the hard systems theories.

We will brief a few soft systems theories which we find promising in terms of the modern life of the 21st century. We will have no room for details, of course. Our choice includes:

1. The Viable Systems Theory, because it sheds a new light on hierarchy of subordination in organizations as organizational systems. It requires cooperation and communication rather than commanding.
2. The Soft Systems Methodology, because it offers concepts and a framework method of working on solving problems tackled by action research.
3. The Dialectical Systems Theory, because it offers concepts and ways of influencing the human beings, who are supposed to work and cooperate in organizations as organizational and business systems. In modern circumstances, they are supposed to attain as much holism and innovation as possible, i.e., the requisite holism.
4. The Critical Systems Thinking, because it tackles management, especially conflict management, and offers a matrix-type combination of a "system of systems methodologies" and "a matrix of conditions and problem situations". It enables selection of a proper methodology when one works on solving a given situation/process/event.
5. Cybernetics, because it does not concentrate on description, but rather on solving the problems tackled. The evolution of cybernetics makes it more and more usable and useful in organizations as organizational and business systems.
6. Dialectical Network Thinking, because it revives the over-neglected Network Thinking Methodology and completes it up with the Dialectical Systems Theory's influence on human attitudes, and vice versa.
7. Control Systems Theory, because it provides a good case of synergy of the Living Systems, Viable Systems, and Dialectical Systems Theory as a new theory.
8. We are adding in Ch. 11.3.10 a presentation of systems thinking in the ancient Maya culture.

The role of information is essential in both, cybernetics and all kinds of systems theories. Sometimes it is conceived too narrowly. Hence, we are adding some comments on it.

Complexity is real, but also stressed a lot by systems theory. It can be hard to handle. We are hence also adding some ideas about some tools usable for simplification of management in/of complex processes and situations.

11.3.1. The Viable Systems Theory (VST)

Stafford Beer, the author of the VST, defines viability as the capability of an entity such as an individual or an organization to survive in a rather demanding environment (Espejo, in Mulej et al., 2000). His viewpoint of research and application is management of complexity, which arises from inside and outside the organization/individual under consideration. Relations expressed in impacts on the organization require this organization to behave according to the law of requisite variety. It means adaptability, which requires a response to be prepared and used against every potentially dangerous impact from the environment.

Towards this end, organizational principles must free creative capabilities of organization's members by enhancing their autonomy along with caring for cohesion of the entire organization. This is made possible by stressing communication/information flows among process participants enabling working and cooperation rather than hierarchical subordination limited to subordination without creativity and decentralization. Complexity namely shows up both, in environment and within the organization. If the lower levels, in terms of the traditional hierarchy of subordination, receive enough autonomy and information to be able to do their jobs efficiently and effectively, the remaining variety does not burden the central top management too much. Knowledge of the information requirement and good information channels, along with diminishing disturbances from the environment, and strengthening the person'/organization's impacts on its environment, are the bases for them to act as viable systems.

A to-be-useful work along the lines of both, research and practical solving of the problems of any kind should well consider these experiences:

- Hierarchy should not be too much of commanding, but rather supply of information top down and bottom up, as well as from one's environment and into the environment.

- Research should provide for knowledge, operation management should use it on the basis of "now and here", top management on the basis "there and then", i.e., from broader and long-term viewpoints.
- Communication channels should help members use the "requisite variety" (as an aspect of requisite holism) in both, blocking off the impacts from environment and strengthening the impacts on the environment.

Democracy replacing the commanding hierarchy with the organizational one is the essential attribute of the VST as a method reaching towards requisite holism based on ethics of interdependence. Our notions do not show up in it, of course (see Schwaninger, 2006). The Soft Systems Methodology can provide another complementary contribution.

11.3.2 The Soft System Methodology (SSM)

Peter Checkland, whom we mentioned above as the author of the distinction between hard and soft systems, developed also a framework methodology of dealing with soft systems called SSM. It consists of seven steps, which are delimited into a group of five steps (1, 2, 5, 6, 7) concerned with the real-life situations in a hands-on style, and a group of two steps (3, 4) concerned with them in a lab-style (Mingers & Uršič, in Mulej, 2000, and earlier).

SSM, as a procedure:

- Allows for permanent learning,
- Includes interdependence by both, the notions of participants' cultural diversity and wish to cooperate.
- Helps its users care for holism by its aim to explain the problem situation (including its context!), not the concrete problem only.

Objectives are hard to define and require humans to work on their definition first. This work includes finding out the initial state of the situation under consideration, its components, structure, processes, clear description of the given state and possible options, contents and influential circumstances of the intervention, contents of the problem and of its solving.

This is the basis for one's transitions to the lab-style (called systemic thinking by Checkland). One should attain a broad and deep enough understanding of both, the situation and trends, of all essential viewpoints of the action under research such as customers, doers, transformation, aspects, owners, and environment. These findings and ideas should enable the modeling of activities for a transformation to follow. In the next phase (number 4) the ideal model (called formal system) and the conceptual model (about the way of solving the problem situation under consideration) are compared from the viewpoints of what and how it should be done.

Models may have to be changed in the lab-style, but they are after that also confronted with the hands-on real situation, in phase 5. In phase 6 one finds out, which changes of the model are desired and makeable, before action takes place in phase 7.

In terms of dealing with requisite holism and ethics/VCEN of interdependence, the MMS model provides for learning and action to be interdependent and thought about quite holistically. Perception, argumentation, comparison, decision-making and action—they all may be covered in a quite holistic style. The result is not simply a solution to a problem, but a change in the broader situation, followed by a new learning. Of course, the subjective judgment of observers/actors/decision-makers is unavoidable, prevalence of a VCEN (i.e., prevailing habits, views, ways of life and working, including the legal and political documents and practices) is important. These prevalences of human attributes are important in both, VST and SSM. This notion brings us to Mulej's Dialectical Systems Theory, which is concentrating on the human personality and work process a lot (see Ch. 12; Mulej *et al.*, 2000 and earlier, since 1974; Mulej & Ženko, 2004a, b; Mulej *et al.*, 2008).

Especially important features in such a process are the conflicts of the parties involved. Here, the Critical Systems Thinking may be helpful.

11.3.3 The Critical Systems Thinking (CST)

CST's authors M. Jackson and R. Flood specialized in application of systems thinking to management, especially to management of conflicts. They accept Checkland's criticism about the over-generalization of the hard systems theory and extend it to his Soft System Methodology, because it avoids working on conflicts (Jackson, in Mulej *et al.*, 2000, and earlier). They require five attributes to be included in a system/methodology of thinking:

1. Critical awareness (rather than satisfaction with the given situation, i.e., routine-loving);

2. Social awareness (to include the broader rather only direct consequences into thinking);
3. Complementarities in theory (to allow all different kinds of knowledge available to be included);
4. Complementarities in practice (to allow the same for all different methods available);
5. Human emancipation (to allow for creativity to flourish rather than subordination to reign).

The tools they produced and use, make a system of systems methodologies for the problem solvers to select the proper one/s from, according to the attributes demonstrated by the matrix of conditions and problems. But attributes of humans as users of CST are not addressed like in DST.

In terms of dealing with requisite holism and ethics/VCEN of interdependence, CST can also be quite useful. All available knowledge can be listed for an overview to replace oversights, so can all attributes of the real situation, which one is able to detect. One is supposed to criticize and to accept criticism, to feel interdependent (under the notion of complementarism, both, in terms of theory and in terms of applied methods), to employ all capacities available, and to think of the broader consequences able to show up, usually as side-effects. These consequences are very much what the modern life of the 21st century is going to be all about (see Affuah, 1998; Wood, 2000). But unlike DST, GST does not address impact on humans and innovation and related human attributes. Still both, DST and GST, deal with management of human lives.

Dealing with management of our lives, not only organizations, brings us closer to cybernetics.

11.3.4 Dialectical Network Thinking

Rosi's Dialectical Network Thinking (Rosi, 2004; Rosi & Mulej, 2005, 2006; Rosi et al., 2006) is also trying to help humans improve, i.e., innovate their management. But it takes more the viewpoint of management processes in activities full of networks and related influences.

The concept of network may apply to any complex feature in the life reality; it is very close to the notion of system consisting of a set of elements and a set of their relations plus relations with their environment in an open system concept. (See Bertalanffy, 1979: 139-154)

This fact offers the basis for a soft-systemic theory called Network Thinking (Vernetztes Denken, NT; Gomez & Probst, 1987, 1997). NT supports holistic problem solving, especially in organizational management. Of course, one should apply NT with no limitation to a fictitious holism resulting from authors' decision to use a single viewpoint when defining the "system's" contents. One should neither strive for the Bertalanffian total holism that cannot be attained. One should rather meet the Mulej/Kajzer law of requisite holism, which offers a middle way between the two extremes. See Table 2 and 8 and related thoughts.

Our question is: Does NT make requisite holism possible and DST unnecessary or vice versa? Originally, both, NT and DST, are about 30 years old and may need a renewal. It is our thesis (proven in Rosi, 2004) that one can develop from DST and NT a new model of systemic thinking, called Dialectical Network Thinking (DNT). Our research showed it makes sense to apply project management in DNT, too, in order to rationally organize work processes.

Attempts to diminish the weak points of both, NT and DST, have led us to creation of DNT as a new methodology—a new systems theory—leading towards innovation of the problem solving process. DNT helps organizations and organizational-business systems (OBS); it:

- Requires and supports team work;
- Supports requisite holism in the human work on development of OBS;
- Supports the organization/OBS structures aimed at innovation and development;
- Shows what kind of OBS suits the modern environment and is capable of using the DNT;
- Stresses the interdependence of development and structuring of OBS with the development of project management;
- Supports change management in OBS;
- Makes easier the detection of the need to restructure the OBS;
- Supports learning and unlearning in OBS;
- Serves as the bottom-line in making of the contemporary organizational culture.

Thus, DNT leads towards more holism and innovation in problem solving, which is a precondition for a more sustained mastering of the modern turbulent business conditions.

11.3.5 Control System Theory

The new Control Systems Theory (CoTS) provides a new basis for a possible innovation of social business-organizational systems such as organizations of the public sector, e.g., the health care sector in Slovenia. CoTS eliminates a lot of the possibility of excessive self-organization, chance synergies, and one-sided intuitive measures, quite often so far, e.g., in the Slovenian public health care system (PHCS). At the same time it allows and supports professional and in particular inter-disciplinary creativity and inhibits bureaucratic rigidity. CoTS results from synergy of Living Systems, Viable Systems, and Dialectical Systems Theories. In CoTS terms, e.g., the Slovenian PHCS may be discussed as a living system and a viable system at the same time, requiring and enabling its participants to behave in line with the Dialectical Systems Theory (DST), both, in PHCS theoretical definition, description, management, and supervision, and their innovation.

11.3.5.1. The Living Systems Theory (LST) in CoST

The LST is very useful for the description of details on the single selected theoretical basis, but not much more (Miller, 1978; Mlakar, 2000, 2007). In applied research, we try to apply CoST to the public health care system (PHCS) of Slovenia.

The Viable Systems Theory (VST) focuses on organizations, which are one of the types of living systems, and on their ability to survive and keep their identity in complex conditions. VST therefore does not deal with description of facts from a selected viewpoint representing a living phenomenon and features common to all of them, but it covers rather an aspect of actions necessary for the organization to survive as a viable system (VS) (Beer, 1979).

The practice, e.g., in PHCS in Slovenia shows, that it is easier to strengthen the ability of the organization, if its internal features are known in detail. Therefore, VST alone is not sufficient for us. We will link it, within CoTS, with LST and DST. In short, this new synergy is illustrated in Table 15, representing a schematic image of a CoST model (Mlakar, 2007). Inside all VSs attributes of living systems can be found, in our case, at least, but with different concrete contents due to prevailing of different viewpoints, found essential.

- VS1 represents the operative execution of tasks, e.g., in the outpatient clinic or another part as a basic unit of a hospital;
- VS2 represents the functioning of a department, or division, linking the operative staff according to short-term optimization standards, e.g., in a hospital ward with several outpatient clinics or other parts of a ward;
- VS3 represents the functioning of the organization—the task performer linking the departments, according to medium-term optimization standards, e.g., in a hospital with several wards;
- VS4 represents the functioning of libraries, and the like, within PHCS, collecting information of various research work, for operative and politically strategic decision-making; medical research work includes, e.g., concerning drugs, their national and international inspection and approval prior to use;
- VS5 represents a political view of the functioning of PHCS, representing long-term guidelines; it optimizes the whole PHCS according to long-term standards, e.g., with decisions taken by the Minister of Health, Government and Parliament, including taking into account international bodies concerning medical care.
- VSs 1–5 as subsystems work within PHCS in activities and cooperation of participants of different specializations (e.g., medicine, economics, law, and finance).

11.3.5.2. Participants of CoTS in e.g., PHCS

The participants of VS1 have the function of operative performance of tasks within the competences as a PHCS subsystem (a doctor or a nurse).

The participants of VS2 have the function of departmental heads and their staff. Their task is to ensure a continual performance of tasks, work organization, analytical monitoring of the tasks carried out in comparison with plans, quality of their work, reducing the effects of the environment, which could cause unwanted errors, organizing procedures for rectifying the effects of unwanted errors, and so forth.

The participants of VS3 have the function of managers of operative staff (they analyze the business results, and make decisions regarding measures).

The participants of VS4 act as information providers, expert and research bodies at the hospital, country and international levels. They cooperate with subsystems VS1, VS2, VS3 and VS5. They define the scientific and specialist bases and proposals concerning the scope of rights and capabilities, which can be realized through the funds available within PHCS.

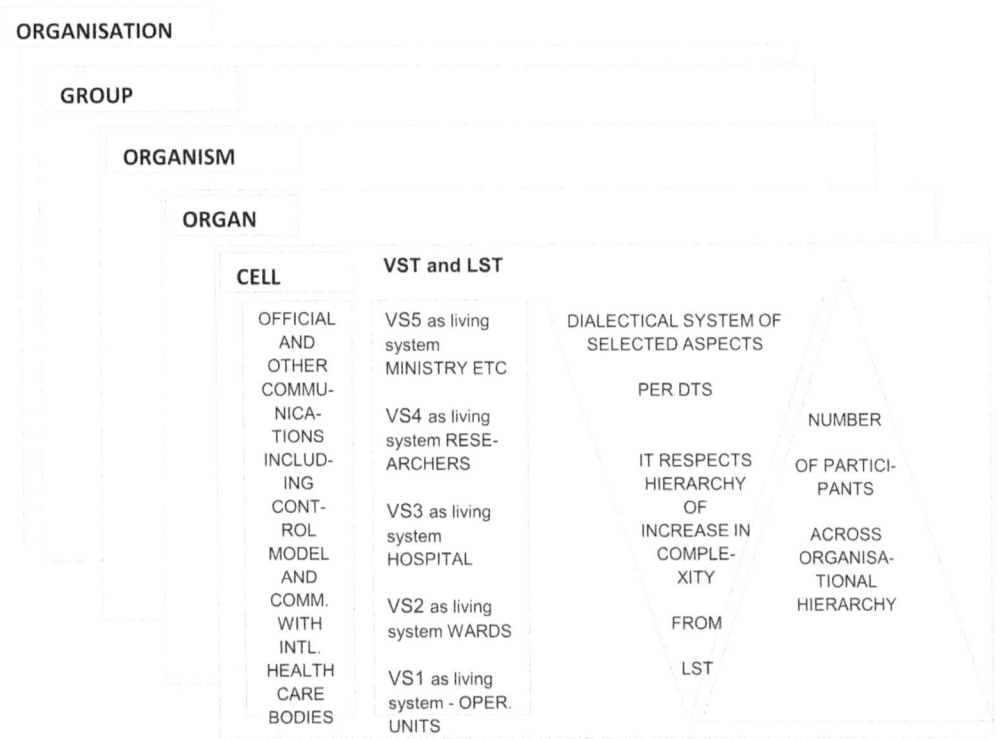

Figure 1 *CoST Model applied to PHCS*

The participants of VS5 work together with VS4, and they form the PHCS strategy of the country based on VS4's proposals. For this purpose, a dialectical system of criteria and aspects must be taken into account, priorities must be decided, and ethical rules respected, which means: the law of requisite holism must be observed.

Table 15 (Mlakar, 2007) illustrates how subsystem levels from LST link with five VS from VST. Table 15 lacks space to demonstrate all forms of VS 1-5 in the final column, i.e., separately within each VS. In the case of e.g., an outpatient clinic the role of VS5 can be assumed by the manager of the outpatient clinic. VS4 consists of consultants giving advice from literature and practice, including internal clinic and external staff, helping managers and practitioners

VS SUBSYSTEM	LST LEVEL	VS 1-5	LEVEL
VS 5/7, e.g., in PHCS: United Nations Organization	LST 7	VS 5 VS 4 VS 3 VS 2 VS 1	Normative – long term Strategic and other info Operative – middle term Operative – short term Operative – immediate term
VS 5/6, e.g., in PHCS: World Health Organization	LST 6	VS 5 VS 4 VS 3 VS 2 VS 1	Normative – long term Strategic and other info Operative – middle term Operative – short term Operative – immediate term
VS 5, e.g., in PHCS: National health authorities, e.g. Parliament, Government and Ministry of Health of the Republic of Slovenia	LST 5	VS 5 VS 4 VS 3 VS 2 VS 1	Normative – long term Strategic and other info Operative – middle term Operative – short term Operative – immediate term
VS 4, e.g., in PHCS: Medical research and development organization and associated activities	LST 4	VS 5 VS 4 VS 3 VS 2 VS 1	Normative – long term Strategic and other info Operative – middle term Operative – short term Operative – immediate term
VS 3, e.g., in PHCS: A hospital as a medical, business, legal, etc. system.	LST 3	VS 5 VS 4 VS 3 VS 2 VS 1	Normative – long term Strategic and other info Operative – middle term Operative – short term Operative – immediate term
VS 2, e.g., in PHCS: Hospital wards.	LST 2	VS 5 VS 4 VS 3 VS 2 VS 1	Normative – long term Strategic and other info Operative – middle term Operative – short term Operative – immediate term
VS 1, e.g., in PHCS: An outpatient clinic within a hospital ward.	LST 1	VS 5 VS 4 VS 3 VS 2 VS 1	Normative – long term Strategic and other info Operative – middle term Operative – short term Operative – immediate term

Table 15 *LST levels and VSs in a medical care system (PHCS)*

run and renew current practices of work in the outpatient clinic. The VS3 role can be assumed by individuals with the task to harmonize the teams in this clinic in the medium-term prospective, e.g., one month; therefore, we deal with work plans and the assurance of conformity of practice with it. VS2 roles are assumed by those who harmonize the teamwork for short-term periods, e.g., by weekly and daily plans. The role of VS1 belongs to operational teams during day-to-day work in the given outpatient clinic.

11.3.5.3. Dialectical Systems Theory in CoST

While LST provides for a very detailed and clear insight, VST provides for a clear way from a commanding to an informing and learning organization.

In Figure 1 and Table 15 the summarized essence of DST means, that people working in VS 1, VS 2, VS 3, VS 4, or VS 5, even if they use all 19 subsystems and all other 5 aspects from VST, will tend to differ in their definitions, what belongs in "the requisite holism". This makes them complementary to each other, and thus they are better able to cover all essential viewpoints, relations, and synergies. The LST enables them to cover details, and VSM enables them to work in an organized way. Thus, all three systems theories are well applied in a new synergy—the CoTS, be it in a medical care or another real life situation.

11.3.6 Cybernetics of the 1st, 2nd, and 3rd Order

In the mid 1940s the idea showed up, that it might be worthwhile to transfer the concept of the nervous system of the living bodies into the machines; this might make machines able to transfer information and make decisions. From this effort automatic machines, including computers have emerged. The thinking machines were very much based on the feedback mechanism, which is a special case of (a hard systemic and rather predictable, hence routine-based) interdependence between parts of the same entity.[1]

This concept was later on called the first order cybernetics. This label was introduced, when the second order cybernetics was discovered. This was a new revolution in science (Umpleby, in Mulej et al., 2000):

- One was no longer dealing with the natural attributes as given facts only.
- One stopped observing passively the natural processes and leaving the observing persons and their process of observing outside the scope due to the selected viewpoint.
- One found that the observers are actually a part of the process. This means that the laws of nature are not objective on a holistic level, but only inside the selected viewpoint and (pre)conditions.

---

1. This belongs to the rather fruitful examples of isomorphisms. But it has a creative interdisciplinary cooperation as its basis.

Hence, there are no objective scientific findings, because they depend on the human approach and selection of viewpoints and interpretation of findings.

This was a revolution in science—from the expectation of a total objectivity to admitting the role of humans. The next step of evolution of cybernetics came to be cybernetics of conceptual systems. It is not important only to introduce cybernetics of cybernetics, as the second order cybernetics has done by including the observer into observation. It is also important to consider and influence the ideas of observers. The relation between the individual ideas and society around them is one of interdependence and interaction. Therefore it is too rough and shallow to consider the ideas in general terms only. It is individuals from whom the impacts start (Umpleby, in Mulej *et al.*, 2000).[2]

A quite natural next step is the finding that it is not realistic enough to talk about human beings as observers only. They actually combine observing, decision-making, and impacting. This finding resulted in third order cybernetics (For some details see Vallée, in Mulej *et al.*, 2000). This is a new revolution in science:

- Science is no longer limited to finding out facts and general laws about them, neither to a passive description of them.
- Science combines finding out facts and technology of using them for an influence.

Cybernetics, due to such an evolution, may no longer be seen as a science and practice of using feedback as its only attribute. There are at least five attributes making a dialectical system: (1) management (in terms of any kind of impacting) of (2) complex, (3) dynamic, and (4) stochastic, probabilistic and even possibilistic, features (as systems), (5) by using information, including feedback as one of its types.

In the case of dealing with requisite holism and ethics of interdependence, we see that life is partly outside the human impact and partly under the human impact. Hence, several kinds of cybernetics may serve investigators and practical actors, be they individuals, businesses, other organizations, or governments, as usable tools of consideration, decision-making, and action.

---

2. But the question how are the involved individuals" attributes, is not asked like as it is in DST.

They all use information, the above definition says.[3]

Let us delimit the four types of cybernetics in Table 16 (Mulej et al., 2007, reworked).

This finding brings us closer to the notion of information and to the need for a few thoughts about it.[4] Let us add a few more findings to the ones in Chapter 6 in this context in which we see information as a general tool of decision-making and impacting based on observation and thinking, which produce information.

11.3.7 A New—4th Order Cybernetics

Our (i.e., Božičnik's and Mulej's) new Universal Dialectical Systems Theory (UDST) is presented in our article: "From Division to Integration of Natural and Social Sciences by the 'Universal Dialectical Systems Theory'" in *Cybernetics and Systems: An International Journal*, 40/4 in 2009. Our tacit application of it is presented in our article "Corporate Social Responsibility and Requisite Holism—Supported by Tradable Permits" in *Systems Research and Behavioral Science*, 2009. Here we want to make a further step in the same stream of theoretical development and suggest a new 4th order cybernetics.

Let us not repeat what we have said about cybernetics so far, but only add the following thoughts.

Even the most frequent applications of systems theory leave many complex problems un-tackled, because a single discipline alone cannot cover all important attributes—it cannot cover all crucial viewpoints, their relations and resulting synergies. Therefore, one should make a synergy of:

---

3. There are at least two crucial types of information: (1) the feed-forward information, such as provided by forecasting, planning, designing a machine etc., and (2) the feed-back information, which provides for (2.1) either the decision maker and actor to see the consequences of their impacts, or (2.2) the blocking off the tendencies for a deviation from the equilibrium to take place. (See: Kralj, 1998, and earlier.).

4. Actually, all knowledge and theories, including systems theories and cybernetics use and produce information, or try to do so, at least. Dealing with e.g., climate change or any other topic tends to attain the same: to define, understand, and solve the problematique by solving a dialectical system of problems with a dialectical system of solutions, which are conceived and undertaken requisitely holistically. (See, e.g., Božičnik et al., 2008; Ečimovič, (ed.), 2008; Ečimovič et al., 2002, 2007). All the briefly summarized systems theories can together make a kind of a "system of systems methodologies", which serves as a toolkit, when the insight into the "matrix of problems and situations" lets us see what must and can be done.

Nature is a whole and holistic. Humans are not capable of total, i.e., real, holism of behavior, but they need holism to survive/succeed. Humans win in their fight/interaction with nature and other humans, if they are more holistic, hence considering their interdependence more than their opponents are.

Humankind and the other nature – both are entangled: complicated and complex. The effort to master entanglement gave birth to human practice and science, including cybernetics of several kinds.

CYBERNETICS of the 0 ORDER	CYBERNETICS of the 1ST ORDER	CYBERNETICS of the 2ND ORDER	CYBERNETICS of the 3RD ORDER
NO THEORY	ONTOLOGY	EPISTEMOLOGY	PRAGMATICS
Practice as it happens to be	Whole, holism and wholeness of practice	Whole, holism and wholeness of findings about practice	Whole holism and wholeness of decisions and consequences
Life with hardly any thought	Observation of attributes of practice	Impact of practice	Decision-making based on observing for impact
Practice without theory	Positivism and materialism	Relativism/constructivism/idealism	Praxeology
Acceptance of given facts as a fate – superficial	Learning of given facts – natural attributes – analysis to find deeper background and essence	Learning of given facts, but of the ones selected by observer in his/her focus of concentration selected viewpoint	Basis for impact based on observation – risk because holism of insight is not possible
Given visible facts	"Eternal truths"	Opinions, estimations	Decisions
Collection of experiences considered logical because they repeat	Generalization of repeated findings by analysis	Hence observation of observers and observation processes, too	There is no observation or impact without decision – analysis and synthesis
Experience-based laws	Proof-based laws	Doubt about laws, refusal of them with new proofs (falsification)	Decision must be professional and rational
There is no science, only experience – response to "what" and "how" rather than to "why" questions	Formation of science and more and more sciences – classification by specializations – response to "what", "how" and "why" inside them in form of models of the selected part of reality	Consideration of aspects to which specialization limits humans. Revolution in science – laws are relative only, valid under selected conditions, neither for ever nor in general	A single profession is only exceptionally enough for the requisite holism rather than a fictitious holism of approach; the latter causes unpleasant and dangerous surprises

Opposition to alien and other novelties, because they do not match the old experiences	Observing, learning, and networking of findings complying with models; limitation to unmentioned selected viewpoints	Consideration of impact of the selected borders of observation and viewpoints of consideration	Rationality means that values and other emotions are not taken in account in decision-making; human nature makes this supposition impossible
One-sided learning – superficial and with no deep engaging in the hidden essence ("supra-natural forces")	Despite analysis, learning is one-sided, since models leave aside the unselected parts of attributes; this causes fictitious insights, leading to unpleasant and dangerous surprises due to oversights of attributes	The human is essential with her network of "knowledge and values and other emotions" – science and practice of impact depend on it; surprises are expectable	Decision based on taking circumstances in account; it depends on selection of attributes to be considered or left aside; hence inter-disciplinary creative cooperation is needed for more holism
Holism is not the selected viewpoint, but adaptation to nature might be	Holism is limited to the selected specialization – science, unconsciously perhaps	Holism depends on one's selection of the network of viewpoints, which might be interdisciplinary	Holism of decisions – up to capacity and will of decisive ones to cooperate with opposing thinkers
"It is as it is." Logic of old experience, including experience concerning impacting and deciding, not observing only	"Laws are objective, but inside single sciences" – the traditional sciences. Details are exposed, links and synergies are overseen. Limitation of entanglement to complicatedness, perhaps complexity inside a single science only.	"Laws are relative, depending on humans and their selection." Along with details links and synergies may be exposed, but not unavoidably. Limitation to complicatedness and to complexity inside a single science is possible, but not unavoidable.	"Humans decide how they understand and apply findings, including laws" – use of science in practice. Limitation to complicatedness and to complexity inside a single science is expectable, but avoidable, if there is a team in decision-making
Natural holisms of humans, who feel, observe, think, decide and impact, while they are able and willing to cooperate with opposing/diverse thinkers. Cooperation makes them more successful than the one-sided ones. It is reflected in e.g. yin-yang, ancient Greek and modern dialectics, Maya, etc.	Descriptive sciences, which offer with analysis inside their specialization precious, but partial findings that they do not link with insights from other sciences.	Descriptive sciences that do not work on nature, but on human impact; their viewpoint covers especially methods and holism limited to them. They offer precious, but partial findings that they only exceptionally link with insights from other sciences.	Descriptive sciences that do not work on nature, but on human impact as an interim phase between observation and impact. Holism is limited to this topic only. They offer precious, but partial findings that they only exceptionally link with insights from other sciences.

Table 16 *Delimitation of cybernetics of the 0, 1st, 2nd and 3rd order (continuned).*

The point is in practice with no theory.	Bertalanffy's statement, that he has created systems theory against over-specialization and hence as a worldview and methodology of holism is not considered a lot.	If human impact is dealt with in an interdisciplinary way, this statement of Bertalanffy is considered more probably.	If human decision-making is dealt with in an interdisciplinary way, this statement of Bertalanffy is considered more probably.
Natural reductionism, often dangerous	Scientific reductionism unavoidable and often dangerous	Scientific reductionism probable and often dangerous	Practical reductionism probable and often dangerous
NO REFLECTION IN SYSTEMS THEORY DUE TO LIMITATION TO PRACTICE ALONE. A KIND OF SYSTEMIC THINKING IS POSSIBLE.	REFLECTION IN DESCRIPTIVE SYSTEMS THEORIES, BUT INSIDE A SINGLE TRADITIONAL SCIENCE, AND IN MATHEMATICAL SYSTEMS THEORY.	REFLECTION IN SYSTEMS THEORIES CONCENTRATING ON HUMAN IMPACT.	REFLECTION IN SYSTEMS THEORIES CONCENTRATING ON HUMAN DECISION-MAKING.
In a global and innovative society, humankind's survival and business success of its organizations depends on requisitely holistic behavior. Systems theory is not applied for a rather deep description of components and relations inside the considered features from a selected single viewpoint, but in an interdisciplinary creative cooperation.			
Even if the most generalized Bertalanffy's definition of the notion system is applied and taken off context, it reads: "system is made of components and relations between them giving it its specific attributes different from attributes of its parts alone"; it leaves open the practical issue: "Which part of the feature dealt with are we considering, since we cannot consider all of them for natural reasons?" If humans were capable of the latter, many less professions and specializations would have shown up over recent centuries and decades. Each and every one of them is precious, but none is self-sufficient or sufficient for holism of behavior and wholeness of outcomes, and none can prevent all unpredictable and frequently unpleasant consequences of one-sidedness.			

Table 16 *Delimitation of cybernetics of the 0, 1st, 2nd and 3rd order (continued)*

1. Transition from the 1st and 2nd order cybernetics to the 3rd order cybernetics;
2. Inclusion of the Cybernetics of Conceptual Systems into this novelty;
3. Application of DST in this synergetic novelty in order to provide for the requisite holism of behavior of influential actors and requisite wholeness of outcomes of their actions; and application of the:
4. The Universal DST as a bridge between the natural and social sciences.

This might transform the suggested novelty into an innovation, i.e., a new and accepted source of new benefit from the new idea for its users, and consequently, also its authors (EU, 2000: 6). This official document links the invention-innovation-diffusion processes very directly with the kind of systems theory Bertalanffy has been striving at[5]. So do we (Mulej, 1974, 1975, 1976, 1979; Mulej et al., 1992, 2000, 2008; Mulej & Ženko, 2004 a, b).

All five interdependent components of DST can be applied also with no formal use of systems theory, if one uses the USOMID methodology.[6] See Ch. 13.

The essence of the Universal Dialectical Systems Theory is published (Božičnik & Mulej, in *Cybernetics and Systems*, 2009).

The resulting suggested new synergy may be found in Table 17.

In any thinking and action, be it supported by systems and cybernetics, information is a crucial ingredient, input, throughput and output. Thus, let us devote some time to it now.

11.3.8 The Role of Information in Systems Thinking

In daily life, the word information tends to be used in many different contexts and contents with a lack of a clear definition (see Rosicky, in Mulej et al., 2000). We were able to see the same reality concerning the word system.

---

5. "'The Action Plan [First Action Plan for Innovation in Europe, 1996, based on Green Paper on Innovation, 1995] was firmly based on the 'systemic' view, in which innovation is seen as arising from complex interactions between many individuals, organizations and environmental factors, rather than as a linear trajectory from new knowledge to new product. Support for this view has deepened in recent years.'"

6. Application of USOMID (Mulej, 1981; Mulej et al., 1982, 1984, 1986, and later in books referenced above) has supported the invention-innovation-diffusion process in several hundred cases in several countries yielding many millions of Euros in innovation-based incomes of its users.

Background of action	Result of application of background
Use of the 0-order cybernetics	Impact with no theoretical background – a big danger of one-sidedness due to the usual limitation of approach to a single viewpoint, but incidentally requisite holism/wholeness is possible due to the involved humans' natural capacity of creative cooperation or due to simplicity of the tackled problem allowing for a single capacity to suffice
Use of the 1st order cybernetics	Insight with a theoretical background – a big danger of one-sidedness due to the usual limitation of approach to a single viewpoint, but incidentally requisite holism/wholeness is possible due to the involved humans' natural and/or learned capacity of creative cooperation or due to simplicity of the tackled problem allowing for a single capacity to suffice
Use of the 2nd order cybernetics	Impact with a theoretical background – a big danger of one-sidedness due to the usual limitation of approach to a single viewpoint, but incidentally requisite holism/wholeness is possible due to the involved humans' natural capacity of creative cooperation or due to simplicity of the tackled problem allowing for a single capacity to suffice
Use of the 3rd order cybernetics	Impact with a theoretical background – a big danger of one-sidedness due to the usual limitation of approach to a single viewpoint, but incidentally requisite holism/wholeness is possible due to the involved humans' natural and/or learned capacity of creative cooperation or due to simplicity of the tackled problem allowing for a single capacity to suffice; decision-making and action are exposed
Use of any systems theory	Insight with a theoretical background – still a big danger of one-sidedness due to the usual limitation of approach to a single viewpoint, but incidentally requisite holism/wholeness is possible due to the involved humans' natural and/or learned capacity of creative cooperation or due to simplicity of the tackled problem allowing for a single capacity to suffice
Use of Cybernetics of Conceptual Systems	Impact with a theoretical background – still a big danger of one-sidedness due to the usual limitation of approach to a single viewpoint, but incidentally requisite holism/wholeness is possible due to the involved humans' natural capacity of creative cooperation or due to simplicity of the tackled problem allowing for a single capacity to suffice; humans are more considered than in Cybernetics of the 0, 1st, and 2nd order as member of society
Use of the Dialectical Systems Theory	Insight and impact with a theoretical background – no danger of one-sidedness due to no limitation of approach to a single viewpoint, but normally requisite holism/wholeness is possible due to the involved humans' natural and learned capacity of creative interdisciplinary cooperation

Use of Cybernetics of Conceptual Systems supported by (U)DST in synergy	Insight and impact with a synergetic theoretical background – a little danger of one-sidedness due to no limitation of approach to a single viewpoint, normally requisite holism/wholeness is possible due to the involved humans' natural and learned capacity of creative interdisciplinary cooperation and consideration of involved humans as members of society and mutually different as humans and as professionals
Use of the new 4th order cybernetics as a synergy of (U)DST, cybernetics of 0, 1st, 2nd, and 3rd order (as appropriate) and cybernetics of conceptual systems	Insight, decision-making, action and impact with a synergetic theoretical background – a little danger of one-sidedness due to no limitation of approach to a single viewpoint, thus regularly rather than incidentally requisite holism/wholeness is possible due to the involved humans' natural and/or learned capacity of interdisciplinary creative cooperation and consideration of involved humans as members of society and mutually different as humans and as professionals, including their ethic of interdependence; Result: less one-sidedness under the name of systems theory and/or cybernetics, and resulting failures, more success in control of complex reality/events/processes

Table 17 *The concept of the 4th order Cybernetics (continued)*

This is far from being the only attribute they may have in common.

Let us repeat: information must, first of all, be delimited from the notions of data and message:

- Data exists when signs are ordered into a syntactic entity, such as a word, an organ of a body, a leaf of a tree, a picture, piece of music, and so forth.
- Message shows up when data is ascribed a meaning, thus receiving its semantic dimension.[7]
- Information shows up when a message makes an impact by coming to be understood, accepted and causing an action. This is called the pragmatic dimension of information.

From such a definition, one can see that it is problematic, if the making and application of computer is called informatics, even less so to let it monopolize the notion.[8] The same would apply to ordering data in book keeping, in libraries, for example. One can find something, which all of them have in common, and so do all other features meeting the mathematical definition of a system:

- Information is an influential relation.
- There is no system, hence, without information.
- There is no entity, hence, without information.
- There is no order, hence, without information.
- Information is a natural phenomenon, which is not limited to humans and their relations and organizations.
- Information is, potentially, but not unavoidably, supportive of holism.[9]
- Information is an expression of interdependence in general.[10]

---

7. The role of the selected viewpoint already is entering the scene, which has not been so critical with data. This is even truer in the case of information. Different viewpoints make different messages and information from the same data. E.g.: "Inflation rate is high." This data may cause information: "Let us spend money quickly." Or "Let us change the interest rate of the National Bank." Or: "Let us acquire foreign currency and sell it again later or save in it."

8. On these terms it is wrong / superficial to speak of "information society" and include the modern communication facilities and tools, only, more or less, such as computers, telephones, internet, etc. They make only a fraction of the innovative business and innovative society. We will come back to them in Chs. 14 and 15.

9. An impact can make people behave on a biased basis, e.g., by political propaganda, etc.

10. Impact is normally a both-way relation: one must adapt to one's listener to attain influence over on him or her.

- Information can be a physical (e.g., in a stone, in a machine), biological (in a living cell, organ, organism), and/or human (in a group, organization, society, humankind) relation.
- Information can be linked with evolution (e.g., of a cell of an embryo, developing into lever, of another cell of the same embryo becoming the eye) and with development (of e.g., a society from a nomadic one to a postindustrial one over many steps in the process).
- Information can be a tool against entropy, a tool of negentropy, because (and if) it induces order, evolution, development, holism, interdependence, and relations, keeping or transforming an identity of an entity under impact of/by information.
- Information can (also, but not only!) be a product of consciousness in terms of knowledge, data interpretation, learning and other experiencing, indeterminism and determinism.
- Information can be insufficient and/or exceed the information requirement/needs.
- Information can be subject to individual subjective understanding of given data and messages. It depends on the selected viewpoints, of course.

The viewpoints in which the traditional sciences were specializing did not focus on information—but rather on energy and matter and their flows. The issue, e.g., was how much energy or food an embryo may need to become able to be born and survive. The issue from the viewpoint of cybernetics and systems theory results from a different viewpoint: Why will an embryo become a dog or an elephant rather than a tiger? The answer is: information.[11]

In dealing with, for example, climate change, humankind of today may still have to come across a similar change of questions put from different viewpoints. As long as only the traditional question was asked, only (!) the basic process (the one of production of products and/or services, its supplies and their sales) was found worth consideration. Cybernetics found the information and management processes to be (interdependent and interactive) preconditions of the basic process (and impacted by it, too, of course). Hence, consequence in the form of e.g., climate change, do not come from the basic processes only,

---

[11]. In this case information is found in the *structure of the system* (= *complex entity* rather than mental picture) of genes, as humankind has learned, not in genes, which are something of matter with little energy, but a lot of impact, as a system (in the mathematically based sense of its definition, i.e., an entity).

even less from them alone or in isolation from anything else, but they also/rather come from the information and management processes, involving both, humans and the other nature, even the Universal one (Myers, 1991).

These two processes govern the basic process—in a way of the choice made by the owners of the management process. Its owners, of course, must consider the basic process very carefully in order to place the right instructions into it, but this is their choice anyway[12]. This set of findings lets us see the growing complexity of managing an organization (of any size) and of its consequences. Let us, hence, take a look at possibilities to simplify management processes, which may let us have more time to deal with the crucial open issues of the dialectical system of all three processes.[13]

11.3.9 Some Tools Usable for Simplification of Management/Impact in Complex Processes and Situations

First of all, we should never forget the sentence by Albert Einstein, the great scientist in physics who has never developed his exceptional findings by experiments, but rather by mathematics (which is the biggest level of abstraction and reduction). He said (Thorpe, 2003):

Let us simplify as much as we can, but no more.

---

12. Which instructions are the right ones? This decision depends on the selected (dialectical system of) viewpoint/s and hence on the subjective (and objective) starting points of the decision makers, not only on circumstances. It depends also on the data receiver, of course.

13. Mulej and Kajzer (in Mulej *et al.*, 1994) published (a dialectical system of) four sets of guidelines aimed at making the basic, information, management processes and the process of their coordination creative and innovation friendly in human activities. Experience proved they were useful, especially in the transition processes from older to innovative society and economy: they help their users to think requisitely holistically about their issues.

What can be simplified? Reality is as it is, it cannot be simplified, this would reach beyond the human scope[14, 15]. The human image of reality can be simplified; this generates dialectical systems, systems, and models. They are used as the bases of the human action, not insight only, of course. If the basis is too simplified, the action will tend to have a too unrealistic background. Hence, this simplification may be helpful and dangerous, both at the same time, even.[16] See Table 7. Object[17] exists and has all attributes it has by its

---

14. Humans are humans, not simple animals, because humankind produced and produces tools aimed at simplification of our life and work. They are technological, organizational, and other kinds of tools. And they do simplify humans' insight and influence rather than reality, frequently, but mostly on an one-sided basis. Earlier we recalled how very complex and complicated the consequences, including the side-effects, of an exaggerated simplification may be. Climate change is a very visible case of such complex and complicated consequences, resulting / emerging from many one-sided "simplifications". See Table 12 again. How important this point is, can be visible seen from the fact that the Nobel Prize for peace 2007 was granted to scientists and politicians working on climate change problems; they are interdisciplinary and international, which helps them attain the requisite holism.

15. For example, over millennia of development of humankind, many humans moved from forests to towns. In old times, in forests, it was good enough to bend down and pick up a berry in order to eat it. In the modern, developed/advanced times and societies, the process has many more steps and requires much more investment, which is called progress and productivity growth. But it includes making cars, trucks, roads, refrigerators, supermarkets, etc.

16. Of course, new and new tools are invented and (in a very small percentage) transformed to innovations, too. From the viewpoint selected by their authors, they may simplify the human dealing with reality. They may diminish the remaining variety, as the VST would tend to say. But reality is as it is, from other viewpoints. If chemicals are used a lot in agriculture etc., it is easier to produce more food. But: what about their impact over on health, nature, etc.? What about the longer-term usability of the same arable land, after it has been poisoned with millions of tons of chemicals a year for long decades? Cars make our life easier—in some respects, such as moving around. What about parks changed to parking lots, about breathing clear air, being in physical shape and hence healthy, etc.?

17. The object (feature, process, event, …) exists and therefore has *all* attributes. Once one starts to deal with it, one selects among those attributes the interesting / important / essential ones. In the case of this book it is humankind, which is considered as an object from the dialectical system of viewpoints-systems that are modeled in chapters.

nature[18]. Dialectical system[19] allows for a requisitely holistic presentation of the object. System[20] allows for a one-sided presentation of the object from a single selected viewpoint. Model[21] allows for a rather understandable presentation of the system, not of the object. In human interaction models are used; hence the basis of human interaction is a very simplified one, compared to the reality that humans try to comprehend and master, with application of traditional sciences, systems theory and cybernetics included; if there is not enough of the creative interdisciplinary cooperation, success is rarely possible.

If simplification is unavoidable, reduction from the object level to the model level is so, too. The reductionism, which Bertalanffy was fighting rightly under the label of over-simplification (and we are as well), is back, for natural reasons. The issue, which shows up reads:

In which ways can one simplify/reduce the total amount of attributes to the requisite one, in order not to exaggerate, but to rather comply with the law of the requisite holism?

---

18. The point of the Table 7 in terms of the real life and the human dealing with it reads: it is hardly possible to deal with anything complex without cooperation of several persons, and hence without communication between them. For communication to be possible, one comes—as an individual—rather unavoidably to the model, while one tries to deal with the object. In other words, all three steps of simplification in Table 7 are normal. The basis for cooperation, thinking, decision making, and action, is therefore quite far away from a totally realistic one. Thus, returning to the level of the dialectical system (of models) is crucial for success, which is unavoidably based on the requisite holism (according to experience we have watched). That's why we said a second ago that *simplification may be helpful and dangerous, both at the same time*.

19. In the case of climate change, which we have taken as an extreme example: the viewpoints of description of the nature, of climate circumstances, of climate change, of systems thinking as a thinking tool, etc., and their relations—this is what makes our dialectical system about climate change (See: Ečimovič et al., 2002, and later). Other authors have different options.

20. In the case of this book: our thoughts in the framework of our single viewpoints make our single systems per chapters (which emerge to the synergy called the dialectical system, when we link them to make the book).

21. In the case of this book: every chapter is a model of a single-viewpoint system, so is every paragraph, sentence, and table. From every selected viewpoint a system as a mental picture of the considered object is introduced and modeled. They all emerge to the synergy called the dialectical system of models and make the book, which is the final model for us to convey our findings to our readers.

It was discovered that it is better to face the topic of complexity and its simplification by a process approach rather than by a structured approach to organization, production, and so forth. (Schiemenz, in Mulej et al., 2000, and earlier; Kajzer, 1982). Oversights tend to include less many important attributes in this way.[22]

Hierarchy, if it is not limited to a commanding hierarchy of subordination with no creative cooperation between bosses and their co/workers, is a useful tool of simplification of management. It allows for parts different from each other in the same process to be considered as relatively independent entities on which specialists can work. The interdependencies and interactions among members of such a sub-entity (e.g., finance department in an organization) are more frequent and important than the ones among different sub-entities. Acknowledging the differences among parts of the process is the basis for division/distribution of labor and even more for coordination of work processes into one entity.[23] Specialization does not make e.g., departments special only, but also interdependent: they are complementary rather than self-sufficient. The same applies to organs of a body, parts of nature, products, and so on.

Recursion is a different way of simplification, although quite closely linked with hierarchy in a number of cases. The point of this simplification of management is not in the differences, but rather in the similarities, which show up again and again. Specialization to a specific profession is such a case: it is easier to become a good boss in the same specialized department and industry than in a different one. Repetition of the same features (= recursion) allows for routine and requires creativity to be employed to the remaining, non-repetitive, non-reoccurring, non-recursive features.[24,25] Standardization is enabled e.g., if attributes of products, or processes, are built into automatic

---

22. In the case of dealing with anything, e.g., the climate change, both the process of its making and the one of solving problems related to it matter more than the issue of who is whose boss. The latter issue shows up as well, of course, but it is the process that makes the structure needed and appropriate rather than vice versa.
23. This is how a process-based structure comes in. Departments of an organization pass their plans, but they coordinate them, too, hopefully, in order not to make more damage to each other than benefit to themselves.
24. This would be the case of the remaining variety in the Law of the Requisite Variety and its application in the Viable Theory/Model. See about it in Ch. 11.3.1.
25. In the case of dealing with the climate change, it may make sense to invite the ones who have already collected knowledge by education and experience (as well as suitable values, emotions, and talents, of course), to work on a new plan concerned with dealing with e.g., the climate change locally elsewhere (Ečimovič et al., 2007).

machines, in decision-making (at least on a framework basis), or somewhere else. Today, recursion is often called fractal structure/attribute.

"Black-box" can also help simplify the management. Car drivers need not to know the functioning; it is enough, if they know only the behavior of their cars, so do TV-viewers, users of kitchen appliances, persons cooking their own tea, or coffee, who are no profound professionals. Frequently it is not necessary to know the inside, the "hidden processes", to manage, these cases say. Sometimes these processes are impossible to know but on the level of behavior, i.e., on a (more superficial) black-box level, such as processes in a brain. In business, democratic bosses may have much less work to do, because they are capable of trusting, hence of considering their subordinates as black-boxes and concentrate on the remaining variety.[26]

Feedback may help such bosses to control the process well enough. But feedback is not only a type of input-output relation between human beings and/or other parts of nature. It is a basic attribute of artifacts based on first order cybernetics, as discussed above. All automata are self-regulating due to feed-back, but the level of temperature of a water-heater, for example, is predetermined by feed-forward information installed. This is called regulation rather than self-regulation, for this reason.[27] Nature applies self-regulation as well.

In nature, there is a lot of self-regulation, if humans do not intervene too much. Harmony arises from interdependence and interaction of different parts making the same eco-system; it is a process of a dynamic stability.[28] A trusting boss with trustworthy coworkers may use a black-box approach and self-regulation much more than other bosses. This can be called autonomy.

---

26. In the case of dealing with the departments of an organization, the coordinators among a number of them should better trust their coworkers and concentrate on what is not covered.
27. In the case of dealing with e.g., production, reports to coordinators are a case of feed-back loops, mutual impacts of parts of nature on each other are another case, measurement with instruments can let us see them better / more easily.
28. The issues of climate change e.g., emerge from both the self-regulation of nature and the many human interventions, which are (more or less unavoidably) one-sided in their character, and their emerging synergies.

Autonomy can be found in nature and in organization. It can be called a way of using the black-box approach, hierarchy and recursion/fractals combined, as well as regulation and self-regulation combined, or even all of them combined.[29]

Standardization is another way of simplification of management. Standardized parts of machines are easier to replace. Standardized rules of conduct are easier to follow. Programoteque is such a case (on a framework level!), as mentioned in Ch. 13.2. Standardization of decision-making is also possible, but it is easier to attain in terms of methods than in terms of contents (Potočan, 1999).[30, 31]

Trustworthiness and reliability of partners in business, private life, working together allows for granting much more autonomy than the opposite attributes do. It can develop to substantial savings, too, and is therefore investigated a lot under the term of creditworthiness (Knez-Riedl, 2000, and earlier). Social responsibility is a new attempt for linking trustworthiness and requisite holism (Hrast *et al.* (eds.), 2006, 2007, 2008); it was reduced to corporate social responsibility in the 1960s, but it is considered in broader terms now, e.g., as a next step in development of methods of (informal) systems thinking about innovation such as Total Quality Management, Business Excellence, Business Reengineering, Innovative Business, Baldrige Award, European Excellence Award, Deming Prize, etc. (for an overview see Gorenak & Mulej, 2009). Some authors even see a chance for worldwide peace in development of social responsibility (Crowther *et al.*, 2004a, b; Crowther & Ortiz Martinez, 2004; Ečimovič *et al.*, 2008). In combination with creativity with a purpose rather than laboring as the most unpleasant part of one's life, ethics of interdependence,

---

29. If the selected viewpoint makes somebody concentrate on an autonomous unit, the impression can result that this is an independent unit. This shows up as a wrong conclusion, if another—broader—viewpoint is selected.

30. In dealing with e.g., climate change, standardization may be seen in the typified framework models how to approach the problematique, such as Local Agenda and Local Agenda 21, in measurement, in adoption of somebody else's experience, etc. In a creative work, the standardization remains on a framework level based on successful experience, of course. It serves as a reminder / check list, etc. Guidelines in DST are such a case. We will come to them in Chs. 12.3.3 and 12.3.4. Creativity concerning the details is required. In a routine-loving way of work, rules of conduct replace the users' creativity. Every step must be fully covered. (In the modern daily life, users of computers face rules all the time, so do users of most other machines.)

31. Application of quantitative methods that are quite close to a hard systemic approach requires some standardization of data and procedures, too. We will come to their support for innovative and systemic thinking in Ch. 12.3.6.

and requisite holism we even see in social responsibility a humankind's chance to find a way out from its pending danger of "side-effects" of affluence (Mulej et al., 2007; Mulej & Prosenak, 2007; Potočan & Mulej, 2006a, b, c, 2008; Prosenak et al., 2008).

11.3.10 Historical Parallelism of Systems Thinking: Maya and the Evolution of Consciousness vs. Contemporary Systems Thinking

Systems thinking is a framework that is based on the belief that the component parts of a system can best be understood in the context of relationships with each other and with other systems, rather than in isolation. The only way to fully understand why a problem or element occurs and persists is to understand the part in relation to the whole (Capra, 1997).[32]

There are some historical facts regarding systems and systems thinking. Systems thinking, as a modern approach to problem solving, was revived after WWII even though it had been an ancient philosophy. We can track systems thinking back to antiquity. Differentiated from Western rationalist traditions of philosophy, C. West Churchman often identified with I Ching (http://en.wikipedia.org/wiki/I_Ching) as a systems approach sharing a frame of reference similar to pre-Socratic philosophy and Heraclitus (Hammond, 2003). The first systems thinkers can be found in the oldest of human societies—the ancient Phoenicians with their cuneiforms, the Egyptians with their pyramids, the Chinese with the yin-yang concept, the Greek philosophers, and the Maya Indians are the earliest ancient societies of system thinkers. We will briefly summarize Maya here.

11.3.10.1 Maya and their Concept of Systemic Thinking

The Mayan numerical system and long count units has been proven as one of the most accurate systems for describing the present and future of the civilization in which we have all evolved. The Mayan calendars Tzolkin and Tun, based on mathematics as a strictly rational factor and enriched by intuition, are examples of an evolutionary system of human consciousness. The calendars and their meaning for sustainable society were completely explained and scientifically proven by the Swedish microbiologist and Professor Carl Johan Calleman. The calendars presented personal intents of individuals and prophetic meanings for civilization (Calleman, 2004). Basically, he deciphered

---

32. Thus, Jere Lazanski calls a complex entity rather than a mental/emotional picture of it a system, in this chapter.

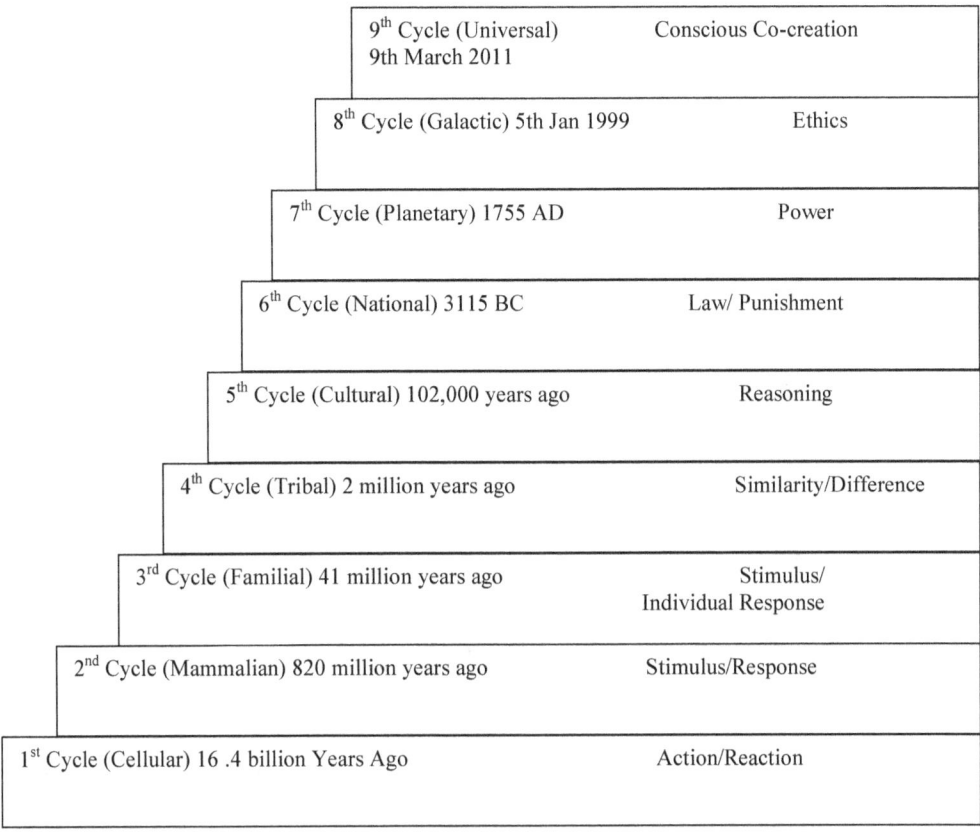

Figure 2 *The evolutionary system: Conscious evolution of each cycle (source: www.mayanmajix.com)*

the purpose of the calendars, what they represented and meant to the Mayans and how they used them. He discovered that the calendars were timing the development and evolution of consciousness (individual, societal, universal).

Maya Indians as ancient system thinkers offer a radical shift of perception in thinking to a modern man. Contemporary society is an interdisciplinary society, an interdependent phenomenon that interrelates with almost all areas of human lives. This fact perfectly describes society as a complex system whose problems are softly defined and phenomena uncertain. Decision-making is difficult and stressful; solutions are short termed. Usually people make their decision in a traditional way: they use conventional/linear thinking. The use of systems thinking offers an individual broader perspectives, long-term solutions, **naturally achieved sustainability** and harmonious elements within the society.

One very interesting fact, as the following history shows, is that systems thinking has not been mentioned among leading philosophers, politicians in the centuries between the breakdown of Mayan civilization and the beginning of World War II. The reason for this can be found by following the **Mayan nine-step system** (see Figure 2) of creation and evolution of human consciousness. "The nine-step Mayan pyramids are thus telling us that consciousness is created in a hierarchical way and that each Underworld stands on the foundation of another" (Calleman, 2004).

Figure 2 shows the Mayan nine-level pyramids system that represents the evolution of consciousness. **The evolutionary system** started with the "big bang," 16.4 billion years ago, when the first living cells appeared. The level of consciousness was at that time only action followed by reaction of organisms (Calleman, 2004). On the other four steps, consciousness evolved from stimulus/response, stimulus/individual response, similarity/difference, reasoning—started by the shamans of the tribal groups when they (the shamans/the learned or wise men) came up with the reasons as to why things are/were the way they are. These shamans then evolved into priests and religions on the basis of cosmology, through the cultural cycle and formed a single pantheon with the religion of Egypt. This is where people developed their own styles for survival according to their reasons for/of life. The last four steps of consciousness as an evolutionary system are the most important ones, since we can find parallels with events that happened and happen today if we carefully observe this system.

The sixth step of consciousness, which began in 3115 BC, was law and punishment and lasted until 1755 AD. The concept of good and evil developed (Adam, Eve and the apple—the idea of retribution), but also the Laws of Nature and Science were discovered (e.g., Laws of Thermodynamics).

The seventh step of consciousness from 1755 to 1999 was a consciousness of power, **where there was no place for integration** but analyzing, separation, creating towers of power, wars and manipulation. This is a reason that no one would think of connection and integration, of systems thinking in its highest meaning; not one philosopher or politician. People came to be above the rule of societal law, the Industrial Revolution happened, followed by the knowledge brought by the IT revolution; society has "horseless carriages with engines that have the power of many horses." Everything depends on the individual viewpoint of the application of power that makes the use of power good or evil.[33]

---

33. The same amount of conscious evolution occurs in each cycle. This means that the amount

The description of Mayan nine step pyramids starts with the 1st cycle and big bang, which ended with the beginning of cellular consciousness, 16.4 billion years ago. The 2nd cycle ends with the beginning of mammalian consciousness 820 million years ago. The 3rd cycle ends with the beginning of familial consciousness 41 million years ago. The 4th cycle ends with the beginning of tribal consciousness, 2 million years ago. The 5th cycle ends with the beginning of cultural consciousness, 10002 years ago. The 6th cycle ends with national consciousness 3115 BC. The 7th cycle ends with the beginning of planetary consciousness, in 1755 AD. Each cycle represents a step on a pyramid. The complete explanation of evolution of consciousness was described in Calleman's works (Calleman, 2001, 2004).

Today we are living on the eight level of consciousness which started in 1999 and will end in 2011. It is a consciousness of ethics, where all the towers of manipulation of negative power are collapsing. Ethics in the largest sense means spontaneous solutions through the application of law and power to the benefit for everyone. It shines from within and is personal, knowing the right thing to do and doing it. It is a refined consciousness. Now, the powerful people who make the laws and lead the nations and societies cannot get away with anything without being exposed, all abuses of power are becoming uncovered.

This consciousness leads towards the last ninth step of Mayan evolutionary system, towards conscious co-creation, which begins at the end of 2011. Here we can talk about the integration of all systems, which leads the planet to **one harmonious system**, which was mentioned by system thinkers in the 20th (Bertalanffy, 1952; Wiener, 1948; Mulej, 1979; Senge, 1994; Ackoff, 1999). Russell Ackoff (1999) in particular clarifies the differences between the conventional, linear thinking and systems thinking. A clear understanding of the difference between analysis and synthesis is crucial for an introduction to the theory of a system.

Ackoff explains that analysis has been the dominant mode of thought in the Western world for 400 years. Here we can compare the sixth, seventh, eighth and ninth level of Mayan nine steps system consciousness with Ackoff's discoveries. Analysis explains how the pieces of a system work. According to

---

of change that happened in the 1st cycle of 16.4 billion years happened in the 2nd cycle 20 times faster in 820 million years.

the Mayan calendar and the evolution of consciousness, there were certain steps in the evolution, which represented the consciousness of analysis: law and punishment (6th step, from 3115 BC-1755 AD) power and fear, division and ruling (7th step, from 1755-1999 AD). We need to synthesize in order to understand the system and the interactions between its parts as they work together. As much as this is valid for living systems, it also represents the 8th and the 9th steps of the Mayan calendar regarding the transformation of human consciousness. At present, we are in a period of ethics (8th step, from 1999-2011), which connects and foresees the integration of the elements into a whole—so called oneness or wholeness (9th step, from 2011 onward).

Understanding the implications of seeing the organization as a system leads to the conclusion that cooperation (integration) is more effective than internal competition (separation) in leading any organization to work more effectively. Cooperation and integration, which systems and systems thinking brought were not welcomed by those rulers who wanted power through wars and separation (the consciousness of power). For example: Napoleon I, rulers in the time of WWI and WWII.

Systems and systems thinking were revived by Ludwig von Bertalanffy with his manifesto of general system theory (Bertalanffy, 1952) and Norbert Wiener with *Cybernetics* (Wiener, 1948) as a methodology for complex phenomena research, theory and cybernetics became an important entity in different fields of scientific research. Although the word "system" denoted a whole consisting of parts and was the axiom for ancient philosophers, General Systems Theory (GST) and cybernetics clearly indicated the relevance of the order and structure of elements within a whole for its behavior. System dynamics (Forester, 1961) and systems thinking (Senge, 2006) are equivalent and can be unified within a system concept (Kljajić, 2008).

In the world of systems, we find five typical characteristics of them and define three categories of systems: hard systems (e.g., Mayan pyramids, contemporary hardware), soft systems (e.g., Mayan civilization, modern organization), and evolutionary systems (e.g., consciousness). The characteristics of systems are connected to the purpose of each system: seeking balance to serve specific purposes within larger systems, combining the parts in a way for the system to carry out its purpose and the fact that every system has feedback. The latter is represented graphically with feedback loops, which connect entities among themselves.

Hard systems involve simulations, often use computers and the techniques of operation research. They are useful for problems that can justifiably be quantified. However, they cannot easily take into account unquantifiable variables (opinions, culture, politics, and so on), and may treat people as being passive, rather than having complex motivations.

Soft systems cannot easily be quantified, especially those involving people holding multiple and conflicting frames of reference. They are useful for understanding motivations, viewpoints, and interactions and addressing qualitative as well as quantitative dimensions of problem situations.

Evolutionary systems, similarly to dynamic systems, are understood as open, complex systems, but with the capacity to evolve over time. Bela Banathy uniquely integrated the interdisciplinary perspectives of systems research (including chaos, complexity, and cybernetics), cultural anthropology, evolutionary theory and evolution of consciousness (Banathy, 2000).

11.3.10.2. Systems Thinking Principle: "The Divine Plan" vs "The Big Picture"

"One of the principal objects of theoretical research in the department of knowledge is to find the point of view from which the subject appears in its greatest simplicity" (J. Willard Gibbs, quoted in Burch 1999.)

Systems thinking emphasizes looking at wholes rather than parts, and stresses the role of interconnections. It is circular and focuses on closed interdependencies, where *x* influences *y*, *y* influences *z*, *z* influences *x*. It has a precise set of rules that reduce the ambiguities and miscommunications that can crop up when we talk with others about complex issues. It offers causal loop diagrams, which are rich in implications and insights. It opens a window on our mental models, translating our individual perceptions into explicit pictures that can reveal subtle yet meaningful differences in viewpoints (Anderson, 1997).

Table 18 presents systems thinking principles of the ancient and modern societies—Mayan and modern systems thinking. Einstein's saying that "problems that are created by our current level of thinking can not be solved by that same level of thinking," leads us towards the big picture principle or the divine plan, or the higher purpose. Systems thinking offers a crucially different way to communicate about the way we see the world, and to work together more productively on understanding and solving complex problems. Long-

MAYA SYSTEMS THINKING PRINCIPLES	MODERN SYSTEMS THINKING PRINCIPLES
The "divine plan"	The "big picture"
Long term, short term perspectives	Long term, short term perspectives
Measurable and non-measurable data	Measurable and non-measurable data
Dynamic, complex and interdependent	Dynamic, Complex and interdependent
We are part of a system	We are part of a system

Table 18 *Principles of systems thinking: Maya versus modern systems thinking*

term and short-term perspectives in Mayan time can be explained by their having equally treated each day's celebration ceremonies as well as long-term perspective wisdom. Modern systems thinking treats equally short- and long-term solutions of the issues, depending on the issue. Mayan numerals were equally important as the data written on "stela,"[34] which is a carved monument (Guernsey, 2006). The fourth principle of complexity is also valid for ancient time as well as for contemporary society. Interdependency and turbulence in everyday society causes fast systems dynamics.

The development of complexity in Mayan civilization has been invigorated by a series of spectacular finds in the lowland regions of Veracruz and Peten. These have demonstrated the gifted development of complexity among lowland relative to highland Mesoamerican societies. The discoveries emphasize the use of mature writing systems and the formation of formal political hierarchies, centuries earlier than once believed (Canuto, 2007). The last principle, which we stand, is that we all are a part of a system—either a small, big or the biggest one. The Maya wrote this statement to a Tun—one of the three calendars they use, the prophetic one.

---

34. La Mojarra Stela 1 is a Mesoamerican carved monument dating from the 2nd century CE. It was discovered in 1986, pulled from the Acula River near La Mojara, Veracruz, Mexico, not far from the Tres Zapotes archaeological site. The 4½ foot wide by 6½ foot high, four-ton basalt slab contains about 535 glyphs of the Isthmian script. One of Mesoamerica's earliest known written records, this Epi-Olmec culture monument not only recorded this ruler's achievements, but placed them within a cosmological framework of calendars and astronomical events. The monument is an early example of the type of stela which later became common commemorating rulers of Maya sites in the Classic era.

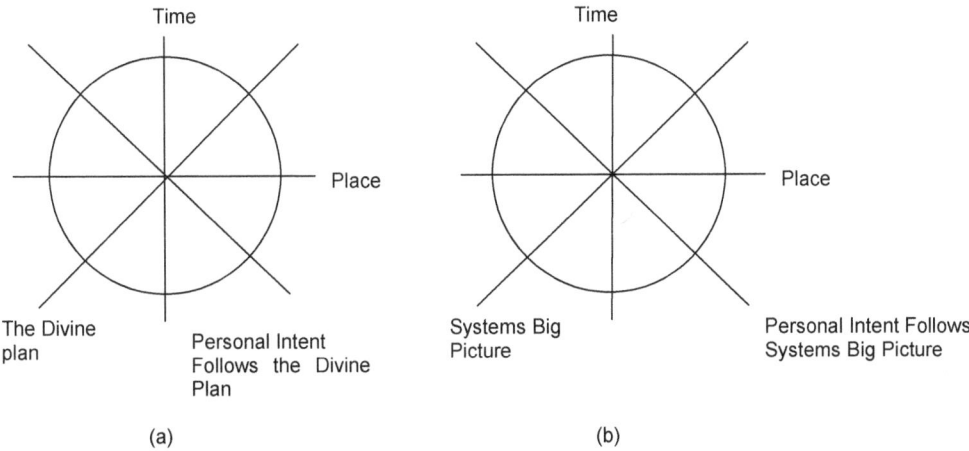

Figure 3 *(a) Mayan Divine plan and (b) Contemporary Big picture*

Let us examine one of the systems thinking principles (Figures 3a and 3b). They have seen the bigger picture, the one that now a days we call "the first principle of system thinking" (Jere Lazanski, 2008). The Maya had a so-called "divine plan," which was actually a big picture they were aware of and were part of. They knew that the civilization will achieve the system of co-creation, but they also knew what many of contemporary individuals do not know. Seeing the world from the big picture point of view is reaching a level of awareness, where linear thinking is replaced by system thinking. First, we can present the Mayan systems thinking of "divine plan or a big picture".

The Maya had a third orientation to their existence called "personal intent"[35] and a 4th orientation called the "divine plan"—"big picture". When considering the personal intent and divine plan in relation to the Mayan Calendar, it is important to understand that the Mayan Calendar is about patterns and cycles and people have to recognize the patterns. They solved problems by having a system, which brought them peace of mind. They used four steps to get them into the peace of mind. First, they were conscious of being a part of a much bigger system, which they called a divine plan. They knew they had to recognize **patterns** in order to achieve certainty. The second step was their recognition that certainty guided them to centeredness. The third step described how centeredness led them to the fourth step, called peace of mind. Today, we achieve the same state of mind when we get to the systemic structure, which drives patterns and events.

---

35. "Personal intent" means knowing what (and where) you are supposed to be doing with your life and how you can develop your own consciousness.

We reach the three levels of understanding: reaction to an event, adaptation of patterns, and creating change affecting the structure. The real power of structural-level thinking comes in: actions taken at the level of event are creative, because they help us to shape a different future, the future that we want. Our ability to influence the future process increases as we move from event-level to pattern level and to structural level thinking. Sometimes, the best action we can take must remain focused on the present. The art of thinking at the systemic structure level comes with knowing when to address a problem at the event, pattern or structural level, and when to use an approach that combines the three.

11.3.10.3. Conclusions from Chapter 11.3.10

For centuries systems thinking was overlooked. The reason for this fact can be found in its main message, following the Mayan development of human consciousness as evolutionary system: the integration of all systems, which leads the planet to one harmonious system. The evolution of human consciousness did not allow the systems thinking to appear as common thinking before the time.

11.3.11 Some Comparative Conclusions about the Selected Soft Systems Theories

Why are all the summarized systems theories so rarely applied, that world wars and other small and terrible troubles and crises, including the one of 2008– happen? For centuries the social rulers had no intention and ability to help people think and act, i.e., behave, in a requisitely holistic way. Liberalism of Adam Smith required requisite holism and/by interdependence under the name of the invisible hand, but using this notion in his style might prevent the power-holders from abusing their impact. Thus, they caused the current and earlier crises. No free, i.e., liberal, society resulted, but a misused version of it. A new way out of them is needed. Versions of it existed millennia ago, but the rulers dismissed them; thus new versions are needed.

One may conclude that the briefly summarized soft system theories can all be useful in dealing with the modern life of the 21st century. They can all build bridges between the interested parties involved; they can all support their cooperation without destroying their actual differences. They, though, differ from each other in their selected viewpoints and specific topics. Thus, it is up to the users to make their choice—from these systems of systems methodologies—to be requisitely holistic in their behavior, i.e., monitoring, perception, thinking, emotional and spiritual life, decision-making, and action.

Since the parties involved may feel opposed rather than complementary to each other, and may have different habits concerning ways of work, we may want to make them aware of four facts regarding their subjective starting points:

- In terms of knowledge on contents (the issue: What?) they are different and need to remain so in order to be complementary/interdependent/needing and benefiting each other.
- In terms of knowledge on methods (the issue: How?) they have a chance to standardize their methods of work on a frame-work level, which does not replace, but rather supports their creativity and transforming it into a common good, e.g., innovation.
- In terms of values (the issue: Why should we care for working on this topic?) and other emotions they have their interdependence as their background resulting in their ethics of interdependence; on this basis, they may cooperate better and more easily once they develop their ethics of interdependence on each other (Mulej & Kajzer, in Mulej *et al.*, 2000, and earlier).
- In terms of their natural talents (the issue: Is this activity for me, am I for it by nature, profession, and so on?) they have their natural capacities of different orientations and can develop them into better or worse knowledge, values, emotions, activities, outcomes. But they have to work on it, and it may be a hard work. Alternatives may be even harder to take.

11.4 SOME CONCLUSIONS ABOUT SYSTEMS THINKING ABOUT THE CONTEMPORARY LIFE PROBLEMS

The problems of business, living and natural environment which are intersecting[36] each other are not only expressed globally, but also locally.

---

36. Due to this fact, the Fuzzy Systems Theory may also apply to the modern business of the 21st century. We do not include it here as it is somewhere between the hard and soft systems

On this level, these problems are no longer a far-away theory or political declaration, but a hands-on system, i.e., (partial, but complex and complicated) entity of life. It can hardly be resolved by declaration rather than by concrete actions undertaken by concrete persons in charge of managing their daily life conditions and results. As long as these persons stick to biased, narrow views, their chance of success is quite poor rather than easy to attain. Once they include into their views:

- Their interdependence;
- Their impact on their starting points and, later on, all phases in the same process (see Table 19);
- The impact of their starting points on their perceptions, preferences, objectives, tasks, processes, technologies and results.

they have a better chance to survive as business organizations, individual parts of nature, and local authorities in charge of quality of life and working life.

Their transition from the biased thinking to a rather/requisitely holistic thinking may be easier, if they accept in their values and knowledge the concept of systems thinking.[37] The modern businesses of the 21st century address complex problems, which must be managed by human beings who are unavoidably narrow specialists and cooperating with each other at the same time, to be requisitely holistic. A common methodology can support them in this effort. Trustworthiness (Knez-Riedl, 2000) and ethics of interdependence may open doors for their cooperation better than mistrust and their obsolete ethics of independence and dependence, in their values and emotions. Social responsibility (Hrast et al. (eds.) 2006, 2007, 2008) may emerge.

If people find this suggestion too demanding, their alternative is even more demanding: end of chance of survival, but realization of entropy by ruining their conditions of life and work. The choice is up to them/you.[38]

Some people, especially the ones who exaggerate in trusting their own way of life (only), as well as quite many also in the less advanced areas/societies,

---

theories, and as we lack room anyway.

37. See also: (Ženko, 1999; Ženko et al., 2002; IDIMT, 1998, 1999; Mulej et al., 2002) for more about the *informal systems thinking*.

38. Because we love our and your children and grandchildren, we beg you not to choose the alternative, but social responsibility.

tend to think that their economic and social environment still allows them to remain one-sided and superficial. They tend to remain so. It is, or seems to be, simpler. Hopefully, the simple current processes will not cause to them too complex and tough consequences, as it frequently happens. The same fear is possible in the case of too narrow specialists, who are not willing to enter interdisciplinary cooperation.[39]

Prevention of many unpleasant and unhelpful consequences of oversights may be called a major aim of the (dialectical) systems thinking.

In many cases DST has worked well, over its more than 30 years of development and application, but not until it was accepted by the decision-making persons in their consciousness or even sub-consciousness[40] as the basis for subjective starting points, e.g., in the form of the individual and organizational culture. Hence:

The (dialectical) systems thinking, based on requisite holism and ethics of interdependence, can help the humankind of today and in future to come closer to the Bertalanffy's rightfully required worldview—be citizens of the entire world, care for the entire biosphere, see it as an organization with many interdependencies, fight over-specialization—by interdisciplinary creative cooperation with transdisciplinary values of the single-disciplinary specialists completing each other up.

This is why the suggestion made by Molander and Sisavic (1994) that systems thinking needs to become a mass movement, like Total Quality Management is doing, is so important.

---

39. The crisis of 2008- shows, though, how complex are the consequences of over-simplification based on over-specialization. The influential economists do not seem to listen to systems science and/or to other professionals from disciplines different from their own. The neo-liberal school of economics, called the Chicago school, taught them to apply extreme one-sidedness under the name of free market that is not really free because of big companies' bargaining power, monopolies, oligopolies, etc. We will make our comment in Ch. 17.5.

40. We might better use the word "pre-consciousness" than the word "sub-consciousness". Psychologists found that the sub-consciousness is able to be too full of data to really help consciousness, if there is no selection. It is the role of pre-consciousness to act as the door-keeper between the two (Kline, 2000).

	Influences, preconditions, circumstances	
	⇒	
	Definition and development of starting points as a requisitely holistic system	
The objective starting points, part 1: objective needs*	The subjective starting points: 1. Values and other emotions** 2. Knowledge on contents (what & why?) 3. Knowledge on methods (how & why?) 4. Talents	The objective starting points, part 2: objective possibilities***
⇒	⇒	
⇒	The dialectical system of essential viewpoints	⇒
⇒	⇒	⇒
⇒	The selected viewpoint/s	⇒
⇒	⇒	⇒
⇒	⇒	⇒
Selection of the perceived = partially objective needs	⇒	Selection of the perceived = partially objective possibilities
⇒	⇒	⇒
Selection of preferential needs		Selection of corresponding possibilities
⇒	⇒	⇒
	Definition of (well, i.e. requisitely holistically grounded, not merely desired!) objectives: What do we want (with good reason/s)?	

Definition of tasks system/s: What do we have to do in order to attain objectives?****
⇒
Definition of work procedures for every task: How must we proceed to perform?
⇒
Operation: performing all the tasks according to procedures prescribed/foreseen
⇒
Results comparable to tasks, each of them contributing to attainment of objectives
⇒
Influence over the foregoing phases of the process where needed
(Returning to the beginning of the entire process, or a phase of it, as appropriate)

Table 19 *The law of hierarchy of succession and interdependence, applied to the work procedure in general*

*   Objective needs may be visible as demand in the market, etc.
**   To psychologists, values and emotions differ, as parts of the human personality. We agree and we anyway put them together because they are both influenced by motivation, a lot (see. Šek, 2007; Škafar, 2006; Trstenjak, 1981). We will come back to them. – In the case of e.g. climate change, values causing damage to climate have been prevailing over values caring for a healthy climate, in the centuries and decades of the industrial society. Short-term effects of human actions have been preferred over the long-term ones. Etc. Growing entropic dangers result from such values and emotions, and an innovative change in values and emotions must become the next step.
***   Objective possibilities may be visible as the available raw material, equipment, labor force, finance, laws, etc.
**** Usually, for every task the procedure differs from the one for other tasks. The responsible persons or teams, hence, limit themselves to their own tasks; this is their own level of requisite holism. Coordination between them is up to their superiors who take care of the requisite holism of a broader type, and so on all way to the very highest level of the organization. – On the worldwide basis, the broadest type of the requisite holism belongs to the General Assembly of the United Nations Organization (UNO). That's why it is important, in the case of the humankind's environment protection, including the issues of the climate change, that it was UNO, which proclaimed the sustainable development as the necessary way toward survival of humankind of today and in future. Now, we even speak of the sustainable future (Ečimovič et al., 2007).

This in turn is why the quality movement is so important, if its notion is conceived in a (dialectically) systemic style (see Peters, 1995), not in a bureaucratic one (see: Uršič et al., 2000; Pivka & Uršič 2001; Pivka & Mulej, 2004).

This is why the social responsibility may enlarge its criteria towards more holism (Gorenak, 2008; Knez-Riedl, 2002, 2003a, b, c, 2004, 2006, 2007a, b, c; Knez-Riedl & Hrast, 2005, 2006; Knez-Riedl & Mulej, 2001; Knez-Riedl et al., 2006; Mulej et al., 2008a, b; Waddock & Bodwell, 2007).

Systemic thinking needs to be made more popular and generally applied, including the chaos and complexity theories (see Chs. 11.1, 11.2).

And this is what demands more of a modern learning (Lessem, 1991; Parsloe 1995; Flood, 1999; Hofkirchner, 2005; Trunk-Širca, 2000).

This is what the consideration of complexity (e.g., of globalization of economy, climate change) supports, but does so better, more effectively and efficiently, if it is conceived and used in a (dialectically) systemic style, rather than e.g., in a bureaucratic, biased, one-sided one.

The bureaucratic one tends to show up, if more short-term, one-sided and superficial business criteria tend to prevail over more long-term, holistic and deep ones, i.e., over systemic ones.

In all given and similar cases of attempts to simplify management, there is one more link to systems thinking, at least: a requisitely holistic preparation and action are prerequisites for such attempts to work rather than to disappoint.

What could and should be done for sufficiently/requisitely many (influential) individuals (and organizations) to accept the (dialectically) systemic style of thinking, decision-making, and action, so that the humankind could solve our issues concerning any real problem? How can we make it all happen/be accepted? See Chs. 12 and 13 for some thoughts about DST and its implementation.

Chapter 12

DIALECTICAL SYSTEMS THINKING: ABOUT COMPLEXITY, INTERDEPENDENCES, WHOLES AND REQUISITE HOLISM/WHOLENESS INSTEAD OF CRUCIAL OVERSIGHTS

12.0 THE SELECTED PROBLEM AND VIEWPOINT OF DEALING WITH IT IN CHAPTER 12

The briefly summarized systems theories did not ask a question, which we have found crucial from the very beginning (Mulej, 1974):

> *What kind of people are defining and realizing objectives; what help might they need to be requisitely holistic in order to fail less often in their social, professional, and personal efforts.*

12.1 DIALECTICAL SYSTEM (DS) VERSUS THE SYSTEM AND THE OBJECT OF CONSIDERATION

We are returning to Tables 1, 2, and 8 to repeat the basics and to go on from there.

The traditional definition reads that a system is a complex or very complex feature made of a set of elements and a set of their relations; it may have an environment and relations with it, as well, or not. Due to relations, and hence interactions[1]—a system is more than a sum: the whole system has properties, which single elements do not have: it has synergies. In mathematical terms, a system is an ordered set and systems form a hierarchy (Schiemenz, 1994).

The point of the original introduction of the system's concept was to put an end to the oversight of synergies that result from interdependencies. Holism of the systems view replaces one-sidedness due to which humankind has suffered terribly for ever and especially in the first half of the 20th century (two world wars, world economic crisis in-between, poorly working composed products, and other conditions of one-sidedness). Decades of application of this concept of the "system" demonstrated that it is a good concept in its very general, philosophical contents, but its aim is hardly attainable, if no education and methodology supports holism well enough.[2] This fact demands human attitudes and behavior to change, to be innovated—a narrow specialization is still unavoidable, but no longer sufficient; it oversimplifies.

What we humans need, is both, a narrow specialization and (the requisite) holism achieved by interdisciplinary cooperation backed by ethics of interdependence.

If the selected approach causes an exaggerated simplification, the usual oversight causes complex consequences: it is not reality, which is simplified, but only human dealing with it. That's why we must be able to be holistic in our monitoring, perception, thinking, feeling, spirituality, decision making, and action, i.e., behavior. But, systems theory has never been able to become

---

1. The terms "interdependence" as the background of interaction was used by the first author of the General Systems Theory, Ludwig von Bertalanffy (Elohim 1998), but does not seem to be frequently used later on until the Dialectical Systems Theory by Matjaz Mulej (1974 and later).
2. This experience lets us see, after half a century, that Bertalanffy has been right when he called GST a "teaching" rather than a theory.

a mass movement instead of being a rather closed-in science worked on and used by a few only (Molander & Sisavic, 1994). There were several attempts to solve the said problem, such as the General Systems Theory, Cybernetics, Soft Systems Methodology, Living Systems Theory, Viable System Model, Fuzzy Systems Theory, Critical Systems Theory, Autopoiesis, to name but a few. Each of them made a different contribution. We summarized some of them in Ch. 11.

Some +35 years ago, a specific systems theory was produced in Slovenia, too—the Dialectical Systems Theory (DST) (Mulej, 1974, 1976, 1979; Mulej et al., 1992, 2000, 2008; Mulej & Ženko, 2004a, b). DST starts from the notion that a system does not exist, but the object does, and the system reflects it as its mental or emotional picture, but partially: it exposes the part of object's characteristics, which is relevant from the selected viewpoint, only. Thus, a system is supposed to support holism, but it is fictitiously holistic in its own traditional definition:

From the formal mathematical viewpoints, a system is an ordered set, hence holistic.

In its contents, a system embraces only a part of the really existing characteristics, hence it is not holistic; its scope is limited to the viewpoint(s)/ aspect(s) selected by the observer(s)/manager(s) of the topic in question. See Table 7 again.

DST defines holism (Table 1) as a system (= ordered set) made of

- *Systemics (= consideration of global characteristics of the event under consideration, i.e., in a synthesis, aggregation, synergy), and*
- *Systematics (= consideration of detailed characteristics of the event under consideration per parts with no synergies, i.e., analytically, in separation), and*
- *Dialectics (= consideration of interdependences among mutually partly equal and partly different or even opposing elements of the object under consideration, and of processes, which are caused by these interdependences and make the transition from systematic to systemic characteristics of the object under consideration), and;*

- *Materialism (= realism, consideration of reality rather than self-bluffing about the characteristics of the object under consideration).*[3]

Holism of human behavior, i.e., monitoring, perception, thinking, emotional and spiritual life, decision-making, communication, and action normally includes consideration of the environment of the event under consideration, of course. This is called an open system concept. The open system concept, on DST terms, does not simply include environment from the specific selected viewpoint only, but also a system of all different viewpoints which might be relevant, essential for achievement of a sufficient/requisite holism (Tables 2, 8)[4].

The problems caused by the lack of systems thinking in the first half of the 20th century and later cannot be solved by a fictitious holism of narrow specialists alone. The solution suggested by the DST is called the dialectical system (DS). Its definition reads (Tables 2, 7, 8):

DS is a system (= network) of all relevant and only relevant viewpoints/ systems (= mental pictures of the object under consideration from the selected viewpoints).

What is relevant? It depends on a subjective decision what is found relevant, and varies in time, conditions, knowledge and values, i.e., starting points. Different characteristics of e.g., globalization are of different relevance to different sciences/practices/viewpoints (aspects) selected; the same holds of an enterprise and anything else. There is no one single truth, unless each and every aspect is made a part of a total aspects system, which is a total synergy. This would reach beyond human capacities. A DS is a reduction still making sense: it does not go all way to a one-sided and superficial approach of one viewpoint/system from the old times.

---

3. See P. Checkland's classical example: water is a real object; it has two basic elements, hydrogen and oxygen, from the chemistry viewpoint; they are interdependent and their synergy makes water's chemical characteristics which are quite different from the ones of its components. See also the example about training a horse by Balle cited earlier, the case of edible salt as a synergy of two poisons, etc.

4. Let us remind again: A total holism of human behavior is actually impossible due to natural limits of human mental capacities. Remaining inside the framework of one / two few single viewpoint/s causes a fictitious holism; this is hardly helpful. Requisite holism is the middle way out of trouble; for it, one uses the dialectical system.

DS makes a requisite holism possible. (DS and requisite holism reduce reductionism: Tables 2 and 8).

To make the DS survive as a concept, the Dialectical Systems Theory was created and employed.

12.2 THE DIALECTICAL SYSTEMS THEORY (DST)

It is no mere coincidence that the Systems Theory surfaced first as the General Systems Theory (GST), called originally General Systems Teaching (which includes impact!) and after World War II. Both world wars and the great depression between them showed extremely clear how dangerous one-sidedness is rather than holism in behavior of influential persons/organizations/countries, such as Fascists and Nazis under Mussolini and Hitler. In addition, the post-industrial and post-world-war period made the human existence increasingly and clearly depend on holism and creativity applied for innovation, which demands holism on its own (Fagerberg *et al.*, 2005 V, and later). But the GST concept, as it has been practiced, did not prove viable for innovation[5]: it did not include enough of interdisciplinary cooperation to act in line with Bertalanffy's words that he had created GST against over-specialization and for wholeness as a worldview rather than as another specialized discipline of science. By his education and practice, Bertalanffy was an interdisciplinary person (Drack & Apfalter, 2007). Norbert Wiener created cybernetics in an interdisciplinary team (Hammond, 2003). Application of both, GST and cybernetics, within a single traditional discipline may be progress, but not really the basic point: it does rarely cover the uncovered area that only an interdisciplinary cooperation can cover. Without the latter wholeness of insight or holism of human behavior is often not requisite for success of the endeavor. Isomorphisms and shared formal language help, but not enough for humans to avoid crucial oversights, as one can now see in the current blind alley of the global economic life, the big number of wars after World War II, and environmental problems of humankind.

The basic ideas—holism of human behavior and wholeness of insight and basis for action—must be revitalized, because GST and Cybernetics are still very accurate in terms of description and isomorphisms, but lack a methodological

---

5. See many conferences on systems and cybernetics, including the ones mentioned earlier: there are very few papers on innovation and very few team-work papers, even less many interdisciplinary coauthored papers (Mulej *et al.*, 2005).

support to interdisciplinary cooperation, creativity and innovation, although GST has been created as a worldview of holism or wholeness and methodology supporting it and against over-specialization, as Bertalanffy (1979: VII) put is. This attribute has become unavoidable in modern conditions.[6]

This means that the starting points of defining a DS must support creativity and holism instead of a routine-loving and one-sided behavior. DS may no way let creativity and holism disappear from the definition of the acting humans' subjective starting points via the definition of objectives, tasks, and procedures all way to the final step—goal attainment. The alternative is clear and very

---

6. According to the Nobel Laureate Lučka Kajfež Bogataj (2009), now, the number of humans is 6 times bigger than in 1800, and every human—on average—uses at least 5 times more energy, making the consumption of energy more than 30 times bigger overall. In terms of GDP every human is 17 times richer, on average, and one thousand times more mobile than in 1800. This trend keeps growing. So far one billion humans were making the consuming society; now five more billions are expected to join it. Thus, now already, the human impact on warming of the Planet Earth that is an undeniable fact, is five times bigger the natural changes such as Sun's activities. Humankind can no longer afford releasing—by burning fossil fuels—four million tons of CO_2 every hour, or cutting down 1.500 hectares of woods every hour, and adding to these causes of troubles 1.7 million tones of nitrogen in the soil—by dunging with mineral fertilizers—as humankind is doing today. A diminishing of 80% is necessary, but not in plans of the developed countries. The rich ones do not want to do it, while the poor ones are not able to do it. Though, such objectives can be attained with the given technologies, but not with the given consumption patterns and without big changes—innovations—in the structures of production and use of energy. Currently, 80% of humans live in un-appropriate housing, 50% are under-nourished, 19% live on less than one Euro a day, 50% on less than two dollars a day (as ex-president Clinton said in Slovenia on 30 October 2009), 17% have no access to healthy drinking water, while six percent own 60% of all wealth/property, and a single percent of all humans have university degrees.

Thus, the requisite holism is needed:

- Instead of the dangerous fictitious one, i.e., one-sidedness of over-specialists, and
- Instead of the unattainable total holism, and is attained, if
- Humans decide to consider all crucial and only crucial viewpoints and their relations/interactions and resulting synergies.

The dilemma reads:

- We may choose a complex way of dealing with the issue at stake, such as sustainable future, and attain simple consequences because nothing essential will be forgotten about.
- We may chose a simple way of dealing with the issue at stake, such as sustainable future, and attain complex consequences because something essential will be forgotten about.

Therefore holism is a better solution than simplification.

bad: the entropy as the permanent natural tendency towards destruction of everything may become reality, not only a threatening tendency any longer[7].

Mulej's DST made the DS concept (instead of a GST's one) methodologically supported by becoming, formally, a dialectical system of three elements and three relations, in general. They are supposed to both, influence humans as observers, thinkers, decision-makers, and decision-implementers, and help them attain the requisite holism of their behavior to enable the best possible results. Success is thus attained with no excessive effort and no crucial lack of insight and influence; this requires a well-managed process in which all crucial steps, interdependences and synergies are considered. In order to attain such a success, human attributes of both, decision-makers and decision-implementers, must support creativity and requisite holism with cooperation of complementary professionals. In this effort, it helps a lot, if application of theory can take place in an informal style. This is why there are three relations and three elements in the DST as a dialectical system: all of them are both

---

[7]. This is how Mulej (1967) defined entropy. Because nothing except the entire Universe is a closed entity in nature, entropy is only a tendency. Nature fights this tendency by laws that Prigogine has described with his principle of order through fluctuation, and economy and human society by flexibility and adaptation, taking the form of innovation (both technological and social) in current practices. If such efforts are one-sided rather than undertaken with requisite holism, entropy is no longer a tendency only. If a person, organization, or society tries to live without innovation, this is a case of one-sidedness in the current conditions of global and local competition by innovation. On the other hand, every created novelty destroys something existing so far, e.g., nature to build towns, raw material, energy, except knowledge, to make products; knowledge undergoes entropy by becoming obsolete. Schumpeterian process of innovation as creative destruction reflects the permanent synergy of entropic and anti-entropic attributes of any natural or work process. The point is in the final equilibrium or destruction: if e.g., costs for maintenance of the natural preconditions for human survival are only postponed rather than covered (Božičzicnik, 2007; Božičzicnik et al., 2008; Ečcimovičc et al., 2002, and later; Stern, 2006, 2007; etc.), the entropic tendency is growing at the expense of the same or generations to come and does so in very dangerous and rapidly growing dimensions: in a best-case scenario it may cost humankind more than both world wars combined, and in the case of further postponing it may diminish the world-wide GDP for as much as twenty percent, according to Stern. Thus, the economic growth of several recent decades is actually fictitious, in a longer-term perspective. This realization of the natural permanent entropic tendency, called entropy, is a clear consequence of human one-sided rather than requisitely holistic behavior. It has the form of natural disasters, diseases, terrorism based on envy against the privileged rather than socially responsible ones, 85% of humankind living on less than six US$ a day, a billion of them under one US$ a day, while the richest ~220 persons would prevent death of four million children, if they donated only 5% of their fortunes, etc. (Crowtheret et al., (editors.), 2004). But entropy is still a tendency: one can work against it, if one attains requisite holism rather than one-sidedness of human behavior and innovates permanently.

essential and sufficient[8,9]; in other words, DST matches criteria of requisite holism.

1. The three relations are:

    i. The law of requisite holism (it demands the author/s of the definition of a system representing the object under their consideration and/or control to clearly state what part of attributes of the object is included into their system; this is the mental picture of the object under consideration and/or control; one must do one's best to fight over-simplification by all available/crucial knowledge and skills as well as by ethics of interdependence)[10].

    ii. The law of entropy (it reflects the reality in which there is a permanent tendency towards destruction, which demands holism and innovations permanently; the latter ones have conditioned survival since the times humankind has given up human's adaptation to nature)[11].

---

8. In 2005 and 2006 we devoted quite some time to support to the qualitative methods of attaining holism with quantitative methods, especially the computer supported ones. (Čančer & Mulej, 2005, 2006). We will come back to them in Ch. 12.3.6.

9. Making new synergies from several systems theories including DST has also proved fruitful (Rosi, 2004; Rosi & Mulej, 2005; Rosi & Mulej, 2006; Mlakar, 2007; Mlakar & Mulej, 2007, 2008). For very brief summaries of them see Chs. 11.3.4 and 11.3.5.

10. Example: a general manager must consider the company with a very broadly defined (dialectical) system of all essential viewpoints. That is why he or she tends to have supporting offices and a team made of best professionals advising him or her. A lower manager is both obliged and entitled to consider the same company from a narrower viewpoint or (dialectical) system of viewpoints. Even narrower is the viewpoint in the position of an assembly-line worker. But each and every one of them must seriously judge the limits/boundaries he or she defines. Otherwise one would oversimplify and cause mistakes. See Tables 1, 2, and 8.

11. Mistakes mentioned above are caused by one-sidedness and result in failures, which result in realization of entropy. Economic experiences demonstrate clearly that the statement is true that entropy becomes reality rather than tendency most easily in closed-in situations (prisoners have no chance of survival without food, etc. delivered from outside; no closed-in economy can have all necessary resources of material and information, etc.). But the entropic danger applies also to a system/entity that is open too much. Namely, openness enables a two-way route of interaction. Nobody can only take from the environment, everybody must expect entities in his or her environments to consider him or her their environment, too, and try to get resources they may need against their own entropy. Simple example: usually, families like visitors, but they still lock their doors against undesirable ones, such as thieves.

iii. The law of hierarchy of succession and interdependence (later events depend on earlier events of the same process crucially; processes and events interact, when and because they are interdependent; interaction is a precondition of survival, too: without it processes stop)[12].

2. The three elements are:
    i. Ten guidelines defining the subjective starting points (values and other emotions, knowledge on contents, and knowledge on methods, as a dialectical system) aimed at making humans go for creativity and holism rather than for routine-loving and one-sided behavior[13].
    ii. Ten guidelines on assuring the agreed policy to survive in later steps of the working process (in which several more narrowly specialized

---

12. Once you have decided to study business, you no longer have much time for books of all other schools. And once you have acquired a certain amount of knowledge in one field inside of business, you can combine it with information in other fields. From an apple seed no fish can grow. Neither can an apple grow in Antarctic nature, because it is interdependent with nature, with bees, warm temperature, long enough spring and summer time. Etc. The Markoff chain expresses the *hierarchy of succession*, but it leaves the law of interdependence aside. The ancient Greek dialectics, as well as Hegelian and Engels's dialectics express the general attributes of all natural processes: they start with interdependence (of e.g., bees and apple flowers needing each other to survive), which is expressed in the language of dialectics as the fight (i.e., mutual impact) of opposites (e.g., bees and apple flowers), which triggers the process of change of the prior quality (i.e., essence) into a new quality, step by step (e.g., of apple flowers to growing apples), which finishes by negation of negation (e.g., the grown-up/ripe apple is still an apple, not a nut, but no longer an apple-flower). Analogical framework applies to making products, running organizations (be them families or international associations), etc. and can be seen in the hidden background of what is called physical, chemical, biological, economic, social or other general rules/laws. In literature on dialectics, one mostly speaks of the three phases of changing without mentioning the background of the process—the interdependence. Even Engels (1953: 66-72) does so, although he calls dialectics the science of mutual relations and mentions three laws. He also mentions Hegel as the author who has developed all three laws as laws of thinking, only, although, according to Engels, Hegel provides on hundred points most persuasive single proofs from nature for his dialectical laws (ibid., 67). Mulej united these laws with their background and speaks of the law of hierarchy of succession and interdependence.

13. In the same situation, different persons can see different needs and possibilities on the basis of their different conscious and subconscious parts of their personalities. Since the process of behavior starts here, we call this the "subjective starting points". E.g., an optimist sees in a glass that it is still half-full, a pessimist sees that it is already half-empty. In the same glass a physicist sees a different content than a chemist or a lawyer, an economist, a cleaning person, etc., neither on the basis of their professional knowledge alone, nor based on their values only, but on the basis of interdependence of knowledge and values. This is their action basis. We will come back to the definition of the guidelines helping humans' starting points to support innovation and holism in Ch. 12.3.3.

and routine-loving persons normally enter the stage)[14].

iii. A methodology of creative cooperation aimed at making DST viable in the daily practice as an informal systems-thinking by a shared framework programming and executing of the human creative activities (e.g., our own method called USOMID in Slovene acronym)[15].

In Table 19 we can see in which way all components and relations of DST work and show up in the case of the human work process (which is what DST addresses):

- The work starts by definition of subjective starting points in interdependence with perception of the objective starting points by involved persons.
- The subjective starting points in turn (should) cause a quite broad definition of the dialectical system of viewpoints, e.g., on the level of the top management of an organization.[16]
- Individual sectors of the organization are represented in the top management, but they tend to limit their own efforts to their own sector-based viewpoints only, once their representatives return from the top management session to their own respective sectors.[17]

---

14. In an organization, the top management and the owners define subjective starting points of the entire organization (e.g., expressed in vision, mission, politics, strategies, and objectives). According to the usual division of labor, other persons are supposed to take care of parts of the process in which these objectives should be realized. Their care for single parts may cause the big picture to be lost quite easily, which may lead to failures. Special attention must be devoted to the care for survival of the starting points, requisite holism and creativity, hence. We will come back to these guidelines as well in Ch. 12.3.4.

15. It is good to know the basic theory, but for many it is enough to know the know-how with not much of know-why. We will come back to USOMID methodology in Ch. 13.

16. Viewpoints of all top management members and their advisors, such as production, financing, human resources, supply, marketing, environmental care, social responsibility, etc. and their synergies make a dialectical system in an organizational management. In a school curriculum, all selected courses/subjects do so. Nothing essential may either be forgotten about or exposed as the only important viewpoint and attribute.

17. See the case of General Motors in the book by Barabba (2004): to make GM viable again, he had first to change the sector-based, i.e., one-sided and limited, behavior of sector representatives in the top management to a requisitely holistic behavior. He had to make them act with more ethics of their interdependence. Then, the synergy of their opposing insights could overcome their opposition to each other in the process of gradual emergence of complementarity of all of them leading to shared synergetic solutions for the problems they were facing. The new troubles of GM show, that Barabba's work has not taken sufficiently deep and long-lasting roots and practices. Too much of the "ivory tower" of departments

- The selected viewpoint defines, what they perceive and find preferential and corresponding (i.e., possibilities interdependent with preferential needs).
- Holism is thus limited to the requisite one, but only now an operational definition of the objective can take place, that leads to meeting of the preferential needs later on.
- Later on steps come, which have been usually considered for long times much more carefully than the foregoing ones: definition of tasks and related procedures, and operational work on them.
- All parallel definitions (components of the starting points; perceived needs and possibilities; preferential needs and corresponding possibilities) as well as all definitions and other activities following each other are interdependent[18].

Let us summarize: Mulej's concept was labeled a DS and his related theory DST in order to demonstrate the difference from other concepts. Dialectics expresses changing based on interactions of the interdependent components of an entity (Engels, 1953). They are interdependent because they are partially equal and partially different. Interdependence is what replaces isolation and therefore one-sidedness, and creates relations, makes them felt and found relevant, therefore also studied and considered. Hence, it is interdependence, which results in synergy towards holism (via the process of emergence). Interdependence exists between both, different and complementary ones (e.g., man and woman; tools and raw material and labor). It is their interdependence which causes them to need each other for their own success and survival, beating their entropy, at least for the time being and to the detriment of others who want/try to consume the same resources; that's why they all permanently both, compete and cooperate.

---

showed up in 1960s division of companies per specialized departments and rooms in which they work in separation rather than daily cooperation (e.g., Balle, 1994; Beer, 1975; Creech, 1994; Drucker, 1987; Grayson & O'Dell, 1998).

18. It makes no sense, e.g., to define objectives that do not meet a part of the preferential needs. It makes no sense to define objectives that cannot be fulfilled in the task-performing phase because there is no equipment, skill, material and other possibilities, etc. Therefore, a synergy of the preferential needs and the corresponding possibilities is a rather realistic basis for definition of objectives. The so-called "desired objectives" tend to be too poorly realistic and hence to lead to failures rather than successes. Due to them, many good professionals may do a lot of good or even excellent work to fulfill the set objectives, but all their effort provides no good results, if objectives are wrong, i.e., the basis for them is not requisitely holistically set and applied.

Thus, DST makes a useful general ground of human interventions, including the ones concerning innovation.

12.3 DIALECTICAL SYSTEMS THEORY: GROUNDS OF HUMAN INTERVENTIONS

12.3.0 The Selected Problem and Viewpoint of Dealing with it in Chapter 12.3

Now we will summarize attributes of DST. As we have said, DST differs from other systems theories in its stress on the attempt to influence human thinking and feeling. Hence, DST follows the aim of GST to be a teaching, and makes more concrete steps towards this end, but no deterministic prescriptions.

12.3.1 The General Ground: Entropy versus Evolution, Human Intervention

Every action that is aimed at survival and/or innovation is undertaken in order to combat difficulties and/or to prevent them from growing and becoming impossible to master. The natural tendency of everything eventually to be destroyed is called entropy. Life is continuously comprised of struggle against entropy; it is evolution, which results, whether it is in society or in other parts of nature. Human intervention tends to support or even direct evolution away from entropy (by having food, shelter, heating, producing and trading goods and services, having friends and other emotional connections, learning). But at the same time we use some resources and thus damage our natural environment, by actions short of requisite holism; they add to entropy (Mulej, 1967 and references in it).

Entropy and evolution are interdependent.

Sometimes we humans succeed in blocking entropy, but sometimes we fail. Innovation process is obviously meant to block entropy, but it fails often; the risk level is estimated about 92%, at least (Nussbaum et al., 2005). Causes may be found in (1) socio-economic and natural environments and their development, (2) in impacts of the environment and e.g., the business system[19]

---

19. Business system is the mental picture of an organization or a person from the viewpoint of working on their business / economic objectives in order to reduce their unpleasant difference between their preferential needs and corresponding possibilities (see Table 19).

under consideration on each other, (3) in the governance, management, and executive work process in e.g., a business system, (4) and in the cost, quality, flexibility, innovativeness, and care for sustainability, and social responsibility that result from the said three types of causes and their synergies (see: Mulej *et al.*, 2000;Ečimovič *et al.*, 2002; Hrast *et al.* (eds.), 2006, 2007, 2008). Thus we see, that a lot of entropy depends on human intervention. In it, the common denominator of entropy is oversights of something crucial. Here lies the cause for the need for requisite holism of human observation, thinking, decision-making, and action. The crisis of 2008- is a clear example.

12.3.2 Grounds for Human Intervention: Objective and Subjective Starting Points

Every human action, taking place, be it routine enhancing or innovative, is undertaken on the basis of some premises, visions, objectives, etc. But there is something, which establishes their background.

First, there is the objective reality as it is, making the objective starting points, like it or not; on the one hand there are needs, on the other hand there are possibilities (like debits and credits in bookkeeping).

Second, there are humans' own characteristics, called subjective starting points; they are made of our sentiments (wishes, desires, values) on the one hand, and our skills (to understand what is going on, and to know how to master processes occurring), on the other hand. In one or another form, the five components of the starting points can always be found (see Table 19).

Moreover, individuals in the same situations have different opinions which means that their subjective starting points let them see and judge equal objective/given reality differently, define different priorities, goals, tasks, and ways of meeting them. Dialogue may change them, so our conclusion is: subjective starting points can be influenced. It is education, experience and other information that exercise such an influence.

12.3.3 DST's Ten Guidelines concerning Influence over the Subjective Starting Points

12.3.3.0 Questions to be answered to define the Starting Points

Consciously or subconsciously, humans tend to address the following ten questions when preparing themselves to undertake an activity, by defining their own subjective starting points:

1. What are we trying to create (= general purpose), because our socio-economic and natural environments require us to be innovative in order to fight entropy?
2. How, in general, should we approach our activity?
3. What is the problem that we are trying to solve, what is the goal that we are trying to attain, and what are the tasks we will have to be able to take care of? All three are interdependent, hence: Are they compatible or do they block each other, and therefore something must change?
4. What are the procedures for each and every task to be performed, on a framework basis at least?
5. Having responded to questions 1-4, new questions arise:
6. How holistic have we been? Have we attained the requisite holism (Tables 1, 2, 8) by a dialectical system of viewpoints considered? What actually makes this dialectical system in the case at stake?
7. Having found out the dialectical system of viewpoints making the requisite holism possible, we usually do not possess all required or requisite skills, knowledge and values. Hence: Have we developed capacity of interdisciplinary cooperation and ethics of interdependence with other, different specialists, by application of modern dialectics, consciously or subconsciously (see Tables 1, 7, 8, 10, 11)?
8. Having developed the capacity mentioned in question 6; do we have the organizational possibilities to use this capacity? If not, how can we find or create them?
9. Having the partners to join us in the organizational possibilities mentioned in question 7, are we sure that both, our and their subjective starting points, are up-to date enough? If not so, what do have to do in order to up-date them requisitely?

10. Having to take care of the up-dating, are we aware of the interdependence of the rational and emotional parts of human personality, i.e., of the knowledge, skills, and values as well as other emotions? In what form is this interdependence surfacing in the given case?

11. Having to take care of the up-dating process in both parts of the human personality, are we aware of the fact, that their attributes have been formed before our efforts, and are we now no blank sheets therefore? In what form is this history showing up in the given case?

Let us now make brief comments to each guideline as responses to these questions.

12.3.3.1 Purpose: To Create Something New against Entropy

Ensuring inventiveness, holism and innovativeness instead of conservatism, one-sidedness, and routine enhancing in the phase of definition of starting points already: we know, that the modern world is a world of rather spoiled, demanding customers. They accept only offerings meeting the systemic quality, which requires excellence, and is best attained by innovation. Keeping to the old criteria (= conservatism), forgetting about holism (= one-sidedness) in the form of out-dated routines instead of ensuring working on new inventions and innovations on the basis of the requisite holism, such a purpose would hardly lead to creation of something new. Of course, in a concrete case this general principle must be defined more concretely (see Tables 3, 4, 5, 6, 7, 9).

12.3.3.2 Approach: To Reach the Purpose in a New Way

Ensuring a methodological, not only methodical knowledge in the phase of definition of starting points already: the methodological knowledge includes formation and deployment of rather new methods, rather than of the merely the established ones. It includes a creative, not just a passive and routine enhancing approach, which is done by a methodical knowledge. Practice changes most, when theory changes, and theory changes when the method changes. The established methods, hence, need criticism and creativity, up-dating, renewal, replacement with new ones, to make the transition from the methodical (= routine enhancing) to the methodological (= creative) approach in terms of methods applied. A new purpose can hardly be achieved with old, obsolete methods (Tables 13 and 14).

12.3.3.3 What, precisely, are the Trouble, the Objective, and the Tasks?

Ensuring a requisitely clear definition of the problem considered, objectives, and tasks in the phase of definition of starting points already: the problem of e.g., a company as a business system is e.g., to extend its existing market share. The objective is to attract new customers by pleasing them with new attributes of products. The tasks have therefore to do with innovation and its management. On this basis, the problem may be seen in the R&D department and other own invention processes, or in buying a patent or a license in order to circumvent the invention process of one's own. In such a case, the operational objectives and tasks may differ from the ones mentioned a few lines ago. All three—trouble, objectives, and tasks—are interdependent and making a dialectical system of their own.

Of course, not everything is fixed in advance; it neither can nor must be, but it helps to have it all as clear as possible as soon as possible, prior to action. That is why companies and other organizations practice planning. The clever ones do it in teamwork of professionals different from each other, because they use systemic thinking to attain requisite holism.

12.3.3.4. How, precisely, does the Procedure go with each Task?

Ensuring a requisitely clear definition of the procedure of performing the tasks defined, and of the problems surfacing, in the phase of definition of starting points already: in general, the procedure was briefed above when we spoke about the law of hierarchy of succession and interdependence (Table 19). Every task has some specific attributes and therefore the work process must be defined specifically.

E.g., all offices of a company, considered as a business system, work towards the same objective—the survival and prosperity of the company. But tasks and related procedures are quite different in production, in marketing, finance, and other fields.

12.3.3.5 Covering Everything Important

Ensuring the inclusion of a dialectical system of viewpoints, in the phase of definition of starting points, already: we are trying to meet this need by bringing many, but only relevant coauthors in our coauthors' team, and by interlinking their contributions into a synergetic one (Tables 2 and 8).

Once goals, tasks, and procedures have been outlined, we should better double-check whether or not we have met the law of requisite holism. All requisite professions and professionals should be taken in account. Their differences are necessary, normally, to make them complementary and attain synergy.

12.3.3.6 Requisite Holism on the Basis of Capability of Creative Cooperation (by Dialectical Thinking)

Ensuring the individuals' capacity for mutual understanding, tolerance, and creative cooperation, especially an interdisciplinary one, in the phase of definition of starting points, already: if we stayed with the medieval metaphysics principles of (1) mutual isolation, (2) full harmony with no essential differences or opposites, (3) no changeability, and (4) total destruction in case of any change, we could perhaps work well in a routine enhancing activity, but not in a creative one (see Tables 10, 13, and 14).

In most schools humans are trained for individual work and competition, rather than for cooperation, even less for creative cooperation. Some persons are lucky to be cooperative by nature. If not, bridges between specialties have to be built. Then, unavoidable holism can be attained, to some extent, at least; it is, hopefully, the requisite one. This does not apply only to schools, but to our entire lives. Different persons need each other because they are different and able to complement each other. In other words: they are and feel interdependent, i.e., dialectical.

12.3.3.7 Dialogue and Organized Cooperation making Holism possible

Ensuring the possibility of a (interdisciplinary) cooperation, in the phase of definition of starting points, already: once the capacity, as a human characteristic, is around, there is a need to have or create a possibility to employ it. This is an objective characteristic. In a usual setting, such a possibility is provided for by meetings and sessions, which are well organized for a real impact of all group members over each other to be achieved. Meetings are supposed to make this possible, as an organized form of dialogue. There are several methods around, ours is called USOMID-SREDIM and was recently completed up with introduction of De Bono's Six Thinking Hats method. We will come back to it in Chs. 13.3 and 13.5.

12.3.3.8 Continuous Up-dating

Ensuring a permanent modernization of the (subjective) starting points in order to avoid their becoming obsolete, in the phase of definition of starting points, already: persons, organizations, regions, countries, which keep staying with starting points lagging behind the more advanced ones, are nowadays losers in the market competition. They need to catch up in order to avoid realization of their entropy. In usual modern settings, this is achieved by a permanent refreshment of knowledge, values and feelings according to changes in one's environment. We quoted Barabba (2004) and Florida (2005) to demonstrate practical consequences of it.

Obsolete behavior causes entropy to become reality rather than a tendency only, in modern conditions of the innovative society. Obsolete team members cause entropy to grow rather than solve problems (see Tables 3, 4, 5, 6, 13, 14, and 19).

12.3.3.9 Interdependence of Knowledge and Emotions

Ensuring the consideration of interdependence of values and feelings, and both, knowledge on contents and the one on methods, in the phase of definition of starting points, already: for several centuries, humans were taught to be rational, i.e., to forget about emotions and feelings as something, which is not scientific and makes objectivity impossible. Now we know that this request is not scientific; it does not consider reality requisitely holistically. In reality, both, rational and irrational components of our personalities, exist and influence each other; they are interdependent.

Education, experience, and other information influence both, knowledge and sentiments, including values at the same time (left and right brain are parts of the same). Knowledge and emotions interact. Learning may e.g., cause interest, and interest may cause learning. For related problems see e.g., (Mulej, 1982; Mulej *et al.*, 1982, 1984, 1986, 1987; 1988; Mulej (ed.), 1995, 1997; Treven, 2005; Treven & Mulej, 2005a, b; Trunk Širca, 2000; Udovičič, 2004; Udovičič & Mulej, 2006).

12.3.3.10 Evolution and Intuition

Ensuring the consideration of path dependence of the current starting points on the foregoing ones, in the phase of definition of starting points, already: before working e.g., on the book (Dyck & Mulej, 1998) or on the book (Ečimovič *et al.*, 2002), or on the book (Božičnik *et al.*, 2008), or on the book (Mulej *et al.*, 2008) we have been experiencing the problems dealt with in these books, in different ways. We lived and worked in different areas, studied different literature and experiences. They resulted in our interest in the topic, in writing these books, and in our knowledge of our need for quite many co-authors. We also could re-employ our experience of arranging cooperation to make it all happen.

No politician is right when he or she says: "we will do everything anew, there was nothing before." Evolution is reality. Revolution only tries to jump over evolutionary phases, which then happen anyway, although under a different label. Nobody can be 50 before 15 years old. Acceleration is possible, but no jumping over the phases is natural.

12.3.4 DST's ten Guidelines concerning Implementation of Starting Points

12.3.4.0 Questions to address after the End of Definition of Starting Points and Objectives

The above logic of DST may be easy to follow and apply because it is plain natural thinking, although its topic is not very frequently found in literature, if at all. It has had a decisive influence. In addition, DST considers another critical situation: making the starting points is one affair, and the implementation of them is another (see Table 19). There are two reasons for this, at least[20]:

1. In later steps of the process when one must concentrate on details, the large image steps back or out of sight completely. Shorter-term horizons may prevail over the longer-term ones.
2. Later steps are frequently done by other persons who are specialists for one or another kind of necessary details.

Thus, it makes sense to remind them, too, of the original starting points of the entire process by guidelines for their subjective starting points concerning their tasks' implementation.

To do this, the following question should be addressed, according to our experience:

---

20. See also the Viable Systems Theory, e.g., in Ch. 11.3.1, and related references.

1. After the division of labor among specialists has taken place, are we still taking care of the entire entity under consideration and/or control requisitely holistically rather than one-sidedly?

2. Can specialists be requisitely holistic, if they are not requisitely open to each other in their daily work life rather than isolated? Are they willing and able to hear opinions that oppose to their own, because they do so and make their team more holistic?

3. When the daily work life on implementation of the starting points prevails, do we sufficiently consider life dynamics rather than static, unchangeability, lack of adaptability of thinking, working and behavior? Does this include adaptation to each other of co-working specialists?

4. If we are trying to be requisitely holistic, open and adaptable in terms of the three questions above, do these attributes cover only contacts with persons of the same profession, emotions, concepts, views, or do they involve interdisciplinary cooperation?

5. If we are trying to be holistic and do our best in terms of all four questions above, may we therefore expect a deterministic reliability of our ideas and actions in our daily lives, or should we rather expect some risk, probability rather than full reliability?[21]

6. In order to meet requirements in terms of all five questions, interdependence of different job performers should better be considered, also after the division of labor has been introduced. Application of materialistic dialectics in terms of Table 10, 13, 14 and 19, is a good solution. How is it practiced, if at all?

7. When using materialistic dialectics, interdependences may tend to hide the necessity of clear limits between the interdependent and cooperating ones. Application of a clear definition, what is meant by a "system" and application of the agreed upon systems and models typology allows for a mathematically precise delimitation. How is it practiced, if at all?[22]

---

21. These five questions address, **what** are we wanting and trying to achieve when we are implementing the defined (more general rather than detailed, adapted to single narrow jobs) starting points in the form of attending partial, specialized jobs, frequently with no links and interdependencies easy to see. In the implementation phase, though, the **how** is equally important as the "what" and "why". Therefore, further five guidelines follow.

22. Tables 1, 2 and 8 provide for instruction and Table 19 supports it: define requisitely exactly from which viewpoint you are considering the object dealt with as a system, and how you present it as a model. Then it will be requisitely clear what part of the really existing many attributes is selected to be in focus, and what parts are outside it.

8. When the entire work to be done is attended by specialists and hence in partial lots, there might be partial and many single outcomes, normally. A generalization may be needed to bring the team to a shared outcome and to a requisite amount of it. Is the generalization done in a way allowing for realism of the out-coming information?

9. When the generalization is being taken care of, and/or new task distribution is under way, there is a chance for a rather one-sided approach, but an approach on the basis of a dialectical system takes a better care of the requisite holism. How is it practiced, if at all?

10. Traditionally, one has tried to prevent too much one-sidedness by putting analysis before synthesis. Analysis is supposed to bring insights, synthesis to create outcomes. But we are putting another question, on the basis of the law of the hierarchy of succession and interdependence: What is the basis of analysis, if analysis is the basis of synthesis? In terms of the above law, definition of the starting points of analysis must take place, i.e., synthesis of them is needed before analysis can take place. How is this process practiced, if at all?

Let us add comments to individual guidelines by responding to these questions.

12.3.4.1 Requisite Holism throughout the entire Process

Ensuring of a requisitely holistic rather than one-sided thinking and behavior in the daily work and life: the authors of a policy/strategy, especially of a broad/requisitely holistic one normally need more narrowly and hence deeply specialized coworkers to work on parts of the entire job needed. A single person may cover a single viewpoint and reduce his or her own requisite holism to it. Without a well-arranged coordination the initial, larger, broader requisite holism may go lost. E.g., every specialist is advised a concept of elaboration of his or her (partial) job, and receives the bosses' framework insight into the entire story.

Total holism is not possible, of course. DST defines holism as a system of both, general and detailed characteristics, their mutual interplay and the interaction with their environment, and realism of thinking (Table 1). It is up to the decision makers to find their way between the (damaging) fictitious and (infeasible) total holism (Tables 2 and 8).

12.3.4.2 Openness

Ensuring an open system concept[23] instead of one's isolation and self-sufficiency in thinking and other behavior in the daily work and life: experience with work division, and its organization in professions and business functions demonstrates the tendency of many persons to avoid the open system concept. This is a consequence of their specialization and feeling that they cannot perceive and digest all influences available, for very natural reasons: we all have limited capacities, if compared to real life information, available nowadays (Tables 13 and 14).

Openness may help the decision-maker. However, it does not cause the same results, if we are open to others within the framework of one single viewpoint, or to other, even, actually, especially, the opposing viewpoints. They are complementary to our insights.

Openness and changing may be two different points. This brings us to the next guideline.

12.3.4.3 Dynamism

Ensuring a dynamic system concept rather than a static one in thinking and other behavior in the daily work and life: for long centuries it was good enough to learn a professional skill once for a lifetime. No change, adaptability, or even flexibility used to be considered good ethics. This is a principle, which is still around, especially with the less advanced part of humankind. This is a part of the sources of their troubles. The innovative society needs reliability, of course, but no obsolescence, especially not under reliability's covering. Dynamism takes place in several steps of mutual contacts (Tables 3, 4, 9, 13, and 14).

This characteristic of human thinking and action is usually reduced to changing in time. Only a part of statistics teaches about interaction. What about flexibility and, sometimes, even adaptation to views of others? If has obviously helped against entropy enhancing results many times in human history. A proverb reflecting this experience: "More heads have more knowledge" or:

---

23. We mentioned the open rather than closed system earlier: it is a precondition of survival because information and other resources enter the entity that behaves as an open system. But we also warned: too much openness may let thieves enter your apartment, not friends only. Every entity is a part of the environment for other entities that also feed themselves on resources from their environment.

"Two heads are better than one" (Ilich (ed.), 2004).

12.3.4.4 Interdisciplinary Cooperation

Ensuring interdisciplinary cooperation rather than single-disciplinary or multi-disciplinary approaches to realize the trans-disciplinary capacity in thinking and behavior in the daily work and life: a single discipline is what an individual can normally cover given the huge amount of knowledge and other information of nowadays. Multi-disciplinary approach makes a set of single disciplines, but no system, i.e., a complex entity, of them because relations and impacts among disciplines are left aside. The third way, a trans-disciplinary approach is important in research of some disciplines such as mathematics, philosophy, or a systems theory of a general orientation; they are aimed at providing for the very general shared vocabulary and insight (see Wierzbicky, 2006). Even with it in mind, nobody comes close to real/requisite holism without inter-disciplinary cooperation of single-disciplinary specialists willing to look beyond limits of their own disciplines (a good way of trans-disciplinary capacity). The first three guidelines which we discussed right now may also in reality happen to be limited and exclude a real holism: one may be (1) holistic inside a single viewpoint, (2) open to persons of an equal orientation rather to others such as one's opponents or complementary specialists of other professions, (3) dynamic without reconciliation of different opinions. Thus, the interdisciplinary approach is worth specific stressing, too. It is the managers' task, first of all, but not exclusively (Tables 4, 9, 19, 20).

The interdisciplinary approach is one of the traditional requests put by traditional systems theory, already, but it has tended to be forgotten about, time and again. Specialists talk more easily to specialists from their field than to specialists from other fields (e.g., economists to economists vs. lawyers to engineers).

(1) The general part or subsystem of interdependent attributes					
(2) Group specific subsystem (1)		(2) Group specific subsystem (n-1)		(2) Group specific subsystem (n)	
(3) Individual subsystem (1)	(3) Individual subsystem (2)	(3) Individual subsystem (3)	(3) Individual subsystem (m-2)	(3) Individual subsystem (m-1)	(3) Individual subsystem (m)

Table 20 *Interdependence of the general, group specific and individual part of attributes*

12.3.4.5 Probability, Risk

Ensuring consideration of probability instead of expectation of a deterministic reliability in thinking and behavior in daily work and life: for quite long times, only the deterministic laws and solutions were called scientific and good enough. One overlooked that in reality every object is a system (= entity) of three interdependent parts of its characteristics:

- The general part of them is equal to all objects of the same kind (animals, persons, products, produces, and so on), and shows up in every case;
- The group-specific part of them is equal only to partial groups of objects of the same kind (female and male animals or persons, mechanical and other products, vegetables and fruits, and so forth);
- The individual part of them is unrepeatable and owned by individual objects of the same group and kind only (first and family name and date and place of birth, the concrete car or refrigerator number, the concrete apple, identity card number, etc.) (see Table 20).

Determinism cannot be fully realistic, but statistically, i.e., with a level of probability, risk.

We humans do not like our cars and other tools to be only probable rather than fully reliable. But they cannot be so. Even less so can nature be, including human beings and society. Some risk remains, even if we try to be very holistic and attain requisite holism. Not everything can be foreseen (Tables 9, 12, 13, 14).

The five statements/guidelines (12.3.4.1-12.3.4.5) say clearly what is needed without saying how this can be done. The next five guidelines provide the answers.

12.3.4.6 Interaction and Flexibility based on Interdependence

Ensuring of employment of materialistic dialectics instead of medieval metaphysics and idealistic dialectics: we debated their difference earlier (see Tables 10, 13, 14, 19): materialistic dialectics has proven to be the closest to modern reality of all three, which are still around (and usable and useful in some situation, too). The medieval metaphysics is still good enough when no cooperation is needed and full isolation and disregarding of changing have to be ensured, e.g., in checking of blood in medical labs. But as quickly as physicians use these labs' results, they have to be linked with other partially gained insights. The transition to materialistic dialectics takes place by linking and by consideration of reality. As quickly as e.g., one does not simply speculate about the possible disease (and a system, i.e., network, of its causes) of the patient under consideration, but checks the patient's real situation, one uses materialistic dialectics, no idealistic dialectics.

In this book, too, chapters are written on the basis of field and desk research and experience; we also tend to detect their interplays and the resulting changes. This is a case of materialistic dialectics. The thesis and antithesis making the coincidental opportunism (Bertalanffy, 1950: 154) and resulting in synergy (Table 10), of which the author of idealistic dialectics Hegel has spoken so splendidly, do not exist as a relation between ideas only. They exist in reality as well in the form of interdependent and therefore interacting parts of any entity, e.g., a machine, a human body, a bee and the flower on which the bee feeds and helps the flower generate its next generation, opponents in a sport competition or legal or political dispute, medical doctors and their patients and nurses, members of the same sport or research team, coworkers in a company or school, inhabitants of the same town, or of the entire world.

This is stressed here again, because in the implementation of single tasks we have to work with a more specialized and narrow concentration than in the case of the general picture from initial formation of starting points. Knowledge and emotion that we need each other (= ethics of interdependence) opens the door to the requisite holism. See Tables 1, 2, 8, and 10.

12.3.4.7 Delimitation of Roles, Jobs, Viewpoints, and Systems

Ensuring of employment of the typology of systems and models in order to clearly delimit the considered characteristics and objects from the others: the problem with dialectics lies in its permanent tendency to consider interdependences and resulting changing (Table 10). This may cause a lack of transparency of the objects under consideration. Mathematically precise delimitation is hence equally needed as linking, even as a precondition for linking. This is enabled by clear definition of the viewpoint/s selected because it says what is taken in account, what is left aside, what is found relevant, what is not, what is put in the environment, what in the system (= mental picture of the object dealt with from the selected viewpoint).

Systems typology delineates what is considered, if e.g., one chooses the open or the closed or the relatively closed system, the dynamic or the static one, the natural or the cybernetic one, the deterministic or the probable one (see Ch. 12.3.7.1).

The same holds of models, which are systems describing systems rather than objects (e.g., engineering design of a machine is a complex thing, but it still is a model, which describes only the engineering viewpoint of the machine under consideration). See Table 7. Every book is a model, too. It expresses the authors' view/s of an object from his or her or their selected viewpoint/s in a way in which finding/s can be expressed, e.g., by text, picture, drawing and/or metaphors of other technical kinds. So is our book, too. Delimitation has the form of chapters and groups of chapters (see Ch. 12.3.7.2).

It is hard to cooperate, if it is not clear what a team member does and what he or she does not. Roles and jobs define viewpoints. In theoretical terms: every different viewpoint introduces a different system to picture the same reality by taking another part of it.

12.3.4.8 Realism in Generalization of Outcomes

Ensuring of a realistic generalization of findings: a business plan, or a book, e.g., especially the one with many coauthors, is in danger of becoming a set (with relations outside consideration), not a system (mental picture with both elements and relation considered, and synergy resulting/emerging from relations). It is also in danger of becoming too huge for a real reading or even deep studying.

Both would cause the business plan or the book, to be useless, even if it was very rich in findings and thoughts conveyed. Therefore, a quick reader summary has become a habit. But the problem surfacing now is: the short version may be loose and lose the essence by becoming an empty phrase of no real contents. That is why it is important to have short generalizations of findings, but to do one's best in order not to lose realism. See Tables 19 and 20.

12.3.4.9 Using a Dialectical System

Using a dialectical system: we defined the dialectical system as the midway between the unattainable full holism and the misleading one-viewpoint-system's fictitious holism (Tables 2 and 8). We did so by consideration of a system (= network) of all essential viewpoints (and resulting systems as one-sided mental pictures of the object dealt with) and only them. Thus, one can be much more realistic than with both alternatives mentioned above again (medieval metaphysics, idealistic dialectics). Thus, before one starts working, and before one starts finishing the work by generalization, one should check which of the three alternatives is used (see Tables 13 and 14).

Neither all viewpoints, nor one single viewpoint, but a system of all essential viewpoints—this applies to the implementation phase even more than to the phase of the definition of the starting points. A narrow specialization is now even more needed, possible, and equally dangerous, if it remains alone.

12.3.4.10 Analysis is based on one Synthesis and leads to another Synthesis

Consideration of interdependence of analysis and synthesis as two phases of work: we have been taught, perhaps all of us, that the first phase of a work is analysis, then synthesis. Two points, at least, are worth consideration: (1) the contents of both words, (2) the basis for analysis, which means its starting points, if it is the first phase.

To (1): In the Webster's School and Office Dictionary of 1978, analysis is said to be (a) separation into constituent parts, (b) summary. Synthesis is said to be (a) combination of parts in a whole, (b) such a whole; this means creation. The definition of analysis under (a) forgets about relations among parts, and so does the one under (b) as well. An obvious consequence is a lack of a (requisitely) holistic approach and a (requisite) wholeness of insight. This matters, because: What makes analysis make sense, if not insight? The real, not the obsolete merely technical definition of analysis is finding the deeper, hidden truth, which happens only partly by separation of the object under consideration into its constitutional parts.

Our answer to (2) will provide the second part of our answer to (1) in a short while. If one searches for truth, and separates the object under consideration into constituent parts, the Dialectical Systems Thinking puts the question: On the basis of which starting points this is done (see Table 19 again, if needed)? The response is obvious: before analysis, a making or synthesis of the subjective starting points occurs. One does some thinking before it, of course, but one decides for it beforehand, too. Both of these phases are interdependent; they only switch their roles of the more or the less stressed one from time to time in the same process.

Analysis covers finding the hidden truth. It takes place inside the framework established by the synthesis of the selected viewpoints. Our coauthors might produce different insights, if viewing the projects from different viewpoints—and find a different hidden truth, and therefore propose a different course of action in their synthesis of conclusions. Hence, a new synthesis is needed that leads towards requisite holism. Therefore, the 2nd order cybernetics is often more realistic than the 0 or 1st order cybernetics, but less realistic than the 3rd or 4th order cybernetics.

12.3.5 The Law of Requisite Holism

12.3.5.0 The Selected Problem and Viewpoint of Dealing with it in Chapter 12.3.5

Let us add some thoughts to Table 2 and 8 and related comments of so far. We believe our definition of this law has been clear enough there. Hence let us add a practical question: How can we make attainment of the requisite holism more operational?

12.3.5.1 The Six Laws of Requisite … and the Law of Requisite Holism

A session at ISSS47 (2003) unveiled six other laws using the notion "requisite" in their names. There are nine laws including the notion requisite quoted in Bausch's book (2003). They are found mostly in the chapter presenting Warfield and his Generic Design. Though, Christakis, in his final speech to the 47th conference of ISSS in Hersonissos, July 11, 2003, mentioned six laws that are not always the same as in the Bausch's book. He combined them by steps into an interesting methodology, called Cogniscope[24].

At the same conference a chapter (by Bausch and Christakis) from a book was made available to participants without saying from which book it is taken. In it (pp. 190-191) the same six laws are briefed as responses to six necessary principles for the structured dialogue to take place (which is the basic principle of the technology of creative cooperation they talk about) and to successfully deal with complex situations. Here they are briefed:

- Ashby's Law of Requisite Variety: Appreciation of the diversity of perspectives of observers is essential in managing complex situations[25].

---

24. For more about Cogniscope see (Christakis & Bausch, 2006).
25. This definition differs from the one we have quoted in Ch. 11.3.1. See Espejo and Harnden (1989).

- Miller's Law of Requisite Parsimony: Structured dialogue is required to avoid the cognitive overload of observers.
- Boulding's Law of Requisite Saliency: The relative importance of observations can only be understood through comparisons within a set of them[26].
- Peirce's Law of Requisite Meaning: Meaning and wisdom are produced in a dialogue only when observers search for a relationship of similarity, priority, influence, within a set of observations.
- Tsivacou's Law of Requisite Autonomy in Decision-Making: During dialogue it is necessary to protect the autonomy and authenticity of each observer in drawing distinctions.
- Dye's Law of Requisite Evolution of Observations: Learning occurs in a dialogue as the observers search for relationships among the members of a set of observations.

In Mulej's DST (including its applied methodology USOMID) we somehow provide for all these laws informally and perhaps uncompletely[27]:

- The Ashby's Law of Requisite Variety (in this version) is covered by interdisciplinary composition of the team, and by consideration of circumstances, conditions, and preconditions, i.e., the objective reality in the team's perception of it and their own work.
- The Miller's Law of Requisite Parsimony is covered by procedure USOMID-SREDIM providing for structured and economical work and cooperation. This is done even better once we have introduced the application of De Bono's Six Thinking Hats methodology.
- The Boulding's Law of Requisite Saliency is covered by creative cooperation in the said procedure using 1) brain-writing, 2) circulation of notes for additional brain-writing, 3) brain-storming, and 4) shared conclusions, in the procedure USOMID-SREDIM, including the very first step (in which the participants select their topic/problem to be worked on; so, it is not imposed over them).

---

26. A set differs from a system, of course: a system is made of a set at components and a set of relations.
27. Our research and their research took place with no knowledge about each other.

- Peirce's Law of Requisite Meaning is covered by studying the system/network of processes and picturing it in the programoteque. The latter is a dialectical system of framework models of processes of creative work and cooperation. It is produced and used, when strict algorithms cannot apply, because creativity is involved, not only routine.
- The Tsivacou's Law of Requisite Autonomy in Decision-Making is covered by the same procedure as mentioned in comment to Boulding's law, and by the roles played by team members for all requisite organizational aspects of the creative process to be covered.
- The Dye's Law of Requisite Evolution of Observations is covered by the same procedure leading through an agreed-upon number of iterations.

Thus, the six laws cover mutually different and complementary/interdependent preconditions for the law of requisite holism to be attained in practice—if used as a synergy rather than individually. It is our experience that we have attained the same without ever mentioning the six laws. The Mulej/Kajzer Law of Requisite Holism showed up as a later reflection of the long-term experience. Its basis was the application of the notion of the Dialectical System, which had started in 1974 as an article (Mulej, 1974) and in 1975 as a lecture (Mulej, 1975) following several years of practice of organizing.

The six laws are helpful. They have their common denominator in attacking the hierarchy of command replacing the hierarchies of organization, process, and increasing complexity (see Mulej *et al.*, 2000, about hierarchies and their distinction), in order to free creativity of the process participants. At the same time they try to avoid the overload in order to free creativity as well, and to enable an organized process of creative cooperation, which they call a structured dialogue. Thus, all of them make the attainment of the requisite holism easier, without losing the real complexity out of sight or getting lost in it.

This is very similar to the experience that we have made over close to five decades of our consulting work with USOMID methodology (especially Mulej and his partners in those times) and its predecessors (developed and used by Mulej and partners in organizing and managing students organizations, students journals, sport events, and other events since the early 1960ies, on the basis of PERT and related teamwork).

12.3.5.2 Information Needs—Framework for Requisite Holism

The briefed seven laws make a dialectical system of preconditions for the work on design of processes and solutions to problems to be more easily able to attain success. They support team's effort to get closer to the Bertalanffian holism/wholeness, in a best-case scenario on the level of the Mulej/Kajzer law of the requisite holism of behavior. Thus, they support more explicitly the effort undertaken in DST and its USOMID[28]. Still, the very practical issue is open to decision and opinion: What is really requisite?

The common denominator of all the (very many) possible cases and examples as well as of all different contents of systems, all of which in one way or another meet criteria of the law of requisite holism, are the information needs or requirements. The latter are addressed by the content of the system(s) as mental/emotional pictures of reality (both, mental and/or physical), which are tackled from the (dialectical systems of) viewpoints that are selected by those, who introduce systems to (re)present the selected attributes of the selected parts of reality. Once authors match information needs with no lack of crucial information and no overburdening with unnecessary data, the law of requisite holism is met.

12.3.5.3 Some Concluding Remarks about Requisite Holism

Our conclusions may therefore read:

1. One faces over-simplification in the definition of the contents of a system, if the observers/controllers introduce system(s), as informational representation of the object(s) under consideration, by their selection of viewpoint(s), which is conceived more narrowly than the information needs. One would be better off with a requisite holism.

2. One faces the other exaggeration, the tendency towards a total system, in the definition of the contents of a system, if the observers/controllers introduce system(s), as an informational representation of the object(s) under consideration, by a selection of system of viewpoints, which includes so many, too many data, that information can no longer be extracted. One would be better off with a requisite holism.

---

28. It might be interesting to make a similar comparison of these laws and Cogniscope technology to others of the close about 30 methodologies / technologies of systems thinking mentioned in a meeting of the 47th conference of ISSS 2003, some other time.

3. One meets the law of requisite holism, if one does not exaggerate either way from (1) or (2), but defines the contents of the system. One does so in the role of its observers/controllers. A requisitely holistic informational representation of the object(s) under consideration is attained, when one selects exactly the dialectical system of viewpoints, which includes exactly that many data and those data, that all essential (not all in general!) information needs are exactly met.

This is, of course, much easier to state and argue theoretically, than to attain practically. Humans tend to deal with complexity, which tends to reach beyond their/our own individual and team capacities. Besides, humans tend, in the contemporary times, at least, and in the so-called modern conditions of life and work, at least, to want to achieve more than time allows for. In addition, changes are permanent, and the daily practices must be adapted to them.

Therefore, information blanks are usual (in the case of the requisite variety, they show up as the residual variety, which is not matched by teams or persons in charge). Blanks may be bigger and more essential, or smaller and less essential, even unimportant (if the decision, what is included in the dialectical system of viewpoints, matches the situation and the trends of changing just right). Thus, one more conclusion may result:

4. There is hardly a finding and/or action that are at the same time exact, realistic and really holistic rather than partial. The actually attained level of holism depends on humans, their values, other emotions, knowledge, talent, skills, creativity, commitment, and perseverance, while searching for information needed by themselves or others for whom they are trying to do something.[29]

In other words, it is always up to the influential ones to decide, which level of holism is the appropriate one. This is a serious responsibility. It is naturally impossible, today, to take this responsibility and meet it with a poor or no trans-disciplinarity, interdisciplinarity, democratic creative cooperation of as many specialists as may be essential in the given case. How far we go towards all these attributes of human behavior, depends essentially on the individual/ prevailing subjective starting points, i.e., the system made of values, emotions, knowledge (on what and on how), talents, and skills. They, of course, reflect

---

29. Given the above definition of information, a product, service, gesture (both good and bad) is/contains also information, as soon as it is accepted by its receiver / customer / user and has impact on him / her / them. So does every part of nature. See Ch. 11.3.6.

the external needs and possibilities, too. All of them are interdependent (see Table 19).

Thus, what is called uncommon sense (Davidson, 1983) opposes over-specialization, and contains a very common sense: we humans must think about our thinking and face the dilemma: be requisitely holistic or fail (including causing the end of humankind on Earth, World Wars, and the like).

The 2008– crisis demonstrates that Bertalanffy, Mulej and others who have required requisite holism and wholeness have been right. Humans (and their organizations) having a chance to conquer their benefits from other's lack of requisite holism and wholeness abused their influential positions to profit—even enormously—at the detriment of others. This could not be considered a serious problem, if it was not exaggerated and hit back—in the form of destruction of nature, market, societal dignity causing terror and other form of riot against abuse of the less influential ones. Once the less influential ones develop into an organized synergetic group of a big size, they become the influential ones. Then, socio-economic orders change towards something new that pays more attention to requisite holism. Now, it is time for people to admit that the industrial paradigm of the recent few centuries has led to a big progress, but also to a terrible destruction of the natural preconditions of humankinds' survival. It did so because its principle of the untouchable market has not worked, but produced monopolies of enterprises and national countries that prohibited the requisite holism of behavior and wholeness of outcomes (Božičnik *et al.*, 2008; Brown, 2009; Taylor, 2008).

So far we have used qualitative analysis to attain requisite holism. Now we will try to complete it with some quantitative support.

12.3.6 Support to Requisite Holism by Contemporary Operations Research Methods

12.3.6.0 The Selected Problem and Viewpoint of Dealing with it in Chapter 12.3.6

Creation of inventions and making innovation from them requires consideration of a network of all essential and only essential viewpoints or factors; it is called a dialectical system (Table 9). It enables requisite holism in both, individual and group thinking, in order to define problems and produce creative and useful ideas by methods of creative thinking and networking in interdisciplinary cooperation. Still, decision-makers in enterprises have to select viewpoints to be considered and networked, including relations in networks to be considered, and to develop, choose, and verify possible solutions in order to develop inventions into innovations, leading to improvements in quality and quantity of output. They can do so by several methods, among which we emphasize, in this chapter, the decision-making ones supported with appropriate computer programs. To help them, we completed Belton and Stewart's (2002) general multiple criteria decision analysis (see Čančer, 2003, 2004). Now, we extend it to the process of generating ideas and developing them to innovations. We demonstrate that the modern Operations Research methods can help managers much more than the traditional ones (Jackson, 1991, 2003).

Innovation is a complicated and complex outcome and process of interdependent and interacting preconditions (see Table 9). Therefore entrepreneurs in order to deal with inventions and turning them to innovations should master interdisciplinary cooperation as a means of requisitely holistic behavior. Their effort can be supported with the discussed multi-criteria decision-making methods. This chapter emphasizes, therefore, the need for the appropriate knowledge about the basics of both, systems theory and the applied computer supported methods in decision-making, which we consider as necessary (but not sufficient) conditions for innovations.

12.3.6.1 Support to Creativity in Invention-Innovation Processes

In innovative society (with information society, society of excellent quality, learning society, knowledge-based society, entrepreneurial society, as its partial characteristics), successfulness depends on competitiveness. The latter depends in turn on requisitely holistic and creative thinking and decision-making—of individuals and organizations, from a family to all humankind. Therefore, creativity is at least as important as professionalism; its use is especially important in creating useful novelties, i.e., for innovating. And so is cooperation capacity and related methodological support for a dialectically systemic thinking by creative cooperation of different professionals.

A wide variety of traditional methods for strengthening individual and group creativity are outlined in literature (see: Cook, 1998; Glor, 1998; Hawkins, 2000; Pečjak, 2001). In practice it is enough to select the most appropriate one(s) according, mainly, to its/their function and the problems' nature. "Why" technique, mind mapping, fishbone diagrams and wishing are the most known techniques among wide variety of the techniques for problem/opportunity definition. This set includes also systems thinking approaches like relevance trees and cognitive mapping.

The most known and widely used technique is brainstorming with its variations, applicable mainly for generating ideas, as well as for their evaluation and problems' identification. Nominal group technique, provocations, forced relationships, brain-writing, attribute listing, morphological analysis, synectics, as well as mind mapping and different types of checklists can be used to support generating ideas. Moreover, attribute listing, morphological analysis and mind mapping can be used for decomposing, whereas mind mapping and W technique can be used for analysis. Checklists are also known as toolkits for verifying and systematization.

Besides generating ideas, synectics can also be used for finding solutions. Idea writing can be used not only for generating, but also for ranging ideas. Star rating matrices, the balance sheet method, paired comparison analysis, and reverse brainstorming can be (among other purposes) used for choosing ideas. For this purpose we can also use the, already mentioned, nominal technique, which can be used for defining and evaluation of ideas, too.

Balance sheet can be used for solution implementation as well; for this purpose we can use stakeholder analysis, implementation checklist, critical path analysis and many other techniques, too.

It has been assessed that two thirds of innovations are results of a demand-pull, whereas one third of them is a result of a discovery push (see Pečjak, 2001). Newer theories mention five phases of innovation management theories (see EU, 2004: 23-25):

1. Innovation derived from science (technology push).
2. Innovation derived from market needs (market pull).
3. Innovation derived from linkages between actors and markets, such as chain-link theories.
4. Innovation derived from technological networks, such as "systems of innovation" on a national, regional or international basis, which cause synergies of ideas and information from both, internal and external sources.

5. Innovation derived from social networks, which enable a lot of exchange of information and make knowledge available very rapidly on a worldwide basis. The point is meeting the need for many kinds of knowledge and their convergence from a variety of actors.

Hence, the theory (5) is closest to (informal) Dialectical Systems Thinking.

Of course, the idea is only the initial stage in the innovation process. It has to pass the invention, developmental, production, and commercial, phases before e.g., a product is innovation and diffused, massively produced and sold. Therefore, modern internationally advanced methods for strengthening creativity like Work Simplification, USOMID, 20 keys, ISO 9000, TQM, EQA, re-engineering of business performance, learning enterprise, knowledge management, Total Systems Intervention, project management, and other methods, supporting the innovation acceleration should be used to improve innovation capacity everywhere. Further, research and development, and unprofessional invention and innovation creation should be used to strengthen creativity in core invention-innovation-diffusion processes.

12.3.6.2 Methods for Requisitely Holistic Developing Ideas into Innovations

12.3.6.2.1 Quantitative Models Supportive of some Systems Theories

The 20th and 21st centuries have brought notable developments in the following tools for systems thinking: system dynamics, management cybernetics, soft systems methodology, dialectical systems theory, cognitive mapping, and models, for example viable system model, to mention only these among several systems thinking approaches. These theories, e.g.,:

- Help understand and handle the increasing complexity (Soft Systems Methodology);
- Enable the study of effective organization, communication and control in social systems (Management Cybernetics, Cybernetics of the 3rd and 4th order, and Control Systems Theory);
- Help make people more capable of creative cooperation for innovation (Dialectical Systems Theory, Control Systems Theory, and Dialectical Network Thinking);
- Ease understanding of how organizations work (Viable System Model), the structure and dynamics of complex systems, the ways in which an organization's performance is related to its internal structure and operating policies, including those of customers, competitors and suppliers; and

- Help them use that understanding to design high leverage policies for success, and
- Help model and simulate system behavior over time (System Dynamics).

In practice, simulation (on a qualitative and/or quantitative basis) is used to verify the produced models because of their complexity. They are increasingly used to design more successful policies in companies, and—e.g., system dynamics (see Sterman, 2000)—to public policy settings, too. There is a number of software packages designed to support quantitative system thinking modeling. These include, e.g., ithink, Powersim, and Vensim (for system dynamics modeling), ViPlan Learning System (for viable system model), and Group Explorer (for cognitive mapping).

Decision analysis can be used to determine the most preferred strategy when a decision-maker faces several decision alternatives and an uncertain or risk-filled pattern of future events. It includes decision-making without or with probabilities, risk analysis, sensitivity analysis, decision analysis with sample information, etc. (see Anderson *et al.*, 2003; Čančer, 2003).

Influence diagrams, payoff tables, and decision trees could be used to structure a decision problem and describe the relationships among decisions, chance events, and consequences. When analyzing complex practical real-life problems, we found decision trees very useful since they provide a graphical representation of the decision-making process with the emphasis on the sequential nature of decision problems. Decision-making with sample information and using decision trees can gain support from appropriate (and popular) computer programs, e.g., Tree Plan and Precision Tree.

Risk management is an integral part of inventions assessment (Likar, 2001; Afuah, 2003). Because of the uncertain or risk-filled patterns of future events, decision analysis with the emphasis on risk analysis can be used to provide probabilities for the payoffs associated with a decision alternative. Risk aversion, analysis of probabilistic dependencies and of uncertainty as well as sequential decision-making can be supported by, for example, @RISK, Analytica, DEA SolverPro, DecisionPro, DecisionScript, DecisionTools Suite, DPL Professional, Equity, Netica, Optimal Manager, Qualrus, QMS, TreeAge Pro Suite, etc. Furthermore, a spreadsheet may be designed for any of the decision analysis approaches mentioned above. For the basic decision analysis computation, we can use one of computer programs that are actually used in enterprises, e.g., Excel.

Presentations of available decision analysis packages contain the information provided by the vendors and surveyed by the Operations Research/Management Science (OR/MS) researchers. Appropriate information can be easily found on world web pages. It includes applications like tradeoffs among multiple objectives, analysis of uncertainty, analysis of probabilistic dependencies, risk aversion, sequential decision making, multiple stakeholders, and specific applications for which software is most widely used. Before buying such a decision support package, experts in enterprises can use trial-free versions of computer programs to find out whether a package offers enough possibilities for a convenient preparation of their decisions. However, when using results in answer and sensitivity reports, decision-makers in enterprises must know the basics of the applied methods. Especially in small and medium sized enterprises where the sphere of action of one employee combines a broader spectrum of working tasks than in large enterprises, each expert's knowledge base and ability to learn is of high importance when evaluating and verifying how useful the selected creative ideas are.

12.3.6.2.2 Multi-Criteria Decision-Making Methods

A set of decision analysis methods that are distinguished by applicability in several social fields, are characterized by different levels of problems that are to be identified, structured and solved (personal, business, economic: micro and macro, political, technical, environmental, ethical…). It is described as a multiple criteria decision analysis (MCDA). Multi-criteria decision-making (MCDM) describes the set of approaches that can help individuals or groups in researching important complex decision-making problems. They should be used when intuitive decision-making is not enough for several reasons: e.g., conflicting criteria or disagreement between decision-makers what criteria are relevant or more important and what alternatives and preferences are acceptable.

Results of MCDM should not be understood as the final ("right") answers in the problem solving process. Multi-criteria analysis cannot be justified within the optimization paradigm frequently adopted in traditional OR/MS (see Belton & Stewart, 2002). The appropriate ("objective") analyses cannot relieve decision-makers of the responsibility of making difficult judgements. It is an aid to decision-making, which seeks to integrate objective measurement with value judgment and to manage subjectivity. The latter is evident particularly in the choice of criteria and in determination of their weights.[30] In this choice we introduce some of the MCDA methods because they have already turned out to be applicable in business practice. The following facts contribute to their applicability in solving complex problems:

- The MCDA methods do not replace intuitive judgment or experience and they do not oppress creative thinking; their role is to complement intuition, and to verify ideas and support problem solving.
- In MCDM we take into account multiple, more or less conflicting criteria, in order to aid decision-making.
- In this type of decision-making process we structure the problem.
- Users can compare different methods and assess their convenience in problem solving. The most useful approaches are conceptually simple, transparent and computer supported.
- The aim of MCDM is to help decision-makers learn about the problem, express their judgments about the criteria importance and preferences concerning alternatives, confront other participants' judgment, understand the final alternatives' values, and use them in the problem solving activities.

When applying MCDA methods to several decision-making problems, we concluded that they should be approached step-by-step. We followed the phases of decision-making processes that are commonly acknowledged in literature (see Belton & Stewart, 2002: 6): from identification of a problem, through problem structuring—model building, its use to inform and challenge thinking, to the creation and analysis of activities plan to solve a problem (e. g. to implement a specific choice, to suggest a recommendation, and to monitor performance), but we adapted and completed them up for the problem's type (Čančer, 2003, 2004; Čančer & Knez-Riedl, 2005). In an innovative society, the decision-making process has to link creative thinking on a requisitely holistic basis.

In some environments, decision-makers are not able to cooperate in their group decision-making, they do not want to seek compromise solutions, neither to express their judgments consistently, or they need ad hoc solution. In such cases they are recommended to express their preferences beforehand and to include their goals in the models for goal programming (Čančer, 2000).

---

30. This is why we have stressed the subjective starting points in foregoing chapters and Table 19 so much.

One of the most widely applied sets of multi-criteria methods is the Multi-Attribute Value (or Utility) Theory (MAVT or MAUT) (for a detailed description see Belton & Stewart, 2002). From the late 1960s on this set of methods has been developed not only by management scientists, mathematicians, psychologists, but also by practitioners in management, economic, environmental and public fields. The need to include different scientific, professional fields in the development of these methods results from the need to manage complexity. It has been improved to SMART (a simplified multi-attribute rating approach) and other approaches (for example SWING, SMARTER). They are supported by several computer programs, e.g., HIPRE 3+, Web-HIPRE and Logical Decisions® for Windows.

One decade later Saaty (see 2001) developed the Analytic Hierarchy Process (AHP) method, together with computer program Expert Choice. This method excels by wide applicability, too, and is distinguished by the scales used, the methods used to express judgments about the criteria importance and preferences to alternatives, and the manner of transforming these judgments into numerical values. A (relatively, perhaps requisitely) holistic approach (as the opposite of a linear and piecemeal approach) is used in this method in which all the problem criteria are structured in advance in a multilevel hierarchy.

Furthermore, it is generalized for neural decision processing—the Neural Network Process (NNP). It is, in addition, completed with the interaction and dependence of higher-level elements on lower-level elements and relations in the form of feedback structure that looks like a network—the Analytic Network Process (ANP), which is supported by Super Decisions. It overcomes the traditional OR/MS approaches in the context of Systems Thinking, because it allows us to include tangible and intangible factors and both, interaction and feedback within clusters of elements (inner dependence) and between clusters (outer dependence).

The use of the discussed methods might lead to over-complexities of insight: decision-makers may not need all details of results they obtain with these methods. Namely, some decision problems do not require alternatives to be ranked with respect to their final values; often it is good enough to find out which of them is the most preferred. Therefore the so-called "outranking" approaches have been developed since the 1970s. The most widely applied are ELECTRE in more variants and PROMETHEE (for details see Vincke, 1992).

Further, interactive methods as another set of multi-criteria approaches emphasize dialogues with the decision maker, who reacts to the first solution provided by the first computation step: he or she gives extra information about his/her preferences. These methods are especially applicable when a complete preference model is not constructed in advance and when alternatives need improvements (for details see Vincke, 1992). An evolution from search-oriented to learning-oriented methods can be noticed.

Since common-practice statistical methods—nowadays very valued and appreciated in some scientific spheres—cannot satisfactorily support many complex decision-making processes, including developing creative ideas into innovations, methods for the approximate specification of preferences are gaining power in enterprises. Moreover, theoretical work has been done to extend methods, such as AHP and value tree analysis, so that the decision-maker can express approximate preference statements through interval judgments. Preference programming (see Mustajoki *et al.*, 2005) describes approaches (like PAIRS, PRIME and RICH) that can be helpful in group decision-making, too. Easy-to-use software has been developed to support the interval techniques: WINPRE supports preference programming, PAIRS and Interval SMART/SWING; PRIME Decisions is a software implementation of the PRIME method; RICH Decisions supports the RICH method.

Table 21 introduces software products for MCDM that have received much attention among experts in different practical business fields (because of user capabilities, availability of graphical elicitation techniques, and the possibility to transform subjective judgments into objective measures). It delineates our findings about their applicability according to our experiences and the upgrades for group decision-making, as well.

12.3.6.2.3 Group Decision Support Systems

Group Decision Support Systems (GDSS)—currently a growth area within Management Science—are particularly helpful during the earliest phases of product development, which are focused on the collection and evaluation of customer requirements. They allow far-away organizations to collaborate on the generation, refinement, and systematic evaluation of new product ideas. Some empirical researches (see Salo & Gustafsson, 2004) suggest that group working is especially useful in large-scale industrial organizations, which have intensive communication needs and seek to shorten their product development cycles. Namely, pooling of resources—one of the main

Software	Applicability (according to our experience)	Upgrades to Group Decision Support Software
HIPRE 3+, Web-HIPRE	Especially applicable for methods based on ordinal and interval scale: SMARTER, SMART, SWING, and for measurement of alternatives' values with respect to each attribute by value functions, although it supports also the AHP method by pair-wise comparisons, and direct measurement of alternatives' values	Web-HIPRE: a possibility to combine individual preferences into group preferences with an weighted arithmetic mean method; the group model collects individual preferences directly via Web allowing for use of a distributed mode; HIPRE 3+ Group Link: a GDSS software, which supports on-line group decision making with preference programming
Expert Choice	Especially applicable for the AHP method that is based on a ratio scale (pair-wise comparisons), although it supports measurement of alternatives' values with respect to each attribute by value functions and direct method	Expert Choice 11: ability to accept judgments from multiple stakeholders using wireless keypads or EC Decision Portal in "same time and place" or "different time and place" decision-making; synthesize judgments from multiple stakeholders; weight team members and evaluate outcomes
Logical Decisions® for Windows	Especially applicable for problems where describing the alternatives is of special value (utility functions, AHP, adjusted AHP), and for weights' assessment with tradeoffs, as well as by direct entry, the SMARTER and the SMART method, weight ratios and the AHP	Logical Decisions for Groups: a tool for groups to collect data, prioritize and compare alternatives to reach consensus more quickly; data can be collected either by direct entry or through button box radio transmitters

Table 21 *Some findings about the most preferred software support for MCDM (Helsinki University of Technology, 2005; Expert Choice, Inc., 2005; Logical Decisions, 2005; own experience of Čančer)*

characteristics of group decisions—results in combining and producing more information and knowledge, and it generates more alternatives. Advantages of several involved stakeholders increase acceptance and legitimacy.

- Scientists are trying to show and to prove advantages of their approaches in comparison to other approaches to group decision-making, for example:
- Decision Conferencing, an approach to group decision support that draws on experience and research from information technology, decision analysis, group processes and behavioral studies of actual decision-making, is computer supported by HIVIEW and EQUITY.

- Alternative methods like SCA (Strategic Choice Approach), computer supported by STRAD, and SODA (Strategic Option Development and Analysis), computer supported by Decision Explorer and Group Explorer, place much more emphasis on problem structuring than on optimization.
- Methods based on interval presentation of approximate preferences (see Mustajoki *et al.*, 2005): experiences suggest that the use of preference intervals can be especially helpful in group-decision making. Decision-makers seek consensus by trying to compromise on their individual judgments and reduce the width of preference intervals. This process helps users to focus the discussion on the key issues and this is expected to increase the efficiency of the negotiation process. Table 21 briefly describes how these methods are computer supported with HIPRE 3+ Group Link.
- AHP uses a mathematically justifiable way for synthesizing individual judgments, which allows the construction of a cardinal group-decision compatible with the individual preferences. Its computer support by Expert Choice 11 with EC Decision Portal is briefed in Table 21.

Recently we have faced the group decision-making upgrades of the computer programs that have been most preferred for individual MCDM in the last two decades (see Table 21). Their common advantages are: graphical support for problem structuring, value and probability evaluation; easy-to-conduct sensitivity analysis; analysis of complex value and probability structures; they facilitate changes in models relatively easily and allow distributed locations. Several groups (Expert Choice, Inc., 2005; Helsinki University of Technology, 2005; Logical Decisions, 2005) are actively developing computer aids for decision making. In their software development they emphasize the possibilities of the latest developments in information technology, such as multimedia and Internet.

We emphasize that software products for group decisions for developing inventions into innovations should offer a secure environment to protect sensitive corporate information.

12.3.6.2.4 Conclusions about Quantitative Support to Requisite Holism in Innovation

Methods and software supportive of creativity can help decision-makers define problems, generate ideas, together with recording and organizing, capturing, manipulating, editing and organizing them, and—to less extent—

solve problems. It can be up to the decision-makers to decide if the idea is of value to them, and it may be in their skill to develop the idea into a solution. However, decision-making methods can help them to choose and verify possible solutions.

Multi-criteria decision-making methods that have already turned out to be very applicable in business practice can be used to complement intuition, to verify ideas, and to support their development into innovations. They support requisite holism without talking the systems theory language. This makes them more acceptable. Methods for the approximate specification of preferences are coming into force in enterprises both because of their theoretical developments and because of failures of common-practice statistical methods.

Many learning organizations encourage the use of adequate computer supported methods in decision-making. This facilitates decision-makers' reading, interpreting and using the results obtained in several simulations when evaluating and verifying how useful the selected creative ideas are, and to develop them into innovations. When using these methods, decision-makers are responsible that no unavoidable cost, over-complexity or over-simplification is caused. Hence, the point is reaching requisite holism with only requisite effort by applied systems thinking and innovation.

Within the USOMID methodology of creative work and cooperation, the SREDIM procedure of work (Select—Record—Evaluate—Determine—Implement—Maintain steps), these methods can be well applied especially in the E and D phases. The next topic of typology of systems and models will be more supportive in the S and R phases. See Ch. 13.

12.3.7 Typology of Systems and Models

12.3.7.0 The Selected Problem and Viewpoint of Dealing with it in Chapter 12.3.7

Understanding of each other is made much easier, if there is an agreement what notions such as types of systems and of models mean. In Ch. 12.3.4.7 we mentioned this typology. Now we will elaborate it a little bit. We will start from Table 7 and the related finding:

- A system presents the object at stake as its mental picture from a selected viewpoint, and a model presents the system to make it clear, available for communication, in a simplified form, which can be qualitative or quantitative.
- Formally, both, systems and models, are made of components and their relations. In contents, they differ, due to the selected viewpoint and intention.
- Therefore one always needs our combined definition: in formality a system and a model are a whole, in content it is a one-sided and partial presentation.[31]

This matters, because the limited human capacity requires humans to make decisions based on models rather than on objects: objects have too many attributes for humans to cover all of them. Thus, systems exist in human heads and are described and reported about in models.

12.3.7.1 Typology of Systems

In Webster's dictionary (1978) there are 15 groups of contents of the word "system", based on various viewpoints (in Webster, 1992, they are nine and per professions). On the basis of other references we make the following classification (Mulej, in Mulej *et al.*, 2000: 163-179):

1. Consideration of reality: *real* and *abstract* systems; both of them can be *closed, relatively closed,* or *open*. The real open systems consider environment, hence inputs and outputs.
2. Depth of consideration: *behavior* is superficial, *functioning/working* considers its background, too, to reach beyond the black-box approach, or structure. Relations between inputs and outputs may be seen as either *deterministic* or *hard*, or *stochastic, probabilistic* or *soft*. Behavior can be seen as *equilibrium, stable, unstable, quasi-stable* or *ergodic*. Input can be either *controlled* or *stochastic*.
3. Reduction of insight to differences between components makes *subsystems* show up. Looking for shared attributes of parts makes *partial systems* be in focus.

---

[31]. Authors, who use the word "system" and say it exists in reality, make the oversights of their own selected viewpoint. If this was not true, there would be no need to appreciate the difference between Cybernetics of the 1st and 2nd order that we have summarized earlier.

4. According to natural attributes of the objects presented in systems one can see *natural, technical* and *organizational* systems; the latter two are man-made to fulfill a function.

5. In terms of showing a status or a process one can introduce *static* or *dynamic* systems. By their type of changing the latter can be *evolutionary* or *revolutionary* (*leap-frogging*).

6. Selection of the science to be applied causes systems to be *traditional* stressing the viewpoints of energy and matter, or *cybernetic* stressing information and resulting impact. Impact can be based on exact methods, intuition or a combination of both of them. Cybernetic systems include, along with dynamics, probability, complexity and influence, the feedback making influence easier; feedback can be homeostatic, automatic or based on human conscious decision.

As we have said in Table 1, one can distinguish *simple, complex* and *complicated* systems. All of them can be dealt with in *synthesis, analysis* or *decomposition*, yet in practice systems cannot be *absolute/totally holistic*, but *reduced* due to the *selected viewpoint, viewpoints, system* (= *network*) *of viewpoints*, or *dialectical system*.

If we take the case of a street crossing as our object, the dialectical system might include its construction, pavement, security of passing, energy lines under it, in network; each of them might be modeled in engineering modeling technology of the related specialty, e.g., in drawing, mathematical equations, and so on. If safety is the selected viewpoint, the system might be the obligation to let the right vehicle pass first. The model in this case might be the green traffic light.

12.3.7.2 Typology of Models

The above case shows how far away one can be from the whole reality of attributes of the object dealt with, but meet the information needs anyway with the model. It also shows that there are models about system's behavior or functioning/working and models for it. The first ones are descriptive, the latter ones prescriptive. Both types are influential as the basis for analysis and synthesis, decision and action. Therefore they may not be oversimplified pictures of systems, which in their turn may not be over-simplified pictures of objects dealt with (Mulej *et al.*, 2000: 180-191).

Models must therefore be usable and useful; this attribute depends on the *analogy and similarity to the system*, which can be made understandable with a legend (like in the practice of geographic maps). The level of simplification must be clear and match the purpose[32]. Otherwise disinformation results and requisite holism of approach and wholeness of outcomes is not attained. Consequences may be fatal (in the above case: green light on all streets at the same time may cause accidents).

As a type of similarity, *isomorphism* is more precise than *homomorphism*. The latter is not requisitely reliable in e.g., engineering blueprints as models for building a house or making a car, bridge or another technical artifact. But isomorphism is hard, if not impossible, to attain in social sciences.

Models bring mathematics and verbal expression closer. In mathematics, a model means coming closer to reality by introduction of concrete data in formulas. In other sciences a model means reduction of concrete data under consideration.

Simulation is another word meaning model building aimed at discovering of attributes of the system, which in its turn is aimed at mastering the object. This means that models provide feed-forward information as the basis for action (in the above case: green light makes people stop, waiting and move). Hence, models are also tools of influence (in the above case: tools of the street crossing managers).

Thus, *information* is a model that serves *management, regulation* as well as *self-regulation*. It can be a *hypothesis*, a *decision*, or a *supportive feed-forward* information (in the above case: they first created insight in the situation in the street crossing and created a hypothesis about solving its problems; then they decided to build traffic lights in it; the passengers use these traffic lights as a support to their decision to stop or go, and the crossing managers use them to provide safety in the crossing).

Due to the influential role of models as *partial* informations, which may meet requisite holism or miss it, one must pay attention to dangers of exaggeration in making and using models. Too much or too little mathematics

---

32. If the purpose is e.g., teaching geography, a map is sometimes good enough, sometimes a relief serves better. In traffic, the map must be more precise and may be limited to roads or railroads. In geographic research other attributes may have to be exposed depending on the purpose.

might cause a lack of requisite holism (market situations and trends are less well expressed with mathematics than attributes of technical artifacts, such as engines, hydro-power stations, airplanes, houses). Too many or too few details are another case of danger. Too much or too little attention to the limitations of the model may make it unrealistic. So, conclusions from models can be over-drawn.

From the viewpoint of the *way of expression* models can be verbal (such as books), physical (such as prototypes of engines), graphic (such as pictures, diagrams, maps), or formal (such as quantitative models, e.g., formulas).

From the viewpoint of *analogy* with the system to be expressed, models can demonstrate *functioning/working* (such as electric network), *structure* (such as models of molecules, hierarchies in organizations), or *behavior* (such as models of inputs and output in black-boxes, mathematical equations).

From the viewpoint of *purpose* model can serve *demonstration* (such as teaching or marketing materials), *experiments* (such as in laboratory research, field experiments in agriculture, practicing in sports and theater), or *decision-making* (such a constitutions and other legislation, decision trees). All three purposes can also be *combined*, of course, e.g., per phases of the same process.

From the viewpoint of *research* models run through several phases: (1) modeling of requirement the systems under research should meet as attributes of objects in real life; (2) modeling of hypotheses about attributes of such systems to be met; (3) development and integration of such systems in tangible and intangible forms; and (4) evaluation of the system in terms of suitability or need to return to phases (1)-(3). Within every phase one needs (a) development of models in several steps, (b) collection of research information, and (c) synthesis of information inside models.

12.3.7.3 Conclusions concerning Typology of Systems and Models

Thus, the crucial point is the systems and models selection. It depends on subjective starting points (see Table 19). Consequences may be crucial: this is a case of the law of hierarchy of succession and interdependence. Authors and users of models, who are better at attainment of requisite holism, are usually better at ethics of interdependence, which provides for more interdisciplinary or other creative cooperation.

Actual circumstances: permanent competition • Between humans for resources for survival and an acceptable quality of life and work life, as well as • Between humans and their natural environment for nature's survival, which is a precondition for humankind's survival	
↕	
Application of the Dialectical Systems Theory for a requisitely holistic approach and hence success in the permanent competition of both types	
↕	
Functioning and consideration of the law of entropy as a permanent natural tendency of everything existing towards changing in something else, i.e., destruction	
↕	
Need for efficient and effective human action ⟷ law of requisite holism by ethics of interdependence. And consideration of this law in both: • Preparation and defining of objectives of human activities • Realization of objectives of human activities	
↕	
Law of hierarchy of succession and interdependence	
Guidelines for formation of the subjective starting points of humans preparing and defining the objectives of human activities	Guidelines for formation of the subjective starting points of humans realizing the objectives of human activities
↕	↕
Informal Dialectical Systems thinking of both groups of humans with the DST's applied methodology USOMID	
↕	
Impact of both groups of humans on the circumstances mentioned above by innovation and routine	

Table 22 *Circumstances of life and application of the Dialectical Systems Theory for innovating them*

We humans may never forget: we decide and we are narrow specialists only. And we think in systems and communicate in models rather than real objects. The latter are too complex for our natural capacities as individuals or small teams.

12.4 DIALECTICAL SYSTEMS THEORY AND ITS APPLICATION BY USOMID

Following the DST principles does not necessarily involve knowing all the theory, but the applied methods may be sufficient. That's why they follow now. Towards this end, we created USOMID and applied it with great success in the real world for beyond two decades.[33] As we have said earlier, USOMID is the sixth of the components and relations making DST. For a summary of relations between them see Table 22.

---

33. USOMID tackles creative work processes. There are routine processes as well, of course. Both types can be innovated. Mulej, Ben Graham and many others have learned from the same basic source, from Allan Mogensen. Back in 1926, as an industrial engineer he was in charge of a "motion study" of a worker in at Kodak, Rochester, New York. This means that the engineer was supposed to think instead of the experienced worker, not together with him, and to find out and prescribe the best way for the worker to do his job. Mogensen produced a film about the worker's procedures. In spite of the habit and intention, the worker saw the film about him and asked the engineer: "Please do not use this film, I can do it better." This was a turning point for Allan Mogensen: he realized that he had been educated wrongly and hence did not expect any creativity, skill, knowledge and experience from the worker. On the basis of this experience, Mogensen developed his method of innovating, which is based on cooperation of the experienced performers of the jobs under consideration and consultants; he called it *Work Simplification* (Mogensen, 1981). This source and several more were the basis for us to develop USOMID by combining it with DST principles, and for Ben Graham Jr. to develop "Paper Work Simplification". His method is very good at innovation of the administrative routine jobs by non-technological innovation. USOMID is good at innovation of the creative work processes that can be supported by *framework routine-based procedures* collected in the *programoteque*.

Chapter 13

USOMID: AN APPLIED METHODOLOGY OF DIALECTICAL SYSTEMS THINKING

13.0 THE SELECTED PROBLEM AND VIEWPOINT OF DEALING WITH IT IN CHAPTER 13

It is not always necessary to know theory in details, but the way of application of it.

13.1 THE USOMID CONCEPT: A REFLECTION OF THE SEVEN COMPONENTS OF THE STARTING POINTS

Our experience with application of the DST in non-academic settings soon demonstrated the need for DST's rather philosophical concepts to be expressed in an organizational technology, i.e., methodology. This is why USOMID came about; the acronym reads in Slovenian: Creative Cooperation of Many for an Innovative Work (Mulej, 1982; Mulej *et al.*, 1982 and later).

Creative cooperation can hardly take place, if, e.g., only hard systems methods are used since they are aimed at finding and exposing mechanical, deterministic kinds of relations. They are a very important achievement, if a routine-enhancing behavior is good enough. It is so, when e.g., one deals with very technical details of production for all products to be fully equal. It is different when creativity is needed. Then, soft systems methods become the right choice exposing probabilistic and non-deterministic features/characteristics. Both types are normally needed, of course, frequently even in combination.

For the soft systems methods' support to a work process to be attainable, the DST has developed (and frequently applied) its USOMID method (Table 22). In DST the human creative work process is modeled: it starts from the starting points—a dialectical system of five elements:

- The objective starting points are the objective[1] "needs" and "possibilities".
- The subjective starting points are "values and other emotions", "knowledge about contents", and "knowledge about methods", and talents.
- All further process depends on them. In USOMID every one of them is reflected:
- Objective needs reflect the law of entropy, which causes the need for the modern innovative society, innovative business and its culture, policy, strategies, tactics, innovation objectives, awards; subjectively, the objective needs are in turn reflected in values and other emotions.
- Values and other emotions are impacted by the objective needs of a modern society in the form of a motivation for a deliberate search for many possible changes aimed at creation of inventions, potential innovations and innovations; the motivation for it is created by a well grounded feeling of appreciation for creative coworkers.
- Knowledge on contents is dug out and activated for both above purposes. More than two thirds of innovations are the incremental ones and have to do with the work processes. They can best be results of inventive and innovative efforts of the work performers. A written insight into processes is usually lacking, especially into the rather creative partial processes. Making this insight in the form of "programoteque" is the visible informational outcome of USOMID deployment. It is done on a framework level first and then proceeds towards more and more detailed levels, all way to a computer support. It is first a description of the given facts; then comes their "causes tree" analysis, later one perhaps also innovating.
- Knowledge about methods has a general and a specific part. The specific part is problem/topic dependent. The general part is made of the USOMID/SREDIM procedure of creative work and cooperation.[2]

---

1. The notion "objective" does not mean, that these two components of the starting points are unchangeable given facts like physical or chemical or biological natural and research results. It means that they are more difficult to influence by a single human being than the subjective starting points are. Both of them are interdependent and therefore interacting.

2. As we have announced earlier, we have recently created a combination of USOMID-SREDIM with De Bono's Six Thinking Hats method. We will come to it in Ch. 14.5.

- The objective possibilities for a creative work and cooperation to take place are made of the USOMID Circles, a version of the Quality Control Circles, with some additions.
- The relation between all the elements quoted is double:
- Learning by an initial course, and learning by doing;
- Working out and deployment of the programoteque by a creative cooperation process of the studied job performers and their consultants.[3]

Then, the innovating process can follow (many thousand cases since 1969). Hence, in practice the application of DST via USOMID starts with working out the insight into the processes under consideration. This is usual, of course; in this case it takes place in the form of working out the programoteque[4].

13.2 PROGRAMOTEQUE

There are hardly any organizations, be them enterprises or others, with a holistic insight into their entire work process, i.e., the basic, information and management aspects, as well as their links, especially interdependencies. The basic process of, e.g., a factory covers supply, manufacturing, and sales, and it transforms in-coming knowledge, ethics, money, material and energy into out-going products (and waste). The information process links the basic and the management processes as well as environment by transformation of data in information. The management process transforms the in-coming information into decisions, instructions, and so forth, for the basic and information processes and other out-going information.

It is quite hard to control, or even innovate, the entire business or its production subsystem at least, without a requisitely holistic insight into the entire process. It is equally hard to innovate or optimally control the entire business or any of its subsystems without the motivation of the employees to be creative towards innovation rather than towards abuse; the same is true of their knowledge and experience. USOMID is a method to meet both needs. How?

---

3. Making a programoteque is not unavoidable for innovation of any type, but only for innovation of the creative work and cooperation process. In our experience innovation as rationalization and optimization of the creative work and cooperation is the most neglected one. Working on it yields enormous results therefore.

4. It may make sense to know that only about 20% of all innovations are technological, although one speaks mostly of them.

1. Programoteque provides for a felt respect and appreciation for knowledge and creativity of the job performers, and involves them. It cannot be produced otherwise. Hence, it motivates.

2. Programoteque provides a framework level of structuring, without causing a rigid structure of the process information. Hence, it supports creativity rather than replaces it (such a replacement is—on the other hand—needed within a mass production of equal products except in its engineering and other preparation!). Thus, programoteque supports optimization of the rather creative parts of the work process, which cannot be optimized by hard systems methods, and hence tend to be neglected causing lots of otherwise avoidable costs (several cases in Mulej (ed.), 1997).

Programoteque is produced step by step. It continuously keeps care for insight into work processes rather than subordination, because it is work processes, which create the outcome: the structure of subordination is only one tool of running the processes. Therefore, one begins, if one is a consultant (and every good manager, especially an innovation manager, is rather a consultant than a commander!), with the application of the simplest and most general possible model of the process (see Table 23).

If you compare Tables 19 and 23 you see that the model in Table 23 is a simplified version of the model in Table 19. They could all be related with the usual triangle form of the structural model of subordination: the top management does the planning of objectives, the middle management the planning of how to implement objectives, which means defining of tasks and procedures, the low/operational management and their teams execute the tasks, the monitoring and checking personnel provides the insight in the equality of plans and outcomes so that interventions can take place where ever and when ever needed.

Once the interlinking program is done, consultants no longer talk to the top management; details are supposed to be better known to their more narrowly specialized coworkers. Each of the four steps in Table 23 can be further elaborated in all details chosen as suitable, needed and sufficient (law of requisite holism!). There are only two preconditions:

1. One goes from the very rough and short description towards detailed ones, but one may never switch from insight in processes to structures of subordination and never lose the whole out of the sight.

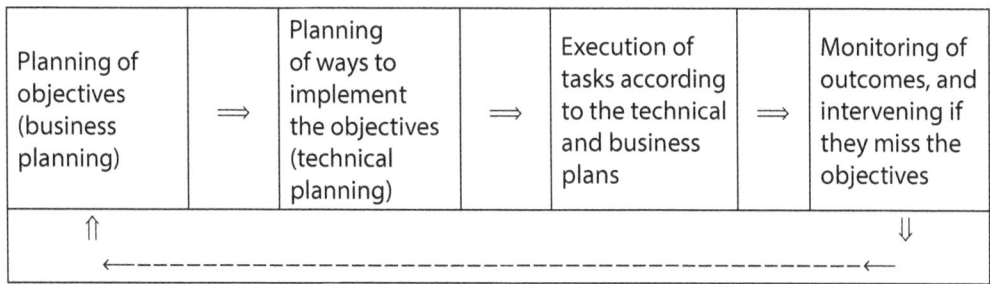

Table 23 *The interlinking general program of the programoteque— shortest version*

2. One never goes so far in details as to lose the framework description in order not to replace the support to creativity with replacement of creativity of the process performers (e.g., by a deterministic prescription which, on the other hand, is crucial in a routine job such as one at the conveyor belt).

In order to meet these two criteria, the principle is that every job performer has information requirements that the forgoing job performer in the same process must fulfill. The technique for making a partial process framework program includes the following steps to meet this principle:

1. The 1st step: consultant talks to the performers of the job under consideration to collectively find out and write facts, how this job is done step by step from its beginning towards its end (the questions put: what, who, when, where, how; no: why!);

2. The 2nd step: the same group talks and writes, on the basis of the notes produced in the 1st step, but now the topic is an analysis of the process under consideration from its end towards its start (the causes tree, mentioned earlier):
   - Is the insight complete enough?
   - Can any activities, which used to be usual so far, be eliminated as avoidable?
   - Is the foregoing step really a precondition for the step under consideration?

3. The 3rd step: the same group talks and writes, on the basis of the notes produced in the 1st and 2nd steps, but the topic is now the influencing environment of the process under consideration:
   - On which other partial processes does the partial process under consideration depend?

- By which documents and information contained in them does the impact of these processes on the process under consideration take place?
- In which way/s and in which step/s of the partial process under consideration is the information received and accepted?

4. The 4th step: the same group talks and writes, on the basis of the notes produced in the 1st, 2nd and 3rd steps, but the topic is now the influenced environment of the process under consideration:
   - To which other partial processes does the partial process under consideration provide their inputs?
   - By which documents and information contained in them does the impact on these processes by the process under consideration take place?
   - In which way/s and in which step/s of the influenced partial process under consideration is the information received and accepted?

Findings can be written in the form modeled in Figure 4 . It can also be computer supported and quantified.

Obviously, the 1st step of this procedure deals with line 4 in Figure 4, and so does the 2nd step. The 3rd step covers lines 1-3, the 4th step lines 5-7. Thus, the requisite holism of the insight is enabled. Again, the level of detail is a matter of a free decision.

From the viewpoint of quality of the outcome, motivation is equally important as professionalism. The job performers are the professionals of the job under consideration, not the consultant; he or she is the professional of the method supporting the professionals of the job in their screening of their job processes. This fact must be recognized, and the job performers must be respected and appreciated for their creative capacity. In other words: the consultant never gives orders or makes decisions, he or she only takes notes and helps in terms of the method. He or she can never know about the process under consideration more than the experienced and trained job performers do; his or her specialty is another one. He knows best the procedure of cooperation.

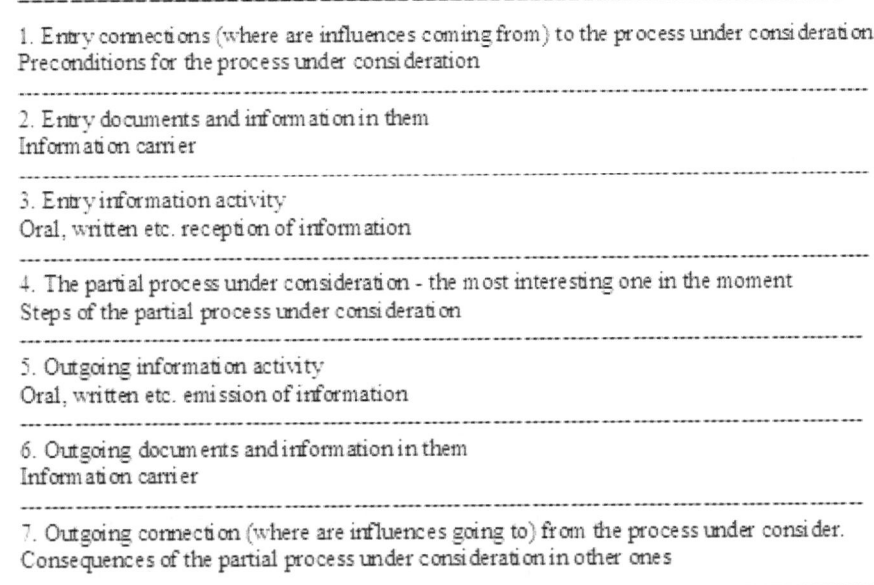

Figure 4 *Form for description of partial processes under consideration*

13.3 THE USOMID-SREDIM AS THE GENERAL METHOD OF CREATIVE WORK AND COOPERATION PROCEDURE

The SREDIM label is an acronym for a procedure of a quite holistic work (see Table 24, vertical steps). Once we had found in working with it, that it did not contain cooperation, we added horizontal steps (see Table 24) to make the procedure closer to the law of requisite holism.

The 6 steps of the SREDIM procedure concentrate on the following issues:

1. Step S (= select a problem) is the step in which the present team members (applying all four steps written horizontally in Table 24) collect their opinions what is the topic for them to work on, what is the broader problematique[5], and what is the problem they select for their work right now (other problems will wait for somebody else or for another opportunity).[6]

---

5. The insight into the broad picture of the reality is equally important as the insight into the details on which we are concentrating right now. Having both pictures at the same time is a crucial part of the dialectically systemic thinking aimed at requisite holism.

6. It makes sense, in our experience, not to limit the collection of the list of topics, problematiques and problems to the team members alone. An *organized criticism and criticizing* is a helpful tactic. It helps to collect many topics, etc. worth working on, potentially. The list of topics, etc.

2. Step R (= record data) is the step in which the present team members (again applying all four steps written horizontally in Table 24) collect data and present them to each other. The question why is not asked now, it will have its turn in the next step. On the basis of experience concerning the way of doing this step, a framework programoteque can be produced and applied when topics are comparable enough.[7]

3. Step E (= evaluate data) is the step in which the present team members (again applying all four steps written horizontally in Table 24) analyze the problem with the collected data and present their findings to each other. The question why is central now, one tries to find out the roots of the problem and the reasons for it to surface[8]. On the basis of experience concerning the way of doing this step, a framework programoteque can be produced and applied when topics are comparable enough.

4. Step D (= determine a solution) is the step in which the present team members (again applying all four steps written horizontally in Table 24) judge the potential solutions and select one on a comparative basis. The question of their realism is central, now. Hence one may tend to evaluate the methods applied earlier as a source of a better or poorer realism.[9] On the basis of experience concerning the way of doing this step, a

---

collected in this way can be publicly exposed as a poster listing them all in one column, and having room in another column where individuals can sign up, if interested in working on the problem in an invention-innovation-diffusion process. If nobody signs up, the problem is going to sit and wait. If they sign up, a team can be made of them to work on the problem as volunteers, but on the company time. Mostly, the team is assigned one hour a week in a period, which makes no serious harm to the daily job obligations.

7. This means, that the programoteque is a dialectically systemic stock of *framework programs* to be applied when appropriate, i.e., when the situation is comparable enough for the old experience to be useful as a reminder on how to deal with the situation. The difference from habits of many well-experienced and successful persons is that their reminder remains in their head (as long as it is not forgotten), and programoteque is written and transferable data, able to become information.

8. Frequently, the mistake occurs that the team members want to prove their possible solution here. This is too early, if one does not know the backgrounds of data collected. Data tend to be superficial. (E.g., a medical case: a higher body temperature is no disease, but its symptom, actually a symptom of many. With no knowledge of the roots and backgrounds one can hardly select the right medicines. Similar is it with economic, business and other topics.)

9. E.g., if one has chosen medieval metaphysics instead of the modern materialistic dialectics, *essentially different findings* may come out (Tables 13, 14). Different kinds of *systems theories* may have a similar effect. So do many different statistical and other quantitative methods, different market research methods, etc. The methods chosen and applied make the framework inside within which findings are achieved; the rest may be out of sight partially or even completely.

framework programoteque can be produced and applied when topics are comparable enough.

5. Step I (= implement solution chosen at the end of the step D) is the step in which the present team members (again applying all four steps written horizontally in Table 24) plan the way of how to change the given reality by application of the invention they have arrived to. The process of its realization, the structure supporting it, the knowledge base required, the potential obstacles, are to be taken into account. On the basis of experience concerning the way of doing this step, a framework programoteque can be produced and applied when topics are comparable enough. Then, the solution should also be applied in reality.

6. Step M (= maintain solution in the real world) is the step in which the present team members (again applying all four steps written horizontally in Table 24) plan the way how to fight the opposition against their innovation, how to persuade their partners, sponsors, champions, apply methods, in order to make the new solution survive. In this step they may also find out what new or unsolved troubles are around and worth starting the entire USOMID-SREDIM procedure again concerning the new issue. On the basis of experience concerning the way of doing this step, a framework programoteque can be produced and applied when topics are comparable enough.

The four steps (horizontal in Table 24) make the USOMID-SREDIM procedure different from the SREDIM procedure: they take care of cooperation in the team. They also make this procedure different from the brain storming methods in important small inventions, such as:

- Team members are invited to think about the selection of the problem, too, not only about its solution; thus, they own the problem rather than feel the problem is imposed over them.
- Team members do not go directly to their oral discussion, because this would make them spend lots of time without actual creativity; time is used more economically and creatively, if the brain writing steps and the brain storming discussion steps are combined.

What is the organizational form like for USOMID-SREDIM procedure to be used?

THE 4 STEPS WITHIN SREDIM STEPS THE 6 STEPS OF SREDIM	1. Individual thinking in a written form (brain writing)	2. Circulation of notes for reading and additional individual brain writing	3. Discussion on the basis of notes aimed at shared findings and conclusions	4. Final minutes on solution agreed upon by the team
1. Select (the problem to be solved)	X =⇒ * =⇒	=⇒ * =⇒	=⇒ * =⇒	=⇒ * =⇒ ⇓
	⇓⇐=========	⇐=========	⇐=========	⇐=========
2. Record (data on the problem)	=⇒ * =⇒	=⇒ * =⇒	=⇒ * =⇒	=⇒ * =⇒ ⇓
	⇓⇐=========	⇐=========	⇐=========	⇐=========
3. Evaluate (data collected)	=⇒ * =⇒	=⇒ * =⇒	=⇒ * =⇒	=⇒ * =⇒ ⇓
	⇓⇐=========	⇐=========	⇐=========	⇐=========
4. Determine + develop (solution of the selected problem)	=⇒ * =⇒	=⇒ * =⇒	=⇒ * =⇒	=⇒ * =⇒ ⇓
	⇓⇐=========	⇐=========	⇐=========	⇐=========
5. Implement (solution)	=⇒ * =⇒	=⇒ * =⇒	=⇒ * =⇒	=⇒ * =⇒ ⇓
	⇓⇐=========	⇐=========	⇐=========	⇐=========
6. Maintain (solution in practice)	=⇒ * =⇒	=⇒ * =⇒	=⇒ * =⇒	=⇒ * =⇒X

Table 24 *The procedure USOMID/SREDIM as the general method of creative work and cooperation*

13.4 THE USOMID CIRCLE—AN ORGANIZATIONAL POSSIBILITY SUPPORTIVE OF CREATIVE COOPERATION

The most usual organizational possibility of creative cooperation is a meeting, session. But most of us have not experienced them as a time freeing a lot of creativity. One reason for this unfortunate experience is that in most meetings they do not apply a method such as the one described here in Ch. 13.3, but they rather follow the logic of one-way communication (the boss speaks, the others ...?). Most meetings also tend to consist of too many individuals for anything else to be possible, too frequently. Therefore, it makes sense to consider the finding of psychology of small groups that creativity is best in groups of 5-12 members. In a bigger group, the number of mutual impacts tends to outgrow the possibility of a real cooperation and creation. In smaller groups the diversity of insights may be too small.

The word USOMID Circle is based on international experience (Quality Circles, Quality Control Circles). Such a circle is an efficient work group on the following basis:

- Members receive training (on methods to be applied);
- Organizational roles are assigned to everybody, so that the burden of organizational tasks is shared rather than completely on the shoulders of the chairperson alone, making this person overburdened and the others irresponsible.

Everybody tries to:

- Achieve the common objective with the results of the shared effort;
- Understand the work method;
- Receive and accept the suitable knowledge, skill and tools;
- Make a contribution towards an encouraging climate in the group;
- Accept responsibility for his or her work, and share acknowledgment for it;
- Receive and accept professional help and support when needed.[10]

Once innovative efforts and results become a wide spread and normal practice, the transition from a random innovating to an "innovative business" can have its turn.

There are 12 roles for circle members to cover all the requisite organizational jobs (Mulej *et al.*, 2000: 529):

Leadership roles include:

- Circle head; he/she is responsible for the general leading of the circle and maintenance of the feeling for the objective and cooperation.
- Secretary; he/she makes minutes of the meeting, while notes of circle members circulate and reach her as well, and final minutes.

---

10. If such help and support is needed, the circle invites other experts to join the circle as members of an "extended circle". They use the same method and share the same rights and duties as the members of the "'original circle'".

- Artist; he/she helps the secretary with notes on a flip chart or in other ways enabling all circle members to see diagrams, tables, notes; thus, she supports overview of the ideas collected by the circle.
- Methodologist; he/she controls the procedure to prevent the circle from deviating from the selected way of work, such as a discussion outside the agenda or jumping over the steps of the USOMID-SREDIM procedure.
- Controller; he/she supervises the division of work to be done by circle members, especially the work between the given and the next meeting, meeting the deadlines, quality of work on the agreed upon tasks.
- Devil's advocate or provocateur; he/she fights the unclear or doubtful statements and the destructive sentences, such as: "We cannot do that." Or: "We would never receive permission to do that." Or: "Don't you talk, you stupid."
- Economist; judges whether or not the use of time in the meeting is rational, other work of circle member is so as well, and tries to collect funds for the circle to work.

Professional roles include:

- Expert for problem picturing; he/she takes the leadership from the leader in phase S of SREDIM, which is when the circle is choosing the requisitely holistic definition and presentation of the problem to work on.
- Expert for data collection; he/she takes the leadership from the leader in phase R of SREDIM, which is when the detailed description and collection of data about the problem worked on takes place; he/she knows methods of data collection well.
- Expert for data evaluation; he/she takes the leadership from the leader in phase E of SREDIM, which is when the collected data are analyzed; he/she knows methods of data evaluation well.
- Expert for decision making; he/she takes the leadership from the leader in the phase D of SREDIM, which is when the selection of one of the collected potential solutions takes place; he/she knows methods for comparative evaluation of versions and decision-making; the leader takes over again in the second part of this phase in which the selected solution is elaborated.
- Expert for brainstorming; he/she takes the leadership from leader in step 3 of all phases of SREDIM, which is when the individual brainwriting and additional individual brainwriting are over and the discussion has its turn.

If the circle has 12 members, everybody takes one of the roles and uses a manual for details. In a circle with less members, the same person can cover several roles. Roles may circulate from member to member, when the circle works on new problems.

The remaining open issue is circle members' capability to cooperate smoothly and creatively. See Ch. 13.5.[11]

13.5 COMBINATION OF USOMID-SREDIM WITH THE SIX THINKING HATS METHOD

The point of this new combination emerged from the insight that the Six Thinking Hats (6TH) method mostly covers the emotional part of the human personality, while the USOMID-SREDIM procedure covers the rational one. The combination means that in every step in Table 24 the appropriate hats are applied.

The USOMID model of creative cooperation enables smooth work covering several professional views and organized procedures, thus leading towards the law of requisite holism. This enables a lot of creativity and a lot of innovation, not invention only. A problem that has remained unsolved over all 25 years is (1) the relative waste of time, (2) fight/arguing and bad feelings. The organizational jobs are supposed to solve this problem, but it does not always work without trouble. This is where the 6TH applies.

The 6TH enters the scene as the third dimension along with SREDIM and the four USOMID steps in everyone of them. The 6TH namely enables all circle members not to argue, but to think in the same direction, and to do so in terms of the exposed part of values rather than of knowledge. Thus, our tendency towards requisite holism is not blocked. The Six Thinking Hats are namely neither used by one person nor by all at the same time, but all circle members use the same hat, and later on another one, at the same time. According to De Bono, this replaces the old western habit that the discussion participants close themselves in their respective viewpoints (like e.g., solicitors or politicians or armies or angry children) and fight for the upper hand rather than for mutual completion and a shared and beneficial new solution (De Bono, 2005). In other words, the 6TH methods supports well the creative cooperation, but from

---

11. There are a number of methods for team work that tend to reflect a similar logic with no explicit link to the (dialectical) systems thinking; they are very useful, too.

different viewpoints than the above-summarized attributes of USOMID do: 6TH points more to the values-and-emotion part of the human personality than to the professional part. Both of them are interdependent anyway. Recall Table19, if necessary.

The difference between DST as we practice it through USOMID, in 6TH is briefed in Table 25. The essential shared point includes the intention and effect of both methodologies—Table 26.

As briefed above, USOMID contains roles for organizational jobs along the shared thinking. With these roles and the USOMID-SREDIM procedure USOMID covers the blue hat, but not the others. The white one may be visible in Phase R (Record data). Procedure USOMID-SREDIM may be better in providing a logical phases order ("vertical thinking"), while the 6TH make more room for "lateral thinking".

DST – basis = dialectical system of viewpoints = all essential professions that complement knowledge of participants towards a shared synergy of knowledge	6TH = essential directions of thinking—focus thinking in turn to human attributes in terms of different emotions and values towards a shared assessment of importance

Table 25 *Comparison of essence of DST and 6TH method*

Both DST and 6TH oppose the practice of un-holistic/un-systemic thinking, e.g.: • Using a set rather than a (dialectical) systems of views, data etc; this leads into a one-sided "argumentative thinking", hence to arguing of "infallible" ones, providing partial rather than requisitely holistic insights and thus causing oversights and troubles. • Analysis provides partial insights, which matter, but are not enough; they oppose each other with no synergy, hence no real solution to considered problems.

Table 26 *Shared points of the essence of DST and 6 Thinking Hats method*

In 6TH all circle members think in the frame of the same hat at the same time. De Bono calls this manner "parallel thinking" that provides for the same orientation, i.e., looking for ideas and proofs. It lets nobody fight each other. Hats enter the scene as phases, ruled by emotional accents of thinking, thus providing the power of focusing, time saving, neutrality and objectivity, removal of "ego": one viewpoint in one moment (by phases—hats). For the essence of hats see Table 27.

Obviously, all Thinking Hats are interdependent and used per phases. With some more detail they can be briefed as follows (details in: De Bono, 2005, or any other edition of the book):

> - White = neutral, objective, facts without interpretation, like a computer;
> - Red = feelings, emotions, intuition, irrationality, unproved feelings, no justification;
> - Black = watching out, caution, pessimism, search for danger, doubt, critique; it all works well against mistakes and weak points of proposals;
> - Yellow = optimism, search for advantages of proposals, search for implementation ways, sensitivity for benefit of the idea, constructive approach;
> - Green = energy, novelty, creation, innovation, in order to be able to overcome all obstacles;
> - Blue = organization, mastering, control over procedure, thinking about thinking.

Table 27 *Essence of each of the 6 Thinking Hats*

White hat:

- Facts, data in the given framework (= law of requisite holism);
- No interpretation—self-discipline (!)—facts, no possibilities/persuasions, verified data;
- An overview (= "map") is made step by step;
- Listening to each other, no prior definitions and/or decisions;
- Practical orientation, all data;
- Like a computer.

Red hat:

- Feelings, emotions, intuition;
- No explanation why something is (dis)-liked;
- Beneficial, although not always precise, correct;
- Intellectual feelings too ("interesting");
- The opposite from the white hat, the irrational aspect of thinking;
- Emotions are unavoidable;
- Intuition that leads to a new view and thus to creativity;
- Opinion = assessment + interpretation + intuition = feeling;
- Emotion to be expressed without delay = background of thinking, values;
- Thinking leads to satisfaction (!!), but:
    - Is it detrimental to others?
    - Short term versus long term?
- Emotions cannot be logical; therefore no justification takes place.

Black hat:

- Pessimism; most frequently—precaution, security, possible dangers, in order to enable survival;
- Critical standpoint, deviation from expectation in order to act against mistakes;
- No exaggeration in order to prevent over-pessimism and abuse of caution;
- Criticism, but all remains logical, although from negative viewpoints;
- No equilibrium; weak points are stressed now, the yellow hat will stress strong ones;
- Doubt about strength of proofs ("Might we better switch to the white hat?") in order to lead to a requisitely holistic insight and assessment of the future situation;
- No limitation to criticism, a contribution is asked for!

Yellow hat:

- Optimism; advantages of the suggestion, positive thinking;
- How to implement the idea in practice?
- Sensitivity for the benefits of the idea;
- Care for making not only black views visible; correction of them, but not in the moment they are being expressed;
- Success; the unstoppable desire to implement the idea;
- Discipline! Conscious search for positive attributes, sometimes in vain, optimism may be exaggerated: "Which action follows now?"
- Assessment of probability that it comes through;
- Backing one's positive assessment with research;
- Constructive approach to strengthen efficiency of realization, but more important changes are not included now, they belong to the green hat.

Green hat:

- Energy, novelty;
- Creativity = the key part of thinking;
- Deliberate;
- Fantasy-based;
- Expose chances—to overcome obstacles that the black hat demonstrates;

- If energy is too abundant, one switches to the red hat to choose the framework of thinking;
- Use it, when experience no longer works well;
- Provocation included, research, risk as well;
- Lateral thinking (= step away with patterns in a new direction);
- Thinking about action, rather than assessment only;
- Logic of nonsense, provocation—"PO" = provocative operation, beyond 'yes' or 'no';
- Alternatives after some results;
- Skill + talent + personality, all of them are needed and interdependent.

Blue hat:

- Thinking about thinking;
- Conducting, control, organizing, double-checking;
- Initial hat/step—to define situation, intention, timetable, sequence of hat application;
- Group head is in charge of this hat all the time, other may intervene;
- At the end—conclusions, summaries;
- Focusing—questions, problem, tasks, procedures, tools;
- Observation, discipline.

The process in Table 24 is completed up in Table 28 (Phases and steps are turned around).

The system (or, network) of all six hats produced many successes in different practical cases (De Bono, 2005). So did the new combination in Table 28, especially among students using it for a semester every week. Why has the emotional part of human personality been so influential? See Tables 11, 19 and Ch. 13.6.

SREDIM Phases / USOMID Steps within SREDIM phases	1. Select problem/opportunity to work on in an USOMID circle	2. Record data about the selected topic (no "Why")	3. Evaluate recorded data on the topic ("Why is central")	4. Determine and develop the chosen solution/s to the topic	5. Implement chosen solution to the topic in reality	6. Maintain implemented solution for a requisitely long term
1. Individual brain-writing by all in the organizational unit/circle	All 6 hats	White hat	All 6 hats, red, black, yellow, green first of all	All 6 hats, red, black, yellow, green first of all	All 6 hats in preparation of implementation	All 6 hats in preparation of maintenance
2. Circulation of notes for additional brain-writing by all	All 6 hats	White hat	All 6 hats, red, black, yellow, green first of all	All 6 hats, red, black, yellow, green first of all	All 6 hats in preparation of implementation	All 6 hats in preparation of maintenance
3. Brain-storming for synergy of ideas/suggestions	All 6 hats	White hat	All 6 hats, red, black, yellow, green first of all	All 6 hats, red, black, yellow, green first of all	All 6 hats in preparation of implementation	All 6 hats in preparation of maintenance
4. Shared conclusions of the circle	All 6 hats	White hat	All 6 hats, red, black, yellow, green first of all	All 6 hats, red, black, yellow, green first of all	All 6 hats in preparation of implementation	All 6 hats in preparation of maintenance

Table 28 *Synergy of USOMID/SREDIM and 6 Thinking Hats methodologies in procedure of USOMID*

13.6 VALUES AND THEIR INFLUENCE

In Table 19 we have only mentioned values with no elaboration. In practice, they are very crucial: they do depend on knowledge, but they also influence knowledge, all the way to the selection for which purpose a given knowledge is applied[12]. This is why we introduced them along with the method to make them influential, in Table 28 through the 6TH.

In general, ethics is a feeling/emotion rather than a part of the left-brain rationality/knowledge/skill. It enables humans to distinguish right from wrong. (For details see: Crane & Matten, 2004; Daft, 2003; Ferrell *et al.*, 2004; Hartley, 2004; Jennings, 2005; Singer, 1999). Empirical researchers consider ethics a synergy of practices, which tend to be preferred in a society or community, as a social group, for long enough periods of time to become codified. (For details see: Jennings, 2005; Shea, 1998; Singer, 1999; Trevino *et al.*, 2000; Trevino & Nelson, 2004; Ulrich, 1997). Moral rules result as a formal next step. They co-create values and culture, of a social sub-group, of SE units, of SEs as complete entities, or a culture of regions, nations, social classes, or professions, and make them sustainable.

Thus, something, which was originally an individual attribute, becomes objectified as a component of the "objective" conditions existing outside the impact of concerned individuals. It becomes a part of broader requirements imposed on individuals, and tends to return, in this way, back to individuals as a part of their (socially obligatory) values. Thus, it enters or re-enters the individual's starting points, which influence perception, definition of preferences, and realization of goals, tasks, and procedures for completing tasks. It means that, for any human activity, ethics is equally essential as professional knowledge and skills, creativity and cooperation capacities are. Moreover, they are interdependent, like both hemispheres of brain are. (For details see: Mulej, 1979; Mulej *et al.*, 2000, 2002, 2004; Mulej & Ženko, 2004a, b; Potočan & Mulej, 2005; Potočan *et al.*, 2005). See Table 19 again.

This means that values—culture—ethics—norms (VCEN, Table 11) have a very crucial impact: they influence the selection of the dialectical system of viewpoints to be considered and all steps of the further process in Table 19

---

12. E.g., it depends on values rather than on knowledge whether or not a person trained in shooting will shoot a person or a paper target only.

including the final attainment of results and returning to the beginning for the next cycle in Table 19, using Table 28.

In order to enhance creativity and cooperation, especially the interdisciplinary one, the common denominator is motivation by consideration of coworkers as creative professionals and humans. This means: most people like working and cooperation, once their peers and bosses and partners let them feel they are acknowledged as creative persons with capacity to cooperate with other, especially the opposing thinkers using viewpoints different from their own. This experience brings us to ethics of interdependence.

13.7 ETHICS OF INTERDEPENDENCE

In Table 11 we mentioned ethics in interdependence with values, culture, and norms. We will keep this notion because in 1998 we introduced the term "ethics of interdependence" (Mulej & Kajzer, 1998). Interdependence is reality in nature as well as in modern human life. This fact must find its reflection in ethics, which teaches humans about the distinction between right and wrong and about defining what is right. Ethics, along with knowledge, helps humans, businesses, nations, and international organizations, to live together, to trust rather than confront each other; this makes them worthy of trust from each other, and thus to attain the requisite holism of thinking, decision-making, and action concerning all crucial topics of our current and future life of humankind (Rozman & Kovač, 2004). Ethics of interdependence, today, is the most realistic form of ethics, for economic reasons, at least.

Perhaps, it has not always been this way, or at least, it has not always been understood this way. According to Wilson (1998), researchers find that the ethical norms have been changing in an evolutionary process based on the interplay of biological and cultural factors (for details see: Shea, 1998; Singer, 1999; Trevino & Nelson, 2004; Ulrich, 1997). This finding may hold true and explain many things, some of which may be quite relevant in the context of this contribution, such as the existence of different ethical principles for each type of society. E.g., Ethical principles of pre-industrial societies were based on their experience that the solidarity of the extended family (and community) helps members to survive, and does better so than the ethics of the individualistic competitiveness of the industrial and post-industrial/market[13] societies/communities. Interdependence is strong within the

---

13. The countries of the Central and Eastern European countries, which used to belong to

(extended) family, which is a business unit, too, and less so outside it, due to the prevalence of a self-sufficient rather than market economy and life-style. Ethics of interdependence is much less so, now.

When one discusses the transformation from pre-industrial to industrial society one must take into account Adam Smith's contribution. Smith is often regarded as the father of economics, and his writings have been enormously influential. In his book "An Inquiry into the Nature and Causes of the Wealth of Nations," Smith (1776) sets out the mechanism by which he felt economic society operated—i.e., the invisible hand. However, this is not all. Smith (1759) presupposed that ethics of altruism would help people overcome their natural selfishness, which was and is making them forget about solidarity and interdependence, once they feel that a narrow individualism would help them better than solidarity. Nowadays, altruism is no more appealing than it used to be, but ethics of interdependence can replace it. (For details see: Mulej, 1979, 2007; Mulej & Kajzer, 1998; Mulej & Ženko, 2004a, b; Potočan, 2000, 2005; Potočan & Mulej, 2005, 2007; Potočan et al., 2005; Rebernik & Mulej, 2000.) Now, again, we see, Smith was right. Interdependence imposes altruism for selfish reasons: nobody can be a specialist in everything and hence self-sufficient.

Democracy expresses ethics of equal legal rights/duties for all men and women, who are included in its definition of the entitled members/participants of the democratic processes. Nowadays, in the advanced parts of the world, at least, political democracy no longer embraces a selected part of society, as it used to when it was normal to own slaves or operate in a feudal society. Economic, organizational, family, local community and similar kinds of democracy compete in order to channel human creativity away from causing too much harm. When ethics of interdependence is omitted, one-sidedness enters the scene once again, and a new kind of problem surfaces (Greer, 2000; Hartley, 2004; Trevino & Nelson, 2004).[14]

Yugoslavia or to the Soviet Bloc and to live in the so- called socialism or communism, gave up that social order in the early 1990s. They are said to move to the market economy. The notion says nothing clear, unless one takes a look at Tables 3 and 4 in this book. The transition, as the process is called, carries them from a non-innovative, routine-loving economy to an innovative one, in which there is a surplus of supply over demand in the market requiring innovation all the time.

14. Let us leave aside the unfortunate fact that political parties understand democracy as their competition with other political parties for their one-sided power rather than their effort to attain requisite holism in dealing with all people's shared issues. They find the number of their members in parliament more important than their proofs / arguments. They should better use what we showed here in Tables 19 and 28. Now, their dialectical systems are often too narrow.

- From ethics of infallible authority (e.g. of feudal masters and company owners) to ethics of individuality (of every person as a specialist in a crucial topic);
- From ethics of inherited differences in wealth (like in pre-industrial times) to ethics of differences created by innovation (in an innovative society);
- From ethics of guilds (caring for quality and allowing no difference between supply and demand) to ethics of the market (causing pressure on those who lack innovation ability) as the power of the innovative ones;
- From ethics of routine and equilibrium (like in times with no/little progress) to ethics of innovation and uniqueness, i.e. sustainable and socially responsible enterprise ethics (in the global market society);
- From ethics of passive obedience including the right of irresponsibility (e.g. "Boss cannot be wrong, so I do not think") to ethics of one's own responsibility;
- From ethics of owning treasures (like in pre-industrial times) via ethics of owning capital (like in early capitalist times) to ethics of owning holistic knowledge and creativity (in the innovative society); and
- From ethics of individual working (before deep and narrow specialization) to ethics of creative cooperation.

Table 29 *Contemporary changes of ethics*

Thus, there is a basis for creation or discovery and definition of ethics of interdependence, which is a precondition for interdisciplinary cooperation, which is a precondition for requisite holism, if the object is considered from a (dialectical) system of viewpoints. This requires more than one discipline/profession/viewpoint as we have shown here so far. People, who are supposed to cooperate, should feel their reality—they need each other in order to succeed, thus, they need ethics of interdependence.

The most modern organizations, only, demonstrate the basic, expected and partly continuous innovating of ethics while they move towards sustainable enterprise ethics (see Tables 3 and 4). It may now depend on innovating of a number of attributes of ethics, along with the economic processes as briefly summarized in Tables 3 and 4, such as the ones briefed in Table 29.

Their common denominator is the innovation of the obsolete feudal and other subjects' culture of dependence, or of the self-sufficient farmer families' ethics of independence, as well as of dependence of the subordinated employees towards ethics of interdependence of specialists needing each other. However, cases that are mentioned at the beginning of this contribution and other cases of power-holders' and subordinates' behavior show narrowness with critical ruining consequences, even though ethics of interdependence's principles are prerequisites of human survival. The 2008- crisis shows this, too.

The very contemporary tendency to enhance requisite holistic thinking is the so-called social responsibility (Hrast *et al.* (eds.), 2006, 2007, 2008; Rozman & Kovač (eds.), 2006).

13.8 SOCIAL RESPONSIBILITY—A WAY OF REQUISITE HOLISM AND ETHICS OF INTERDEPENDENCE

For about one hundred thousand years there were no serious differences in the standard of living in various parts of the world. Then, a 15-20%- of humans became the humankind's advanced part over the very recent centuries and decades. The advanced humans made room for entrepreneurship and innovation by freeing the individual initiative and competition instead of the previous monopolies of guilds, clergy and nobles. The other 80-85% of the world did not do so or did not do so then (Rosenberg & Birdzell, 1986). According to the World Bank data the span between extremes of national per capita incomes has grown from 3:1 before freeing the individual initiative to 150:1 in one century before 1970, and to +500:1 now.

This process changed both, the prevailing knowledge and ethics. Along with the transition from the pre-industrial to the industrial and post-industrial societies and the increasing standard of living of more and more people, individualism has been becoming more and more the prevailing ethics. Narrow specialization has been growing along with the growing huge amount of humankind's knowledge, which is extremely heavily concentrated in the most advanced 15-20% of the world. Besides their good consequences, such as rapid economic development and growing standard of living, one-sided individualism and specialization are endangering the human capability of requisite holism. Over the recent six decades this danger has caused the creation and establishment of systems theory as a tool to fight the one-sidedness and resulting failures, although its practice is still too limited for humankind to avoid current problems well enough. Now, holistic behavior shows up in the form of knowledge and ethics of sustainable development and ethics of interdependence and social responsibility as three further and related ways of applied systemic thinking in both, knowledge and ethics, which are interdependent, too.

One-sidedness of knowledge and ethics of humankind reached a very dangerous triple peak in the 20th century: two world wars and the worldwide economic crisis between them (in 1914-1945). People have kept dying in wars in millions later on as well, based on the one-sided decisions of the influential ones. Scientists, who lived through these terrible decades, felt the need to offer humankind a new capability to prevent such crises from re-appearing.

At first, right after 1914-1945, it was Cybernetics and the General Systems Theory only, now there are many systems theories and many kinds of cybernetics (François, 2004). Many new scientific disciplines that have emerged from them now tend to forget their roots (Umpleby, 2005). Not all systems theorists and cyberneticians keep in mind that L. v. Bertalanffy (1979: VII) has said that he had created systems theory against the current exaggerated over-specialization rather than as one of many narrowly specialized disciplines. Systems theory is to him the culture of wholeness and methodology supportive of it, which includes both, ethics and knowledge. This is in line with other grandfathers of systems theory: Hammond (2003) captures their intentions with the title of her book about their work as "the science of synthesis". Elohim (1999) quoted Bertalanffy saying the reason for it: "the humankind is in danger, it may even disappear, if people do not start behaving as citizens of the entire world and consider the entire biosphere; the world is full of interdependences." Since 1992, this attitude-value-culture-ethics-norm (Table 11) is demanded in the documents of United Nations as the highest political body of the entire humankind. It is called sustainable development, which links care for economic and social development with the care for natural preconditions of humankind's survival (Ečimovič et al., 2002; UN, 1992; Vezjak et al., 1997; WBCSD, 2004, 2005; WCED, 1987, (ed.) 1998).

The topic of systemic thinking and the one of sustainable development keep surfacing as unsolved problems and related needs of humankind at the same time. We have briefly tackled the crucial reasons for it above. The crucial consequences of the unsolved problems include:

- Humankind's history reached, some two to three centuries ago, a point of disappointment of subordinates with their rulers' one-sided ethics, knowledge, and behavior of the foregoing four millennia. Then the oligarchic society management had replaced the previous solidarity of one hundred millennia. Instead, in recent centuries the societal management gave the upper hand to another method for reaching the common benefit—the liberalistic free market and democracy. But the worldwide crisis in 1914-1945 showed that expectations were too optimistic; thus, the influence of government returned to the stage (Reich, 1984). It produced another type of market, with many more interventions than the first authors such as Adam Smith (1759, 1776) had expected, but these interventions are often one-sided to the benefit of the big capital rather than the entire humankind. Democracy shows up more as the changing of one-sided parties in power than a society-wide holism and creativity enhancer. Fighting each other under the name of power of one-sided arguments (rather than by USOMID and the 6 Thinking Hats and similar methods) gave the power-holders, whoever they have been and/or are, the right to impose solutions, which do not always meet needs of the entire society, but the ones of the ruling majority or minority.

- Humankind's history after the World War II produced a speed of development and innovation, being its tool for survival in the competition that exceeds human capability to adapt to newer and newer social and natural conditions. See Tables 1-4 again. Besides, focusing on technological innovation alone produces what Einstein has expressed well: "Excellent tool for unclear objectives." The non-technological innovation needs more attention (IBM, 2006). The 2008– crisis proves Einstein has been right.

Consequently, with full right, humankind needs the development level of sustainable enterprise and the one of the responsible enterprise (in Table 4, esp. decades of 2000- and 2010). It requires requisitely holistic understanding of the current reality and of the role and importance of all humans in that reality, especially of the critical entities such as enterprises. This means that humans must use requisitely holistic thinking in their perception, thinking, decision-making and action for humankind to survive as we have demonstrated here.

What do the experiences briefly described so far have in common? It is the need for ethics of interdependence to replace the ethics of individualism and social irresponsibility of individuals, companies, and countries including their bosses, first of all. Why do we allow ourselves such a conclusion? Over many millennia people have survived on the basis of interdependence (Capra, 2002). But the prevailing ethic changes in historic and geographic conditions. Over the recent decades, interdependence and solidarity have been losing ground, because the most entrepreneurial people were creating values of one-sidedness and the right of abuse that were supposed to prevail and frequently succeeded to do so and benefited the "upper classes". This might be OK, if it caused no crucial side-effects of one-sidedness. An ethics that makes people forget about the broader social consequences of their actions helped the most entrepreneurial and individualistic ones, but caused much trouble to others. The latter are many more, and may cause trouble to the most entrepreneurial and narrowly selfish people. (Unrests of the unhappy people in several countries provide clear evidence, so does international terrorism).

For selfish reasons one may not be too selfish.

Enterprises have a crucial role in this emerging process. Acting in global economy demands from them a new/innovated view on the world, a more holistic combination of managers' competencies and higher sensibility for problems that in the past have been found less important for business. One mentions corporate citizenship (McIntosh et al., 1998: 4, 61). Two main views connected with it emerge:

- We all live in a more open world, where investigation is hard to escape, and
- So is responsibility to a large number of stakeholders.

According to Chomsky (1997: 294) humankind has, for the first time in modern history of industrial society, a strong and broadly accepted feeling that things will not turn better and humans have no exit. Humankind must confront itself not only with its own individual mortality, but also with possibility of collective death of humankind (Keane, 2000: 134).

Over-bravery and communism (with too much solidarity and centralism) would also damage the society—similarly as the egoism of the western contemporaries and the individualism of western laws. In Mauss's opinion (Mauss, 1996: 141) there is no need to wish that a citizen would be either too good or too subjective, nor too senseless or too realistic. He or she must have a smart feeling for him- or herself, for others and for social reality: "He or she must act so, that he or she considers him or her-self, subgroups and society. This moral is eternal; it is common for most developed societies, for societies of near future, and less developed societies, as much we can imagine them. Those are the bases." We find them close to what we call ethics of interdependence. Globalization keeps requiring it.

Globalization is not limited to sharing a global economic development (McIntosh et al., 1998: 35), although this is its main viewpoint; it includes also ideas about connectedness through understanding of manners, interdependence of our local and global communities and environment. A new—shared (corporate)—citizenship is beginning, which is a citizenship with strategic planning based on global interdependence. Profit is its tool, not goal.

That's why businesses must create a new brand of managers, sensitive for new manners of operation in a less compliant world with less organizational limits and higher entanglement. Abilities of global managers of this new kind are crucial for the impact on the necessary changes, so that responsible common citizenship would become real and beneficial. Managers, who consider the new reality, have (McIntosh et al., 1998: 39-40):

- Ability to think in terms of global citizens,
- Real interest for different ideas and opinions,
- Ability to work with people with different experiences and different prospects about the world,
- Ability to create relations and construct new societal and organizational structures,
- Ability to predict other societal realities,
- Ability to lead in complicated and confused surroundings,
- Ability to lead over geographical limits,
- Understanding of the individual's own values and understanding that we are all involved in business with values.

If a company wants to attract and keep consumers (Embley, 1993: xiii), it will have to build a solid relationship between company and consumer. Companies, who already have recognized this fact, enjoy more consumers' support. This is why Table 4 shows sustainable and responsible enterprises.

This is where the notion of social responsibility enters the scene. It may be called an invention and supposed to become innovation in order for the danger to be avoided, which led A. Smith to stress moral sentiments before wealth of nations and as preconditions for the latter. The same danger led Bertalanffy and his partners to make systems theory, and many others to keep following his and their steps. But beware: promotion of interdependence, including in the form of social responsibility, faces serious difficulties (Midgley, 2004; Mulej et al., 2004) despite of old roots of the notion of interdependence (Smith, 1759, 1776)[15]. Social responsibility may not be fictitious or old-fashioned, which

---

15. The ancient Chinese Yin-Yang, Greek dialectics, and Maya culture should be mentioned again.

means writing many nice codices of conduct for many professions without acting along their lines, forgetting about the tacit rules of behavior in any profession or situation, facing the short-term cost problems in an action such as publishing a book on recycled paper.

Social responsibility requires decisive persons in government and other public offices, enterprises, at home and in local communities to "think globally, while acting locally" and "to do good in order to do well". One should do so, because the consequences of one-sidedness are oversights, too. They are able to hit the one-sided person back with bad consequences, which can be prevented with more systemic thinking, covered under social responsibility. (Knez-Riedl *et al.*, 2006; Mulej *et al.*, 2006; Božičnik *et al.*, 2008; Hrast *et al.*, (eds.), 2006, 2007, 2008).

The European Union found social responsibility necessary and worth official support. All member countries, including Slovenia, are starting some action plans, but they seem to still struggle with the notion in terms of its concrete contents (Štoka Debevc, 2006). Social responsibility namely requires quite some balancing of several alternatives (Mulej, 2006c) such as:

- Individuals who do not buy cloths, cars etc., that they do not need, do good for nature, but less so for employment.
- Individuals, who do good by subsidizing fire-brigades, sport clubs, amateur and other culture associations, do good for the quality of leisure-time activities of people, but may endanger their own investment capacity, thus their innovation and competition capability. This, in turn, may endanger their capability to support the said activities later on.
- It is not good, if youngsters must work in e.g., factories in less advanced countries for a rather small pay. But it is even worth, if they have no sources to live on at all.

Practical actions that have been undertaken in e.g., Slovenia right now, include innovation of legislation and rules that civil servants are supposed to pursue on all levels of governing in order to contribute to making the action of the society and its bodies responsible, i.e., requisitely holistic to be efficient, effective and serving clients of public services very well (Korade Purg, 2006).

The medical, journalist and several other professions with a deeper social response have their codices of correct behavior, which is socially responsible behavior, if it is requisitely holistic for the patient, journal reader, television watcher, or radio listener, rather than to mistreat and misinform them (Flis, 2006; Petelinšek, 2006).

It is an interesting game of words, that the notions "message" and "consciousness" are expressed with same word—"vest"—in Slovenian. Thus, journalists and other authors of information must be well aware of their potential impact: literally, information means forming of the internal attributes—the personality of a person.

Smaller enterprises can most easily contribute to ethics of social responsibility and a sustainable society in the local communities in which they run their business. They spend their owners', managers' and co-workers' free time in helping others run sport clubs, singing quires and other cultural organizations, they support events of such contents financially. Local development partnerships in which they tend to help may resonate quite far beyond their own companies' borders and legal duties (Kurent, 2006).

There are organizations trying to develop and practice fair trade. This practice is aimed at helping marginalized producers in the developing countries, including improving business and economic conditions, gender equality, exclusion of child labor, and striving for environment friendly production practices (Dremelj *et al.*, 2006).

Social responsibility belongs to ethics of communication as well, including public relations of enterprises and institutions. With efficient communication about social responsibility and interdependence we could help each other create local, regional, national and global awareness, which would lead to a more holistic behavior of all stakeholders and so to common more/requisitely holistic actions. But: if used in a wrong manner, communication could mislead not only potential customers, but all stakeholders too (Hrast, 2006).

Care for social responsibility is not limited to the European Union as mentioned earlier. In the early 1990s, in the U.S. and in Western Europe, the growing interest of consumers in working conditions all around the world led to the establishment of the so-called Codes of Ethics or Ethics of Conduct in implementing corporate business policies and strategies. As these codes were inconsistent, a next step was to establish Social Accountability International. This is the first international standard defining the basic social principles to be followed by an organization, which wants to have a contemporary ethical relationship with its employees. In 1997 the SA 8000 (Social Accountability 8000 Standard) was published. Its regulations are universally applicable regardless of geographical position, industry, or company size. It is based on the conventions of the International Labour Organization, the Universal Declaration of Human Rights and the UN convention of the Rights of the Child. Companies can be granted certificates of compliance, there are 13 bodies accredited to perform verification in accordance with SAI regulations (Mlakar & Korosec Lajovic, 2006).

Education is very influential for human values—culture—ethics—norms. Therefore it is critical that educators from kindergarten to university comprehend the concept of social responsibility, and introduce it in the renewed/innovated university/education practice, research and consulting (Knez-Riedl, 2006).

Cases of consulting demonstrate that in any organization there are various interest groups. If innovation attempts are imposed on them, they have a very poor chance of success. A more holistic and therefore more promising approach is the one stressing social responsibility indirectly—in form of stakeholder management making innovation indigenous (Steiner, 2006a).

Regarding the non-profit sector, it is important to stress that the topic of social responsibility surfaced under the notion of the corporate social responsibility. Awareness building, education, demonstration of good practices, redefining managerial roles, competences, knowledge, and values can lead to "altruistic egoism" (Vrana, 2006).

There are not yet many empirical studies on experiences of social responsibility in Slovenia, but the first ones were conducted. One found that the span of motives is reaching from economic, legal, ethical, to philanthropic backgrounds. Individuals have two main channels to fight socially irresponsible behavior:

- As consumers they can make their choices,
- As citizens they can conduct and/or join protests, boycotts, etc.

This action can help the influential persons become more requisitely holistic in their perception, thinking, decision-making, and action; this means more social responsibility (Knez-Riedl, 2003a, 2003b, 2003c, 2004; Podnar & Golob, 2006).

Even the unpleasant press, which tries to detect and publicize bad behavior, abuse and misuse of influence and power, is able to help humankind develop social responsibility (Rant, 2006). They can investigate in a variety of misuses. There is even an assessment that the concept of corporate social responsibility shows up as an excuse for one-sided behavior of the influential ones (Tavcar, 2006), rather than as a way to better life of all (Potočan & Mulej, 2006).

This means, that the concept of social responsibility must become a crucial part of the values—culture—ethics—norms in all organizations to make them sustainable/responsible organizations or enterprises. These requirements ask all organizations to comply with the law of requisite holism and hence to (informally) apply Dialectical Systems thinking, which reaches beyond a nice theoretical description of a topic within a single selected viewpoint and applies interdisciplinary creative cooperation with a dialectical system of various viewpoints that differ from each other and are therefore complementary to each other.

Individuals and organizations, which do not feel interdependent with others, but independent from them in terms of nature and economics, not law only, tend to behave with less social responsibility. Their monopolistic position allows them for more one-sided behavior than an exposure to a competitive situation does. Crises, including world wars and worldwide economic crises, including the 2008- one, have demonstrated that such one-sidedness is dangerous to the entire society, including the one-sidedly behaving ones. Thus, interdependence as a practice, and/or perceived need, leads to ethics of interdependence and lets it show up in the form of social responsibility.

Social responsibility may therefore be called a crucial case of practical application of systemic thinking and feeling, quite probably an implicit one. But this implicitness is less important than the fact that it may lead to a sustainable development and sustainable enterprises, later on even to a sustainable society (Gregory, 2006). Humankind's survival may be a crucial positive consequence of social responsibility.

The list of real social problems seems to be endless. Some authors think (Embley, 1993: xvii) that we have no possibilities. One should be more optimistic, although not euphoric. To solve the daily societal problems is not only the problem of someone else or only a problem of government. Each among us should respond to problems in our society!

That's why plural institutions shall include in their visions, and VCEN, their care and responsibility for common welfare (Drucker, 1989: 93-94) and so they would also assume political responsibility. The same should be done in companies. If they don't consider society's needs in the future, they will lose their public support, first of all. The same holds of citizens as individuals.

Social responsibility may be called a new way of informal (dialectical) systems thinking, which indirectly revives the Bertalanffian concepts to help humankind find its way out of the current blind alley. This became especially visible when the International Standards Organization published ISO 26000 in November 2010 (ISO, 2010). See Figure 5 and pay special attention to its two concepts that link all seven central themes.

All of us would be better off with more of requisite holism and ethics of interdependence.

13.9 SUSTAINABLE ENTERPRISE ETHICS (SEE)— ANOTHER WAY OF REQUISITE HOLISM AND ETHICS OF INTERDEPENDENCE

Let us return to Tables 4-6 to develop in some more detail ethics of interdependence under the label of SEE. So far we saw, how crucial the impact of ethics is. Equally crucial is the impact of enterprises. The most holistic is the impact of the sustainable and SR enterprises.

A requisitely holistic system of criteria of SEE can only partly result from the action of legal institutions, although they are essential. Humans create them and use them based on a more or less holistic perception of right and wrong. Therefore, there is also a domain of free choice in addition to a domain of law and one of ethics such as SEE for social standards of right and wrong. (For details see: Ackoff & Rovin, 2003; Crane & Matten, 2004; Daft, 2003; Eriksson, 2003; Laurent, 2003; Ferrell *et al.*, 2004; Hartley, 2004; Jennings, 2005; Singer, 1999.)

Figure 5 *ISO 26000*

Ethical dilemmas always show up because situations in human life neither can be fully foreseen for the framework of existing law to capture them fully holistically, nor are humans able to live on Bertalanffian principles of total holism as discussed here earlier. Therefore, criteria for SEE-based decision-making are hard to establish and maintain. In addition, the socio-economic development changes circumstances; thus, suitability of an accepted ethical system changes (see Tables 1-4). Even today, there are several approaches to define, what is acceptable ethics, e.g.:

- The utilitarian approach was foreseen to be holistic, but its practical simplification to financial criteria made it one-sided (see: Daft, 2003; Greer, 2000; Kohlberg, 1976; and Sutherland & Canwell, 2004).

- The individualist approach stressed the long-term interest of individuals, but many have a hard time making accommodations for the long-term rather than just the short-term (see: Cornell, 2005; Daft, 2000; Jennings, 2005; Kekes, 1988; Ulrich, 1997).

- It is similar to the moral-rights approach, which stresses the decision-makers' obligation to maintain the rights of people affected by their decisions; quite a number of decision-makers around the world tend to forget, that they do not have the right to decide, but the obligation to provide requisitely holistic information to the others to make their work and life easier and more effective (see: Crane & Matten, 2004; Daft, 2003; Kohlberg, 1976; Lunati, 1997; Schermerhorn & Chappell, 2000).

- The justice approach is another good concept, in principle. It requires moral decisions to be based on standards of equity, fairness, and impartiality. Humans, due to their specialization in a profession and a culture of living, are very rarely able to meet this criterion, as soon as they acquire a position of power and lack cooperation with others (see: Jennings, 2005; Kekes, 1988; Kohlberg, 1976; Shea, 1998; Singer, 1999; Trevino et al., 2000).

Therefore, it is interesting to see which factors affect ethical choices (in terms of Tables 5 and 6). Different authors speak of three levels of the personal moral development scale and discuss factors impacting moral development although these are not the only ones (for details see: Daft, 2000; Jennings, 2005; Kohlberg, 1976; Shea, 1998; Singer, 1999; Trevino et al., 2000; and Ulrich, 1997):

- On the first level, the pre-conventional one, individuals face external rewards and punishments and obey authority to avoid detrimental personal consequences; this level may be associated with an autocratic and coercive managerial style in which interdependence is poorly considered.

- On the second level, the conventional one, people learn to conform to the expectations of good behavior as defined by colleagues, family, friends, and society. Law is upheld, and it is hoped that this law is requisitely holistically conceived and executed.

- On the third level, one follows self-chosen principles of justice and righteousness. One is aware that people hold different values and seek creative solutions to ethical dilemmas. One balances concern for the individual with concern for the common good.

We must add that the poorer-and-younger market-and-democracy societies make up 80% of humankind and have not yet had time to learn from their own experience that level three is the best for economic and social long-term effectiveness.

As long as company owners and their governors/managers, with all their major impact on the countries' governments and international relations in the global economy, do not accept SEE as a modern version of ethics of interdependence, prospects are rather grave. For that reason, different researchers quite rightly require a redesigning of society. (For details see: Beauchamp & Bowie, 2004; Daft, 2000, 2003; Ferrell *et al.*, 2004; Goerner, 2004; Hartley, 2004; Koch, 1998; Sutherland & Canwell, 2004.) Others aid humankind by suggesting in which direction of development of the SEE principles could go (see: Buckel & Fridrich, 2001; Cooper & Vargas, 2004; Edwards & Orr, 2005; Freeman *et al.*, 1995; Koch, 1998; Laurent, 2003) from:

- The stage of the legal approach satisfying legal requirements via;
- The stage of market approach responding to customers, and via;
- The stakeholder approach addressing the multiple stakeholders concerns, to;
- The activist approach actively concerned with the environment—making a true SE.

One could call this an important growth of requisite holism; it displays a currently visible interdependence of the economic, legal, ethical, and discretionary responsibilities in the corporate and countries' social responsibility resulting from ethics of interdependence.

13.10 FROM CORPORATE SOCIAL RESPONSIBILITY TO CORPORATE SOCIAL INNOVATION?

13.10.1 Introduction

Stakeholder consideration and integration together with CSR and innovation orientation are not just action alternatives for systems or supportive means for public relation or possibilities for short time profit maximization. Instead, based on the idea of CSI they interact with each other and constitute a common system that is—within a turbulent environment—aiming towards sustainable development. In order to be viable, such multifaceted systems need to be based on integrative systems thinking in order to synergistically enable the (sub)system's (e.g., company's) sustainable development in conjunction with other (sub)systems.

At his 100 days press conference on April 29, 2009 Barack Obama stressed that the big surprising issue so far has been that action is needed simultaneously at various different levels within all systems (i.e., entities) comprising not only the U.S., but the whole world. By that he characterized indirectly what today's world—not only the U.S. is all about: in the attempt to find their future development paths, neither can individuals, organizations, states, countries, cultures, and religions be seen uncoupled from each other anymore, nor can an action alternative for one (sub)system be generated sequentially, but require simultaneous consideration within a broader context instead (this rises the questions where system boarders are or where they have to be set, a question that is also touched in Ch. 13.10.4 where Mulej's philosophy of requisite holism is introduced). Furthermore, it becomes obvious that most systems we have to deal with today show a highly complex system behavior with chaotic patterns; since very many scientific as well as practical approaches and methods are nevertheless still based on the assumption of deterministic (system) patterns, a critical reflection of those is more than needed together with the introduction of more appropriate theories and methodologies—this unavoidably calls for interdisciplinary collaborative efforts based on systems thinking.

The peculiarity of these days is that the occurring patterns of change are increasingly indeterminable. As shown in the ongoing economic crisis, changes are showing highly chaotic behavior, where small changes within the system can lead to all-embracing changes of enormous magnitude. To deal with change, concepts such as restructuring, down-sizing, and re-engineering can be useful to some degree, but are limited insofar as they adopt an inward-looking perspective and tend to be inherently negative; this strategy is not very useful for future oriented companies that depend on innovation, but instead an effective utilization of the organizations' internal as well as external creative potential is needed to cope with highly complex systems instead of relying purely on cutting-down strategies (Steiner, 2008). In order to make use of creative potential and to design promising future strategies, organizations and individuals need new forms of stability in uncertain system behaviors and dynamically changing structures. In this chapter we suggest to implement strategies based on social and ethical considerations that are anchored within the value system of the own system. By that the concept of corporate social responsibility (CSR) can be further developed to attain corporate social innovation (CSI) that is based on sustainability considerations. CSI as innovation- and collaboration-based specification of CSR can become a powerful driver for sustainable development and future.

The chapter is divided into the following sections. First, the characteristics of CSR in comparison to CSI are clarified and underlying system mechanisms are discussed; hence, innovation and social responsibility can fruitfully interact with each other. Collaborative creative problem solving is found to be a crucial prerequisite for the generation of CSI and consequently for sustainable innovation. After this, the peculiarities of stakeholder management appropriate within a CSI approach are discussed. This includes stakeholder identification and involvement and considers the various preference profiles and active or passive roles together with the appropriate means for communication. Following this, the implications of CSI on sustainable innovation are pointed out.

13.10.2 From CSR to CSI?

13.10.2.1 General objectives of CSR

There are uncountable definitions of CSR available, together with some crucial differences between European and U.S. understandings. A definition of CSR by the World Business Council on Sustainable Development (WBCSD) goes back to the year 1998: "Corporate social responsibility is the continuing commitment by business to behave ethically and contribute to economic development while improving the quality of life of the workforce and their families as well as of the local community and society at large" (WBCSD, 1998). On the other hand the European Community defines CSR as "a concept whereby companies integrate social and environmental concerns in their business operations and in their interaction with their stakeholders on a voluntary basis" (European Commission, 2009b). Within this definition both, social and environmental concerns, are included in spite of most U.S. terminology or the definition by the WBCD before; besides, the integration of CSR as voluntary concept within business activities is stressed, together with the interaction of internal as well as external stakeholders. Nevertheless neither of these definitions say anything about how stakeholders should be integrated or what the objectives for this intended collaboration between the company and stakeholders are.

13.10.2.2 Critical Remarks on CSR

CSR has gained increased importance within practice and academics. Nevertheless, also based on its complex character, a stronger practical impact as well as the empirical CSR research are hampered by the lack of a consistent definition together with inconsistent operationalization and measurement (Williams & Aguilera, 2008; Aguilera *et al.*, 2007, McWilliams *et al.*, 2006; Rodríguez *et al.*, 2006). Based on Parsons' (1961) general characterization of social systems (according Parsons' adaptation to the environment, goal attainment, social integration, and pattern maintenance or latency can be observed in any social system), a fruitful classification is given by Garriga and Mele (2004) who divided CSR approaches and related theories in four groups: (1) instrumental theories, "in which social activities are only a means to achieve economic results;" (2) political theories, which are concerned with the responsible use of a corporation's power in the political field; (3) integrative theories, in which the corporation's focus is on the satisfaction of social demands; and (4) ethical theories, in which the focus is on the corporation's ethical responsibility to society. This chapter tries to point out that CSI can be the approach that integrates all four dimensions.

As further aggravating circumstance, too many names are aiming at a similar direction as shown by expressions similar or related to CSR such as corporate citizenship together with corporate giving and corporate volunteering, corporate accountability, corporate social responsiveness, corporate governance, community engagement, business ethics, philanthropy, global citizenship, corporate environmental responsibility, social innovation, and social sustainability. As this is confusing for most people in CSR research, this picture will probably not be clearer for practitioners.

Another critical issue is given for every strategic concept, as long as it is not fully integrated in the overall organization. Nevertheless, it is integration that makes obvious how a concept such as CSR effects the organization as system with accompanying economic, social, ethical, and ecological implications at various organizational levels. Only when the organizational system and its relevant environment are being considered, potential trade-off effects and arising synergies can be understood. However, some attempts have been made to build awareness for those issues such as by the European Commission

(2009a) within the "European Competitiveness Report 2008"; accordingly CSR can have potential influence on (1) cost structure, (2) human resources, (3) customer perspective, (4) innovation, (5) risk and reputation management, and (6) financial markets on a macro and micro level as well. With regard to the core research interest of this chapter, a closer look on the interplay of CSR and innovation will be taken in Ch. 13.10.3.

With regard to a broadened and integrative systems perspective, a crucial question to be answered is what the next enhancements after the previous CSR conception(s) will be. Is it CSI as joint effort between business and society to create innovation based mutual benefits? Is it a more concrete form of social sustainability? Is it sustainable innovation? Or is it an interplay between them? And how can the single approaches be differentiated?

Since CSR is embedded in a more extensive system, its integration in the (business) organization needs to be discussed further. CSR is characterized by a strategic purpose, it needs to be integrated within strategic management with a clear set of objectives, clearly described responsibilities and competences of persons as well as groups in charge, its interconnected inner-organizational as well as its external partners. If and how CSR is integrated within strategic management is strongly influenced by the distinct expectations and attitudes individuals and responsible managers have towards CSR and its potential influence on industry (Williams & Aguilera, 2008; Strike *et al.*, 2006) or societal culture (Waldman *et al.*, 2006).

13.10.3 Corporate Social Innovation (CSI)

A missing link for a better and more purposeful integration of CSR within an organization's strategy may be provided by CSI as a specific and innovation focused form of CSR. CSI can be understood as an adoption of the previous CSR philosophy that has a high potential to be integrated more easily in the organizations' strategies, since it can become a highly supportive means in facing today's turbulent world as traditional CSR has been (also because of its potentially beneficiary innovation orientation). It is the basis of CSI that a CSR philosophy not only can cause cost-savings, but also enhance the innovation capacity and performance that can cause further potential new value creation together with the generation of new revenue streams, hence (European Commission, 2009a):

- Innovation may result from the engagement with other stakeholders;
- Business opportunities may arise from addressing societal challenges; and
- A stakeholder-oriented organizational behavior creates better workplaces, which can be more conductive to innovation.

13.10.3.1 Systems Mechanism

Whereas CSR is mainly understood as a concept in which a company is taking responsibility for its impact on society, CSI moves a step further by extending this one-sided actions to include both-sided actions. This means that not only the company has an impact on society, but also society has an impact on the company by using the society's inherent problem solving capabilities and its creativity (see Figure 6).

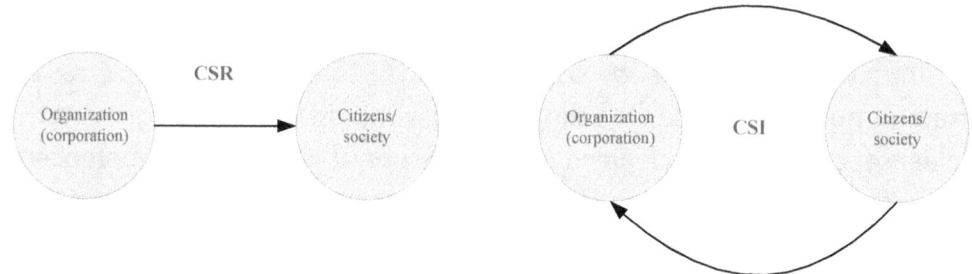

Figure 6 *CSR and CSI: two distinctive system mechanisms*

CSI can be understood as co-creation of values for business and society together with the emergence of new opportunities for cross-fertilization between commercial and social efforts (Esposito, 2008; Esposito & Henderson, 2008). Some most recent but already successfully operating examples of CSI initiatives, such as the "Corporate Social Innovation Program" at Stanford University (Stanford, 2009) or the establishment of a "Leadership Lab for Corporate Social Innovation" at the Massachusetts Institute of Technology (MIT, 2009), underline the meaningfulness of this newly emerging concept.

With regard to CSI, the authors further stress the need for inner-organizational integration, cross-boarder collaboration, "fitting" communication, a shared vision or at least complementary visions of the involved parties, and symbiotic strategic goals. Whereas within CSR society (or certain parts of society) is the beneficiary of a specific action taken by the company for the society, CSI combines collaborative action of both sides for the benefit of the company and the society as well. Hence, CSI is a strategy-based dynamic mutual process between the organization (such as a company) and society to synergistically create benefits (for all involved parties) based on a collaborative creative endeavor with the objective to generate sustainable innovation that is beneficiary for all system stakeholder taken into consideration (although probably to different degrees, according their individual backgrounds, preference profiles and expectations).

To clarify the difference between CSR and CSI a specification of the kinds of benefits is useful. A meaningful distinction regarding the nature of benefits can be made, whether society (or certain citizens) is purely in the position of obtaining something beneficiary without any further need to take one's own action, or its support is more a help for self-help and self-responsible acting such as within CSI. This leads to a big difference from the one-sided "giving"-based approaches, since here the concerned society and its stakeholders are treated not just as pauper(s), but as important person(s) who are in the role of an appreciated partner. Higher degrees of self-esteem, motivation for self-responsible acting, and optimism may be the fruitful result, if the strategy is meant and applied honestly, wisely and with intentions of integrity.

Furthermore, benefits can be of either direct or indirect manner. As already mentioned, stakeholders' benefits from CSR are often more or less charity like and the benefits for the company are gained by indirect effects such as an improved image, a better stakeholder commitment, and an enhanced acceptance within society. With regard to CSI,

- Firstly, external (but also internal) stakeholders' benefits not only depend on the goodwill of the company, but an improvement of the stakeholders' situation is attained by making the company's decisions based on a more accurate knowledge of stakeholders' wants and needs; stakeholders are not anymore just considered as passive actors without being actively involved, but instead stakeholders are integrated actively in communication and problem solving processes;
- Secondly, the company gains direct benefits for its problem solving and innovation processes by getting access to the stakeholders' problem solving capacity and creativity.

Here, the connection to innovation conceptions such as von Hippel's (2007) lead user approach, the open innovation approach by Chesbrough (2006), the open creativity approach (Steiner, 2009) or approaches such as society-driven innovation, stakeholder-driven innovation, and customer-driven innovation becomes obvious. All concepts depend on stakeholder management and a broadened system (i.e., holistic) perspective in order to gain access to often hidden inner- and outer-organization creativity and innovation potential. This is a trend that can be interpreted as the need to go beyond organizational boarders and establish more forms of cross-boarder collaboration with heterogeneous stakeholders.

13.10.3.2 Stakeholder Management within CSI

CSI requires the interaction between the organization (e.g., corporation) and the society (respectively relevant system-related parts of society) not based on accident, but on a systemic consideration of a process-based stakeholder management that besides internal stakeholders additionally integrates the organization's external stakeholders active in a collaborative problem solving process. The stakeholder management scheme as suggested by Steiner is divided into the following phases (for a comprehensive investigation and design of stakeholder management see Steiner, 2008):

1. Stakeholder identification;
2. Stakeholder analysis;
3. Stakeholder classification;
4. Stakeholder action plan (including the phasing of the stakeholders' involvement within the collaborative process).

It is the objective of an effective stakeholder management to build awareness for the heterogeneity of the involved stakeholders in order to enable a fruitful creative collaboration of internal and external stakeholders as prerequisite of CSI. This heterogeneity of the stakeholders involved, probably, cannot be dealt with based on a resource-based view, since this is not only about material and immaterial resources (including knowledge), but it is also about different patterns of behavior based on heterogeneous preferences, heterogeneous value systems, heterogeneous backgrounds and moods, different cultures, variations in predominant thinking styles (about convergent and divergent thinking), heterogeneous competences, heterogeneous expectations regarding the present and future development alternatives that are typically embedded in different social and cultural milieus (Steiner, 2008).

Usually, based on the complex situation CSI is faced with, already available problem solving routines and solutions alone are not satisfying, but instead call for creative solutions, which require creative problem solving processes. A crucial mechanism of CSI is the extension of inner-organizational collaborative creative sources by including external creative stakeholders—on individual and collaborative levels as well—as characterization of an "open creativity" system: "in addition to a system's internal creativity (such as of an organization or a region), the synergetic interplay between internal and external sources of creativity at the individual and collaborative levels also needs to be utilized in the attempt to create innovations" (Steiner, 2009). This is especially valid if the objective is to generate sustainable innovation or, generally speaking, to attain a social, ethical, ecological, and economic sustainable development and future.

13.10.4 Implications for Corporate Social Innovation

The answer to "what's next after CSR?" should probably not be a replacement of CSR by CSI (see Figure 7). A critical systems perspective is needed in any case in order to find out what is appropriate: some cases call for actions such as corporate giving or corporate volunteering (if we just think about earthquake victims or famine); others allow the integration of various stakeholder groups in a mutual problem solving process based on collaborate creative efforts in order to generate innovation appropriate for sustainable system developments. Some stakeholders want to be integrated, others not; specific phases of the innovation process call for the involvement of specific stakeholders equipped with certain competencies and certain characteristics, where a broader form of integration can be hampering (valid for the organization as well as for the society) (for stakeholder management see Steiner, 2008); and every case calls for specific forms of communication in order to allow for mutual collaboration, in which organizations' and society's collaborators need to establish a fitting communication basis (this also calls for certain persons at each side).

To decide, which concept should be applied, a system perspective is needed. As Mulej (2007) pointed out, it does not make sense to try to consider a system totally holistically, nor is it possible, since overwhelming information does not allow a total insight into a system (i.e., complex entity) anymore. Instead of a not attainable and not manageable holism, the focus on "all and only essential viewpoints, professions, sciences and experiences" as proposed by the Dialectical Systems Theory (DST) and its underlying principle of requisite holism, can provide a fruitful strategy for building awareness of the relevant system (Mulej, 2007). Consequently, it is system knowledge that is needed before one can decide whether a traditional form of CSR or CSI is the appropriate approach to be applied.

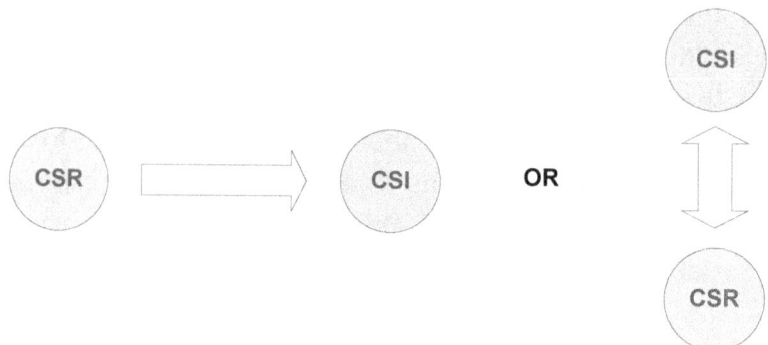

Figure 7 *Development paths of CSR*

The peculiarities of both concepts—CSR and CSI—show that for future developments it will be of interest to integrate both of them in a way that each concept is applied in such a system environment, where viable outcomes can be attained for the system stakeholders. CSI requires conditions precedent to engaging mutually within the innovation process and is therefore a specific form of CSR; it allows for a more intense integration in the organization's strategy based on its far reaching positive implications on the whole organization and its relevant environment (society).

13.10.5 Conclusions and Implications for Research

Firstly, the discussion within this chapter pointed out the need for further clarification, operationalization and measurement of the CSR concept. Secondly, it has been stressed that the one-sidedness of CSR can be fruitfully expanded by considering additionally potential mutual effects for the organization, but also for external stakeholder groups as well as proposed within CSI. This chapter not only shows the need for subsequent research regarding system based research for mutual effects among various stakeholders, but also calls for the consideration of the concepts of open creativity, open innovation and stakeholder management besides the obvious need to consider other forms of sustainability research. Nevertheless, CSI will probably not be able to replace CSR. It should rather be considered as a reasonable extension of CSR for those cases where stakeholders can take an active and self-responsible role within a mutual problem solving process that can lead to a win-win situation for the organization and society as well.

13.11 SOME CONCLUSIONS FROM CHAPTER 13 CONCERNING ETHICS

Requisite holism and ethics of interdependence are showing up in several crucial forms, although they are no general practice yet. They contribute to human and economy's welfare.

In other words: sustainable development (SD) is critical for enterprises in modern society. United Nations (UN, 1992) proclaimed SD in order to make a requisitely holistic synergy of the human care for both, the economic development and the natural preconditions of survival on the planet Earth, a felt need and daily practice. Thus, UN required holistic thinking and action under the label of SD. The basic reasons that make SD critical for enterprises today include:

Market conditions of business operations have demanded more and more innovation and holism over the last few decades, when there was no more room for the random and suppliers markets in the most influential areas of the world (see Tables 1-4). Economic development processes split the world into the well advancing innovative 20% and the increasingly lagging non-innovative 80% divisions of humankind (Brandon & Lombardi, 2005; Dyck & Mulej, 1998; Ečimovič et al., 2002; Koch, 1998; Mulej, 1979). SD/SR presents a possible solution for dangerous and expensive consequences of one-sided development in the industrial era: it offers more holistic and hence balanced development worldwide. This requires innovation beyond technology (Bolwijn & Kumpe, 1990; Cooper & Vargas, 2004; IBM, 2006; Mulej, (ed.), 1984; Mulej et al., 1987; Rosenberg & Birdzell, 1986). SR, including SD, is such a case.

So far, globalization does not seem to diminish the 80/20 divide of the world. Moreover, it is the approach to innovation that causes this 80/20 divide—the innovativeness need paradox: the ones needing innovation the most, like it the least, as well as support entrepreneurial behavior the least (Rogers, 1995: 275). The SD/SR concept can provide for requisite holism by both, technological and non-technological innovations, if humans use more modern—requisitely holistic—behavior as individuals and in local and international communities (Dees & Emerson, 2002; Dyck & Mulej, 1998; Greer, 2000; Laurent, 2003; Lunati, 1997; Potočan, 2004; and WBCSD, 2005).

When SD became an officially declared precondition for humankind's survival, more holism in general and especially in business became an official requirement worldwide. Modern customers prefer suppliers' attributes that are briefed in Table 4—"decades of 2000–and 2010", and tend to avoid other suppliers. (For details see: Dees & Emerson, 2002; Goerner, 2004; OECD, 2000; Potočan, 2002; and WBCSD, 2004, 2005.)

Decisions and actions of all organizations (especially enterprises) must include SD/SR. Otherwise, the profit as the financial outcome of one-sided decisions and actions kills profit as the longer-term benefit: their lack of requisite holism causes oversights and neglects costs to cover e.g., eco-remediation and/or other renewal of preconditions for survival; it causes the need for fixing otherwise avoidable problems, such as health in contaminated areas, instead of moving forward. (For details see: Ečimovič et al., 2002; EU, 2005; OECD, 2000; Umpleby, 2002; WBCSD, 2005; and WCED, 1987, 1998.)

Working conditions of all members of society, especially enterprises, have changed rapidly and critically. In Tables 3 and 4 we briefly comment the development of the market relations over centuries and of resulting basic characteristics of enterprises over recent decades. Due to them and the global impact of enterprises—allowed to be one-sided rather than requisitely holistic in liberal economics, and out of supervision in neo-liberal economics—the old Roman-law definition of ownership is no longer viable. It includes the right of use and abuse of property—and related humans, too. It is not a small and local owner's property that is used and abused, but the entire planet Earth and the entire current human civilization and its economy. Abuse of all of them ruins the life on the planet Earth. SR/SD must replace the industrial mentality

SR tends to be a concept closer to the requisite holism and wholeness, including SD as well for a sustainable and thus viable future of the current civilization of humankind.

13.12 SOME CONCLUSIONS ABOUT USOMID AS APPLIED DIALECTICAL SYSTEMS THINKING

USOMID makes informal dialectical systems thinking possible. Its users need no theory, like users of medical prescriptions. Consultants, managers, professionals of support to requisitely holistic thinking, of course, must know backgrounds of their advices and suggestions, like medical doctors must know the ones of their prescriptions.

Most soft systems theories that we have briefly summarized above, and even more so the hard systems theories, do not pay much attention to human attributes of their users. This makes them complementary with DST: it covers their weak points, they cover DST's ones. USOMID makes them applicable in practice. Innovation is its main application area. Innovation of values-culture-ethics-norms (VCEN) is the most crucial precondition for invention and innovation to flourish, making room for knowledge about innovation and entrepreneurship, and so forth (Table 9) to be created, diffused, and used as well as supported with possibilities (Table 19). A summary of how DST is reflected in USOMID is presented in Table 30.

Now, we can turn to another contemporary attempt to master current humankind's problems—the knowledge management—in order to find out, how well this concept does meet both, the law of requisite holism and enable application of ethics of interdependence. It is namely considered a type of systems thinking (*Journal of Knowledge Management and Systems Science*; Gu & Tang (eds.), 2006). Knowledge society is supposed to enhance innovation, on this basis (Wierzbicky (ed.), 2005).

Law of entropy, visible as the economic development towards the contemporary "state supported buyers' market"; supply is bigger than demand (Chs. 0, 12.3.4 and 14, Tables 3–7 and 25)	→	Objective starting points: needs, visible as demand in the market (Table 19)	→	Innovative society, including innovative business – norms, policies, strategies, tactics, innovation objectives, innovation based rewarding (Ch. 14)
		/ /		
Education and training for requisitely holistic thinking by interdisciplinary creative cooperation making inventions – suggestions, potential innovations and innovations (Ch. 10.3, especially)	→	Subjective starting points as a whole (Table 19)	⊢→ ⊢→ /	Deliberate search for possible inventions in order to make inventions – suggestions, potential innovations and innovations (Ch. 14)
Programoteque = a dialectical system of framework-modeled procedures of work, especially the creative work (Ch. 13.2, Table 23 and Figure 4)		Values and other emotions (Table 19)	←	Justified feeling of being considered as a creative and cooperative person on the part of bosses and peers = motivation (Chs. 13.6-10)
	\			
Optional quantification of and computer support to programoteque	→	Knowledge about 'what' (Table 19)	←	Gaining insight in the work process – from a rough one to details (Table 23 and Figure 4)
			\	
Procedure USOMID-SREDIM-6TH supporting creative work and (interdisciplinary) cooperation (Table 24; Chs. 13.3 and 13.5)	→	Knowledge about 'how', the general part (Table 19)		Causes tree, showing the following order and interdependence of steps in procedure from the end to the start
USOMID Circles (Ch. 13.4) regular composition extended composition organizational roles methodologist co-authors	→	Objective starting points: possibilities, organizational part (Table 19)		

Table 30 *Relationship between starting points according to DST, USOMID and the innovative business and society*

Chapter 14

A CASE OF DIALECTICAL SYSTEMS THINKING: REQUISITELY HOLISTIC MANAGEMENT OF THE INVENTION-INNOVATION-DIFFUSION PROCESS

14.0 THE SELECTED TOPIC/PROBLEM AND VIEWPOINT OF OUR DEALING WITH IT HERE

Let us now turn all the collected insights into a framework action program aimed at enhancing the invention-innovation-diffusion process (IIDP). In it one needs requisite holism unavoidably. Therefore one needs many cooperating professionals different from each other and hence complementary and co-working on the basis of ethics of interdependence. This makes them able to use many types of knowledge and knowledge management (KM), which are quite different from each other in their selected viewpoints.[1] This might help humans improve the outcomes of their IIDP[2] (Chen & Lai, 2005;

---
1. For a current overview see: (Gu & Tang (eds.), 2006).
2. According to the statistics of the US Patent office one single percent of patented inventions become innovations, i.e., yield benefit to their users and hence to their owners. I thank Prof. Dr. Jure Marn for these data. There are further estimations that about 6-7 percent of incremental

Edwards, 2005; De Bono, 2005; Gu & Chroust, 2005; Gu & Tang, 2005; Hays & Wasilewski, 2005; Kameoka & Wierzbicki, 2005; Meyer & Medeni, 2005; Minati, 2005; Nakamori, 2005; Nakamori & Wirzbicki, 2005; Osawa & Mizayaki, 2005; Pavlin, 2005; Steiner, 2006; Wierzbicki & Nakamori, 2006).

14.1 THE DIALECTICAL SYSTEM (DS) IN CHAPTER 14

In order to attain requisite holism, in this chapter our selected DS is networking:

- The process from the making of knowledge and invention, all the way to a well diffused innovation, growing into a beneficial daily practice;
- Invention/innovation typology;
- Dialectical system of preconditions for knowledge to give birth to innovation;
- Requisite holism as a way to diminishing the high risk level of the IIDP;
- Application of the combined application of our USOMID and de Bono's 6 Thinking Hats method as a tool for requisite holism in practice, to which
- Co-laboratory of democracy, and (later on)
- Diffusion of innovation, and
- Heart-storming is added anew as well.
- Creation of absorption capacity for novelties rounds DS off to support innovative business and society.
- Duties of influential people result from the model.

Requisite holism (Tables 2, 8; Ch. 12.3.5) results from consideration of all ten components and relations of the DS at stake and their networking into this DS. The other nine will receive some brief comments.

14.2 THE INVENTION-INNOVATION-DIFFUSION PROCESS (IIDP)

A requisitely holistic—DS—approach to IIDP, according to our starting points for this book, is depicted in Table 37.

It shows that many kinds of knowledge and KM are needed for the IIDP to be mastered requisitely holistically and hence yield success. The vast

---

inventions become innovations, at best (Likar et al., 2002: 11).

amount and variety of knowledge (and values!) requires transformation of bosses from a one-way commanding style to cooperative leadership, even several transformations of leadership (Rooke & Torbert, 2005). It is obviously worth to pay attention to KM in the IIDP (Basadur & Gelade, 2006). Human resources management must be adapted to the innovation rather than to a routine-loving way of handling processes (Bukovec, 2006; Svetlik & Ilič, 2004). Leadership plays an essential role in the creation of the innovative culture (Pucelj & Likar, 2006).

The hard side of innovation management is visible (Sirkin *et al.*, 2005). Effects can only sometimes be measured (Palacios-Marqués and Garrigós-Simón, 2005; Fatur, 2006). Lots of stress may have to be mastered (Treven, 2005), but less than in a neocolonial subordination to the more innovative ones[3].

14.3 INNOVATION TYPOLOGY—AN ADDITIONAL FACTOR OF COMPLEXITY OF REQUISITELY HOLISTIC IIP AND KM IN IT, AS WELL AS MASTERING OF IT

The official international definition of innovation is very general:

Innovation is the renewal and enlargement of the range of products and services and the associated markets; the establishment of new methods of production, supply and distribution; the introduction of changes in management, work organization, and the working conditions and skills of the workforce (EU, 2000: 4).

Economic historians (Rosenberg & Birdzell, 1986) found that it was the:

Innovation of management, which has been the most crucial one in humankind's modern history, going in the direction towards democracy as the relation fostering holism and creativity (Mulej et al., 1987) rather than merely limiting the power (Jež, 2006).

---

[3]. It is still found worth warning, that innovation is not a technological topic only, as one used to think on the basis of old perceptions, and many still need awareness building that business style innovation is even more crucial than technology innovation (IBM, 2006).

IIDP Phases	Crucial inputs	Usual outputs	Usual creators of outputs	Usual economic situation
Creation of ideas, especially invention from knowledge and values of authors	Creative thinking, expertise, ambition, values, time and research conditions	Promising ideas resulting from a part of research, tacit knowledge and values such as interest, etc.	Inventive and professional humans and groups, and (non-)professional researchers	Costs of work and research conditions, no revenue or profit from market
Creation of suggestions from inventions	Writing, etc.—ones' expression of invention	Promising idea written, etc. (made available)	Inventors and advisors about writing, etc. the idea	Costs of message preparation, no revenue/ profit
Optional diffusion of suggestion/s	Offer to the market of inventions –suggestions in or outside "the home organization"	Partly sale, partly giving up, partly transition to development of suggestions	Owners of suggestions (authors and/or others) and their co-workers	Costs of offering; revenue and/or profit from sold suggestions
Creation of potential innovations from suggestions	Creative thinking, expertise, ambition, values, time and conditions for development of a suggestion to a potential innovation	Usable new product, method, procedure, managerial style, potential market, and/or business item	Inventive and professional humans and groups, (non)-professional researchers and developers (incl. market developers)	Costs of work and conditions of development, no revenue and/or profit
Optional diffusion of potential innovations	Offer to the market of potential innovations inside or out-side organization	Partly sale, partly giving up partly transition to application	Owners of potential innovations (authors and/or others) and co-workers	Costs of offering; revenue and/or profit from sold potential innovations

Creation of innovations from potential innovations	Creative thinking, expertise, ambition, values, time and conditions for development of potential innovation to innovation, incl. the entire business operation	Beneficially used new market/product/method/procedure/management style/organization/business item—with users decision and experience	Inventive/innovative and professional humans/groups developers of the novelty and its market, (non-) professional, including entire business operation	Costs of offering; revenue and/or profit from sold innovations (inside or outside organization)
Optional diffusion of innovations	Offer to the market of innovations inside or outside organization, especially to additional customers (after the first ones)	Beneficially used new market, product, method, procedure, management style, organization, business item—with users in broader circles	Owners of innovations (authors and/or others) and co-workers, especially in marketing and sales, but the entire business process matters equally	Costs of offering; revenue and/or profit from additionally sold innovations (inside or outside organization)

Table 31 *Simplified picture of IIP from several (networked) essential viewpoints*

Thus one can structure the quoted definition of innovation in Table 32 (Mulej *et al.*, 1992, adapted in 2009). Its summary reads:

Innovation is every (!) novelty, once its users (!) find it beneficial (!) in practice (!).

In another language (Afuah, 1998): innovation is equal to invention plus commercialization. The latter includes the entire business process, of course, not the marketing phase only.

For success, each of the 40 types of inventions, suggestions, potential innovations, and innovations depends on a specific type of knowledge, values, and KM. But, first of all, it depends on making preconditions for IIDP to flourish, which depends on innovation of management (4.1-4.4 in Table 32) and culture:

From the style "Division in the thinking bosses and the working subordinates" to the style "We all think and we all work and create, invent and innovate, being bosses and their coworkers with all requisite various and different specializations".

Three networked criteria of inventions, suggestions, potential innovations, and innovations	(2) Consequences of innovations		(3) On-job-duty to create inventions, suggestions, potential innovations, and innovations	
(1) Content of inventions, suggestions, potential innovations, and innovations	1. Radical	2. Incremental	1. Duty exists	2. No duty
1. Business program items	1.1.	1.2.	1.3.	1.4.
2. Technology (products, processes ...)	2.1.	2.2.	2.3.	2.4.
3. Organization	3.1.	3.2.	3.3.	3.4.
4. Managerial style	4.1.	4.2.	4.3.	4.4.
5. Methods of leading, working and co-working	5.1.	5.2.	5.3.	5.4.
6. Business style	6.1	6.2	6.3	6.4
7. Management process	7.1	7.2	7.3	7.4
8. Values, culture, ethics, and norms (VCEN)	8.1	8.2	8.3	8.4
9. Our habits	9.1	9.2	9.3	9.4
10. Their habits	10.1	10.2	10.3	10.4

Table 32 *40 basic types of inventions, suggestions, potential innovations and innovations*

This is the management part of the DS of preconditions for IIDP and KM in it. Preconditions are very well summarized in Five Pillars of Total Quality (Creech, 1994) in Table 33.

Perfect product	⟷	Perfect processes
↕		↕
O R G A N I Z A T I O N		
↕		↕
Leading (= managing by example and cooperation)	⟷	Commitment

Table 33 *Five pillars of total quality*

But there is also a business part to IIDP and KM concerning IIDP.

14.4 DS OF BASIC PRECONDITIONS FOR INVENTION TO BECOME INNOVATION

From the business and organization viewpoint, IIDP can be seen as a process dependent on all factors in Table 9, including their interdependences. The sign X denotes that for success none of them may be zero (for details see: Mulej *et al.*, 1997; Mulej & Ženko, 2004a, b).

Now it is time to make a short comment on Table 9.

Invention includes all new ideas that might become innovations. But only one per cent of the inventions that already have become suggestion in terms of Table 31 are able to become a radical innovation, and seven per cent incremental innovations. They are not necessarily results of R&D processes, especially not only of R&TD (technological development) ones, but many are (Likar *et al.*, 2006).

Entrepreneurship is a human characteristic of persons working continuously on the conversion of inventions into innovations on both, the internal and external as well as on local and global markets by new combinations of business factors (material resources, knowledge, business opportunities, creativity, markets, and other factors) (Rebernik & Repovž, 2000).

Requisite holism is a precondition meaning consideration of all essential elements and relations in the process of making inventions and transforming them into innovations. It is based on dialectical (or other contemporary) systems thinking and hence on networked interdisciplinary creative cooperation to meet the law of requisite holism (Mulej, 1979, 2006, 2007; Mulej *et al.*, 1992, 2000; Mulej & Kajzer, 1998; Rebernik & Mulej, 2000).

Management is different from entrepreneurship. It involves efforts at a total and profitable exploitation of all resources that have earlier been created by entrepreneurship. Both, managers and entrepreneurs, need requisite holism (Rebernik in Dyck & Mulej, 1998).

Coworkers matter because neither entrepreneurs nor managers can know or time-wise physically do it all alone. They need co-worker's creativity and routine (without allowing routine to turn into routinism, i.e., refusing of everything new in the name of established skills and habits) (Peters, 1995; Treven & Mulej, 2005; Udovičič, 2004; Udovičič & Mulej, 2006).

Culture (values-culture-ethics-norms) is embodied in the sub-consciousness and tells us rules of right and wrong (Mesner Andoljsek, 1995). If the culture does not support invention and innovation, it is an obstacle in the contemporary market situation (Albach, 1990).

Customers used to be overlooked when the notion still prevailed that innovation is creation of something new. Then, they oversaw the second part of its contemporary definition saying that the novelty also must be accepted in the market, to deserve the label of innovation. Therefore, knowledge of marketing, is as important as knowledge of production. This is true also for internal customers (Zahn, 1995).

Competitors used hardly to exist in pre-entrepreneurship times. Guilds, usual associations of handicraft producers in medieval Europe used to limit the common supply of their members to the quantities and qualities of demand. Thus, innovation was neither necessary nor making sense in economic terms (Table 3). Free entrepreneurship changed this situation (Rosenberg & Birdzell, 1986). Suppliers influence costs and quality due to their own good or poor level of innovating. In a monopolistic position, they may lack care for innovating (Linden, 1990).

Natural environment has been neglected in economic calculations and policies for very long; especially in terms of the sustainable development it is to be considered equally important as other factors of innovation: survival in a poisoned natural environment is not possible (Ečimovič *et al.*, 2002; Ečimovič, 2006; Vezjak *et al.*, (eds.), 1997).

Socioeconomic and other outer-objective preconditions include what traditional economic theory used to consider most or even exclusively: investment, money, allocation of resources, and other economic viewpoints, R&D institutions, legal norms, etc. These remain important, but far from being enough (Mulej, 1995). Therefore traditional economics, especially the neoclassical economics, is no longer good enough (Dyck in Dyck & Mulej, 1998; Mulej, 2006a).

Unforeseeable, random features might be considered a part of objective preconditions, but not necessarily. They cause probability rather than determinism (Hwang, 1997). Good or bad luck, unforeseen synergies, etc., remind us that even with the best information system one only can do a better or poorer informed estimate once it comes to an innovation effort (Knez-Riedl *et al.*, in Dyck & Mulej, 1998).

Innovation is obviously a topic which belongs neither to any single science nor to any single profession or person, but requires all of them to share the effort of innovating and its yield (after a success in the market). Therefore innovation must be integrated in the entire business process as well as in the support provided by the measures of the national economic system and national policy (not only the economic one, but also education, etc.). After all, in the modern situation, not only companies, but also regions and countries compete. They will, hopefully, cooperate more and compete less in a postmodern society (Dyck in Dyck & Mulej, 1998).

Every component in Table 9 requires its specific knowledge, values, and KM, as well as their managerial coordination making a networking synergy emerge from interdependent requisitely different specialists. It is obviously no an easy task, but there are methods around making it easier to master. There are very many methods supportive of creativity and KM, including values management (Mulej, 2006a; Rosi, 2004; Rosi & Mulej, 2006; Udovičič, 2004; Udovičič & Mulej, 2006). A very new method supportive of creative cooperation of requisite and mutually different and hence interdependent specialists, e.g., from different units/sectors of an organization, or different organizations, etc., surfaced in our research recently; we used it in several workshops with very satisfactory responses from participants (Mulej & Mulej, 2006). See Chs. 13.5 and 14.5.

14.5 APPLICATION OF USOMID-SREDIM/6 THINKING HATS METHOD

USOMID (Mulej, 1982) and the Six Thinking Hats (De Bono, 1985, 2005) have been applied in separation for +2 decades, before (Mulej, M. and N., 2006) a synergy was made—see Table 28 and Ch. 13.5. We will not repeat it here.

The USOMID-SREDIM/6 Thinking Hats method of creative cooperation can be completed by another method of creative cooperation, which is called "co-laboratories of democracy". It namely requires dialogue. It is not the only one; there are about thirty of them, according to a discussion during the ISSS conference in 2003 in Greece.

14.6 APPLICATION OF CO-LABORATORIES OF DEMOCRACY

Christakis and Bausch (2006) have created, over decades, a very interesting methodology of creative cooperation based on their experience that (a) dialogue is important; (b) dialogue is difficult; and (c) the only way to overcome the difficulties is through an adequate methodology. Their book is about it along with stressing the need for: (a) an attitude; (b) a philosophy; (c) a call for action; and (d) methodology to be put in synergy in order to apply the collective wisdom of the group.

This approach can support the one in Table 28.

14.7 ACTION RATHER THAN THINKING ALONE MAKES INNOVATION FROM KNOWLEDGE AND INVENTION—DIFFUSION OF NOVELTIES

14.7.0 The Selected Problem and Viewpoint of Dealing with it in Chapter 14.7

If a single user appreciates an invention as innovation, the situation meets the international definition, but does not really cover the costs and bring revenue and profit to authors. This fact makes us think of diffusion of novelties.

14.7.1 Action Based on the Theory of Diffusion of Innovation

In our experience, the weak point of the Six Thinking Hats method is its limitation to thinking; action is not included. This seems true of Co-Laboratories as well. In Co-Laboratories, the computer support reaches far beyond USOMID. Action is included, if one uses USOMID. On the other hand, USOMID covers well enough only the blue hat, while the others complete USOMID very well. Among them, the phases of Implementation and Maintenance change inventions/suggestions/potential innovations into innovations. Here, thinking is far from being enough, while it is very helpful in the preparation part. In the action part, change agents aiming at innovation receive a good advice from Rogers (1995). See Table 34.

14.7.2 Need for Spreading of the Novelty

Innovation management, following invention management (see Table 31) includes also the process of spreading/diffusion new solutions or novelties among potential users. Invention suggestions become innovations, when consumers perceive them as useful and actually use them. The main measures of success are not only the dollars earned, but increasingly the benefit for the humanity. Rogers (1995) studied on the case of agriculture, and many others later on, the whole process of diffusion of novelties from the research laboratories or companies to their potential and end users—farmers and others. There are several organizations worldwide, on which a similar case study could be made. A society is so convinced of the greater benefit for everyone, that governmental agencies are acting as promoters of inventions—in the areas of public health, safety, education, agriculture.

14.7.3 Diffusion and its Preconditions

14.7.3.1 Innovation Process as a Communication Process

Theory on diffusion of novelties is dealing with the difficulties when trying to implement new solutions among people as well as with methods to support activities of the owners and authors of the novelty. Diffusion does not tackle the technical contents of the novelty only, but it is at the same time:

- Communication process among people who know a novelty well (inventor, producer, their agent) and act as change agents, and its potential users;
- Exchange of information and preparation for decision-making;

- Including uncertainties and risks;
- Including a process of changing the society.

14.7.3.1.1 Novelty

Novelty is a new idea, service, process, solution or method that the possible user considers something new and worth considering as potentially useful. Some novelties are easier accepted by the possible users when:

- Users can see the possible advantages;
- Novelty can be applied together with the established methods (similar technology, same equipment can be used);
- If it appears less complicated, it is more likely to be accepted;
- Verification stage. Possible user can study the novelty—samples, testers, model plants;
- Results of the novelty are easily seen, measured, already installed somewhere.

14.7.3.1.2 Decision-Making Process about the Novelty

Decision about the novelty can be of three basic types and their combinations:

- Individual decision;
- Group decision, and;
- Governmental decision.

Every decision making process includes dealing with uncertainties and risks and includes several stages:

14.7.3.1.2.1 Awareness about the Novelty

This is a process of gathering data and information about the novelty at stake. It is a social process of spreading the news about something new, what kind of problems it solves or it is designed for, and how it works or why it can work.

14.7.3.1.2.2 Convincement

In this step the potential users make an informed position, assessment about the novelty. They are prepared to make a stand—accept or reject the novelty at stake.

14.7.3.1.2.3 Decision-Making

Potential users are selecting among the possible options. The novelty with many advances is easier to accept. Change agents like to prepare informative meetings or presentations of the novelty to support the decision-making process.

14.7.3.1.2.4 Use of Novelty

In this stage users have the working novelty. They are observing how it works for them or in their organization. Data are being collected to assess the advantages and the difficulties connected to the novelty at stake.

This is also a stage to adapt the novelty for one's use, supplement it with one's knowledge, and changes may be called re-invention.[4]

14.7.3.1.2.5 Reaffirming of Decision

Once a novelty is well used its users still analyze and think about possible improvement or replacement with a newer novelty or even a rejection of the applied novelty. Disappointment with the quality of the novelty, e.g., encourages the search for new solutions and other novelties.

14.7.3.2 Communication Channels

Communication channels in innovation processes have a few specific characteristics:

- They are about novelty;
- Involved parties are at least two groups: one well informed about the novelty and with clear subjective views, and the other person or group without such information or standpoint;
- There are established ways, paths or means of communications among them.

Change agents can use:

- The channels for mass communication or;
- Interpersonal channels.

---

[4]. This seems to be the main area of the invention-innovation processes known as "Suggestion System" or "Submit Your Idea System" (Mulej, 2006a).

For some types of novelties specialized trade fairs or magazines are more appropriate.

14.7.3.3 Time

Not only it takes time from the problem identification to the idea and its development to a working solution. Also the process of diffusion takes time to accomplish at least the following activities:

- Information about the existence of the novelty;
- Process of preparing the possibilities (confirming, gathering more information, verifications …);
- Decision-making process (yes, no, wait);
- Use of the accepted novelty;
- Re-confirmation of the accepted solution (keep in use, change, reject).

14.7.3.4 Social System

This system (= entity) comprises members of certain social group or community with their relations and connections including their influences. Important are social structure, borders among certain sub-groups (higher class, medium, lower; per gender, profession, age, etc.). One can often observe a formal and beside it an informal social structure. Norms of the social system inform us about the rules or prevailing culture within the community.

14.7.3.4.1 Possible Users of Novelties

Several statistical methods are being used to calculate the number of early users, later users and the median ones.

ADVENTURERS—the earliest users (2.5% of the total population) have:

- Enough financial funds,
- Enough knowledge to understand and use the early complicated novelties,
- More ability to manage great risk,
- More cosmopolitan habits,
- More active practice in information gathering,
- Many connections,
- More years of formal education,
- Higher social status,

- Higher ambitions,
- Higher (possible) mobility—including promotions,
- Larger units (companies, farms, hospitals, schools).

EARLY USERS (13.5% of the total population) are:

- Less different from others,
- More local,
- Having enough knowledge to understand and use novelties,
- Respected in community,
- Having a higher level of empathy,
- Less dogmatic,
- More rational,
- More intelligent,
- More ambitious,
- More in contacts with change agents.
- Other people often observe them and ask them for an advice and opinion—they are opinion leaders.

EARLY MAJORITY (34% of the total population) members are:

- Just before the majority of 50%
- Having many contacts with their coworkers,
- Seldom opinion leaders,
- Important as network,
- The largest early group.

LATE MAJORITY (34% of the total population) members are:

- Just after the majority,
- Cautious, have doubts,
- Observing the social norms of their system,
- Having limited financial possibilities,
- Unable to take great risk,
- In need to feel safe.

LATECOMERS (the last 16% of the total population) are:

- Traditional,
- Having many doubts,
- Having bad experiences from the past,
- Seldom opinion leaders,
- Having very limited financial means,
- Sometimes isolated.

14.7.3.4.2 Social System is no Political Term here

As you see, it is a community that is called a social system here.

14.7.4 Change Agents

Change agents have many contacts with members of the community—social system, but are usually not members of it. They are well informed of the novelty and are economically motivated to have people implement the novelty. They have a high social status. They accept novelties before others do, and have a lot of formal knowledge; they are innovative. They enjoy respect from others.

Most of them are heterophilic (= different in knowledge, etc.). They need a balance of heterophily and of homophily (= equality in knowledge, etc.) with potential users. That is why they have good contacts with opinion leaders.

14.7.5 Opinion Leaders

Opinion leaders are very important for the diffusion of novelties. They are members of a social system (= community with potential users; see Ch. 14.7.3.4). Their opinion is important for other members, they act as role models; they are often asked about their opinion, position. They are more similar to the majority and are more homophilic than change agents. They have good contacts with change agents and become able to understand and use novelties. Also in the group of opinion leaders we could find common characteristics such as:

- Higher formal education,
- Higher social status,
- Reading more magazines and newspapers,

- Being more open for novelties,
- Being more cosmopolitan than the majority,
- Being more innovative users or customers,
- Being not too different to be strange to the majority.

14.7.6 Concluding Comments on Diffusion of Innovation

It is far from enough to detect talents, neither is it enough to collect their ideas as inventions—suggestions and to develop them to potential innovations (see Table 32). But it is usually neither enough to sell a novelty to a single customer, even if she/he is satisfied. This act allows the novelty to be called innovation, but it most probably does not cover the costs of and the yield profit to novelty's owners and authors.

Innovation management includes therefore the diffusion process. Many different people are involved. It includes efforts, financial means and risk. It takes time to make an innovation and it is at the same time also a process of changing the society.

We tackled only a few factors important in the process of diffusion of novelties. While striving to make an innovation, change agents try to reach a critical mass of 5-20% of all possible users. Once their activities have convinced this percentage of their market, the convinced ones will become opinion leaders and contribute to the change agents' efforts to change the social system/community with improved new products, services, solutions, organizational methods, management methods = all types of innovations (for a summary see Table 32). Table 32 includes factors that Rogers originally not has covered.

IIP as a communication process for the potential innovation to be accepted by its potential users and thus to become innovation, turns out to be very complex and to require another long set, or even DS, of specific types of knowledge, values, and KM for IIP to succeed (see also Risopoulos, 2005).

In the communication process, the change agents and their partners (opinion leaders, para-professional aides) try to persuade their potential customers of the customers' potential benefit. This can be called a case of KM. This KM is not requisitely holistic, if one forgets about human emotions/

	VIEWPOINTS TO BE CONSIDERED	Phases of users' decision making about a novelty				
		1 Awareness	2 Persuasion	3 Decision	4 Application	5 Reconfirmation
Novelty customers (potential)	Customers—innovators					
	Early customers					
	Early majority					
	Late majority					
	Laggards					
Potential customers' absorption capacity for the introduced novelty—to-be innovation						
Requisite holism of potential suppliers/authors of novelty—to-be innovation						
Requisite holism of potential customers of novelty—to-be innovation						
Requisite holism of pressure of market, government and bosses concerning novelty—to-be innovation						
Requisite holism of information system concerning novelty—to-be innovation for suppliers and customers to know enough						
Systemic quality of novelty—to-be innovation (based on requisitely perfect products, processes, leadership and commitment, linked in a synergy by organization, and expressed in the system (= network) price, quality, range, uniqueness, and environmental care)						
Requisitely holistic vision, mission, policy, strategy, tactic, operation, and control of the entire process with suppliers (and users)						
Opinion leaders						

Attributes of novelty						
Relative advantage						
Compatibility						
Complexity						
Testability						
Visibility						
Communication channels						
Interpersonal						
Public						
Nature of the culture of customers						
Decision type about novelty						
Optional						
Group						
Authority						
Consequences of novelty						
Desired						
Undesired						
Indirect						
Direct						
Anticipated						
Unanticipated						

Table 34 *Matrix of essential attributes of the diffusion process from the viewpoint of change agents (the darker the area, the more change agents' effort is needed)*

values. We therefore turn to Jensen (2003), who developed the notion of heart-storming to be added to brainstorming. See Table 35 for our very brief summary of it.

Drama is more persuasive than an advertisement text or a lecture with no involvement.			
Story wheel depicts the protagonists' moving from order to chaos and back, but changed.			
Order: Peace as at home. Quiet despair. Nothing new. No new sources to cover current and new needs, poor chances for survival.	Separation: Leaving home to explore the wide world in search for the Grail (= the value/goal worth fight and risk), following the Call to attain something (Grail). But there is a price to it.—The driving idea and story supporting it for persuasion and success come on stage.	Chaos: Unknown, dangerous situations and many battles before one eventually wins one's Grail (if). To enter chaos, one must pass the Dragon Gate: "Are you ready, aware of risk, and determined; did you double-check it all?"	Return: Changed by the initiation following separation and experience of chaos. Perhaps with the foreseen Grail, or with another, such as new experience from challenge, danger, fear, battle, and exploration.

Table 35 *Knowledge and values management in heart-storming*

Jensen's cases relate to advertising. On the contemporary market and technology level, e.g., watches of very different prices serve comparably well, technically. Thus, the hearstorming issue reads: "What will you be, if you have that and that brand of wristwatch?"

After millennia of preferring order to chaos, including innovation and the resulting new benefit from it, most people around the world are poorly capable of accepting and realizing the new innovation-based model of life, including its attributes summarized and networked here. One must become able to practice innovative business and innovative society rather than a routine-based, even routine-loving one, for the first time in history. We can only briefly tackle innovative business and society here. This phase of history requires its specific definition of requisite holism and ethics of interdependence.

14.8 INNOVATIVE BUSINESS

14.8.0 The Selected Problem and the Viewpoint of its Consideration

This is the alternative to the traditional business style, which used to be based on self-sufficiency and a routine-loving way of doing business, when conditions still did not require much innovation (Tables 3 and 4). Until recently we had quite hard times, when we suggested that innovation of business and management style matters more than the technological innovation, although the latter is crucial as well (Mulej et al., 1987, 1994; as disciples of Alan Mogensen who worked on it since 1926). IBM (2006) surveyed the world top companies to find and publish this: innovation of business is at least as crucial as the technological innovation.

14.8.1 Definition of the Innovative Business as a Dialectical System

Innovative business can be simply defined in the following ten interdependent sentences:

1. In principle, every cost is unnecessary. In reality it is so, if we work smarter, not harder, and produce innovations. See Table 30.

2. Today, every product and process becomes obsolete, sooner or later. That's why we must know their life cycles, do research, development (connecting research results with the daily needs and practices), create other inventions and create from them innovations as a new, useful/beneficial basis of survival, on a continuous basis. See Table 30.

3. Survival, and therefore both, good and poor work, is everybody's business. Nobody, neither the superiors nor the subordinates, are entitled to be irresponsible and oppose or disregard innovation in their own life reality. Everybody must be a co-worker. See Tables 13 and 14.

4. Therefore let us continuously, all the time and everywhere, search for possible novelties! Only a small portion of them can become inventions. Some of them will be registered as suggestions. From some of them, by the "research and development" or the "connect and develop" concept (Huston & Sakkab, 2006), sometimes something both, usable and new, might be created, a potential innovation. Customers will accept only a fragment of them as useful/beneficial and worth paying for it, hence making a benefit

to both, customers and suppliers, and therefore deserving the name of innovation. They can be diffused, too. See Table 31.

5. The entire business policy and practice is innovation oriented, not just a fragment of it. See Table 22.
6. Results pay, not efforts. Hence, let us work like the clever ones, not like fools. Diligent stupid humans are dangerous: they do it wrong all the time. So are clever criminals.
7. These six sentences no longer apply to the producing part of the organizations only, but to all activities and all parts of life in all organizations. See Table 9.
8. The effort must be broadly disseminated and permanent, because the pressure of the competitors is permanent. See Table 5.
9. For competitiveness the quality must be systemic, which is impossible without continuous innovation. See Tables 31 and 5.
10. Innovative business is a regular practice of sustainable enterprises with sustainable and social responsibility ethics and systemic quality. See Table 5.

How can one make the effort rational?

14.8.2 Framework Model for Continuous Implementation of Innovative Business

Once we have identified the system of preconditions, which seems to be quite complex, we may try to offer also a framework model about how to meet them. We will brief the model for an organization as it usually functions nowadays. Cases we know of, seem to confirm our model.

A framework model may in general be the same, while details require specialized analysis and planning in every organization. The model always includes vision, mission, policy, strategies, and tactics, before operations (Mulej et al., 2000, and earlier). We must consider that implementation of a strategy is at least as complex as its making; it is here that we confront the established old habits (Feucht, 1995).

The phases are briefed in a logical sequential order, but in reality the process is not linear at all, but dialectical, i.e., full of interdependencies and interactions as follows (Table 36).

MANAGEMENT PHASES	PREPARATION PHASES
Definition of vision ⇓ ⇐	Drafting of vision, mission, policy, strategy, tactics, operation
Definition of mission ⇓	⇑
	Definition of starting points for drafts
Definition of policy/ies ⇓	⇑
	Consideration of experiences, learning
Definition of strategies ⇓ ⇐==	⇑
	Intervening when and where needed in all management phases
Definition of tactics ⇓	⇑
Running the operations ⇒	Checking the results of operation

Table 36 *The cybernetic circle of the preparation and implementation of the management process (a simple model).*

Consideration of the environment—business, political, natural, etc.—is not visible directly in Table 42 36?, but it is part of the consideration of experiences and every other information entering the process in Table 42 36?, as well as information going out of this process.

All phases in the model are linked by information and information requirement: the foregoing phase produces, in a best case scenario, all/requisite information required by the following phase, and only that information. In the case of innovation management this precondition is even harder to meet than in the case of a routine-loving management because decision must be made while all information cannot be available: the future market/customers' reaction to the supply offered will determine what is a failure and what is an innovation (see Peters 1997).

Vision may be briefed as "survival on the basis of competitiveness by holistic creative work and cooperation aimed at a systemic quality in accord with customers' requirement."

Mission: "delight customers with an excellent systemic quality and attract them as sustainable customers."

Policy: "implement innovative business as a source of a continuous systemic quality in all parts of the business process and all units."

Strategy towards the implementation of such a policy may include continuous self-assessment of one's own quality in terms of the Deming Prize of Japan, the European Quality/Excellence Award, or the Baldrige Award of USA, or (as a first phase) the attainment and maintaining of the International Standards Organization's rule ISO 9000 certificate or something similar[5].

---

5. Japan lost the 2nd World War and had a USA control over their its government for a while. Americans were not happy with the quality of Japanese supplies and brought experts to

Tactics for the implementation of such an innovation strategy include organized criticism, followed by teams' and task-forces' work on a solution of the selected problems (on a free-will basis and on company time, one hour a week) with awards for inventions (symbolic in value, but with no delay) and innovations. An innovation reward is foreseen for everybody in the innovative team, all members of their own organizational units, every organizational member including managers, while one half of the value created by innovation enters the company business funds (Mulej, 2003a, b, c).

In daily practice therefore the first step towards continuous innovation is the entrepreneurial and managerial capability and readiness to see their subordinates as coworkers capable of innovating, and both sides of the coin: cost and waste on one hand, and their causes (which are frequently hidden), on the other hand. Cost is namely, according to the older part of the authors' +30 year experience of consultancy and field research concerning innovation, the easiest topic to handle by innovating and a first step towards an innovative culture and innovative business. Once this is usual, work on better quality follows more easily than before, then range, and then uniqueness, and then sustainability and social responsibility, finally, in general (see Table 4). Again, this may be no linear paradigm. As we have said, they are all interdependent.

And we must not rely on traditional economic and business data only, they tend to leave about half of the factors of competitiveness out of sight (Grayson & O'Dell, 1988).

---

teach Japanese about *quality control*. Edward Demming wrote his instructions in the form of 14 sentences; the first sentence required continuous innovation as the source of excellent quality. In 1951, the Japanese government introduced a nation-wide reward for companies achieving excellent quality (the Deming Prize). This was a crucial step towards the Japanese conquering the world markets in of many products due to their permanent link between quality and innovation. A similar quality award was introduced in the U.S.A 36 years later (the Baldrige Award), in the European Union even later (European Quality Award, EQA). In Slovenia, the EQA was made a law in 1998. Another incentive is provided by the ISO 9000 certificates for total quality management, which is less demanding than EQA, but useful as a first step. Unfortunately, quite often, the organizations try to acquire the ISO 9000 certificate without using the incentive for innovation behind its scene. ISO 9000 was therefore updated in 2000, but with a bureaucratic approach, one can meet criteria technically and formally, and keep a lot of the old *non-innovative culture*. (See: Mulej et al., 1997, Pivka & Ursic, 1998; Pivka & Mulej, 2004; etc.)

14.8.3 Total/Excellent Quality and Hence Success in Market—Consequences of Innovation Process Based on Requisite Holism and Ethics of Interdependence

Times of the surplus of demand over supply are over, in the advanced world. The basis of competitiveness has even changed in periods of only decades; see Tables 3 and 4.

In the modern global market no single basis of competitiveness in Table 4 is sufficient anymore if alone. How can their synergy be achieved? By innovating as a continuous activity. The model is briefly presented in Table 36 and Ch. 14.8.2. Cases in the book (Dyck & Mulej, 1998) and other experiences (Mulej, 2006a) tell us clearly the results:

Business must result in excellence to be accepted by the market. Excellence depends on five pillars of total quality which are interdependent and each of them must be excellent (Creech, 1994). They are (see Table 33):

- The product must be perfect to be accepted by the market.
- Processes both, in production and other parts, must be perfect for the product to be perfect.
- Organization is the central precondition for perfection in all parts and processes of the entire business.
- Management must be leadership, not managership, i.e., cooperative rather than commanding for organization, processes, and products to be excellent.
- Commitment of coworkers can then be reached and their capabilities activated as needed.

Methods which change the current way of management and organization may be many, not only the way of Total Quality Management; so we will not discuss them here (references abound; many methods are collected in: EU, 2004).

A product is perfect, if it meets the criteria of "systemic quality" (attribute of sustainable/SR enterprises in Table 4) as found acceptable by customers. No criterion can be missing; they are interdependent. This is a clear experience in advanced countries. If we take a closer look at "systemic quality", we can see two things, at least:

- Attainment of systemic quality depends on innovation in all business functions and by all or most, at least, coworkers.[6]
- A given level of systemic quality can be continuously improved all the way to top world standards of excellence—by continuous technological and non-technological innovation.

Most probably, every activity—managerial, informational and basic (supply, production, sales)—can be (1) made cheaper; (2) improved in terms of reliability, and other features of products and services supplied to internal and external customers; (3) adapted to their requirements in range offered; (4) done in a way making the supply unique for the customer and (5) preserving humans' natural environment (see Tables 4-6). Thus, it provides for (C)SR, RH and RW. In the West, therefore, continuous innovation is the contemporary culture and necessary for businesses to survive (including the jobs in them). With the forgotten four-fifths (= 80%) of humankind it can and must yet be developed (as cases prove in: Dyck & Mulej, 1998)[7].

If innovation is the precondition of excellence, which is a precondition of systemic quality as a precondition of the market success, let us consider the preconditions of innovation. We took a look at the management model in Table 42, but the most essential precondition is the innovation of the management style (IBM, 2006; McGregor, 2006; Mulej et al., 1987; Rosenberg & Birdzell, 1987).

14.8.4 Innovation of Management Style

Management practice is millennia old, since people lived in communities with a division of roles and jobs to be organized and coordinated, of course, although the science of management is about a century old in U.S. and much younger in Europe (Ženko, 1999). It went through several phases of innovation in order to match the ever-new difficulties of bosses with their subordinates, who were no coworkers. It seems that bosses have not been able to perceive that their problem results more from their imposing than from trying to have a novelty taken care of.

---

6. We do not speak of subordinates, because this notion implies the expectation of bosses, that they are not capable of creativity, even less so of innovation. Coworkers accept more, and are granted more, responsibility.

7. As we have exposed earlier, the innovation must result from a more/requisitely holistic IIDP than in the times of the industrial paradigm of the recent few centuries, in order to solve the problems of the current civilization's survival.

The oldest innovation of management may well have been the hierarchical subordination of the less qualified members of an organization to the more qualified and experienced ones (Schmidt, 1993). It is said to have been introduced, when they were building the Egyptian Pyramids. It made and makes sense as long as (1) the superiors do not abuse their position; (2) the subordinates would not be able to perform well enough in their jobs with no supervision and instruction, due to their lack of expertise and interest. It survived all way to the 20th century and is still around.

In the course of the 20th century, an equipment has been more and more introduced that required more skill, education and training. This has made the traditional hierarchical subordination less and less useful and needed, with the extremely different level of education and culture in the 20th and 21st centuries.

Though, in the 19th century, when factories were introduced and employed many people with a poor education, the model of hierarchical subordination was still acceptable—under the same precondition of no abuse of the subordinates by the bosses. To power lovers this precondition has usually been a hardly acceptable one for all millennia since the building of the Egyptian Pyramids until today.[8] In the course of the 20th century the general level of education has been growing all the time in the advanced areas of the world, at least.

The next innovation of management may hence have been the

- Human relations, between the two world wars, and
- Human resources, in recent decades.

Under the human resources model the main difference from the old tradition is the supposition that subordinates are capable of creativity and responsibility. This finding has had a lot to do with the practical experience as well as with the growing need that the companies and other organizations both develop and activate as much of their personnel's creativity, inventiveness

---

8. In November 2006, in USA republicans lost the elections to the parliament in the U.S. The minister who was the first one to be dismissed was publicly said to be unable to listen to his professional advisers; hence he made wrong decisions, which have caused many troubles to his country. This is a clear case of a lack of requisite holism and ethics of interdependence. See public press for many more cases.

and innovativeness as possible.[9]

Gradually, a next innovation of management entered the scene, which we may call innovative business (Mulej *et al.*, 1987), which we have discussed briefly here.

All four steps demonstrate a trend towards more holism and ethics of interdependence in the bosses' views at their subordinates, who become hence coworkers increasingly.

The next step is the spreading of the innovative business from a few companies' model to a prevailing one is society, turning such a society into an innovative society.

14.9 INNOVATIVE SOCIETY

The processes in Tables 3 and 4 brought the socio-economic development to prevalence of the innovative society. In the advanced countries, the innovative business has become the prevailing culture; others are mostly trying to catch up with them (in our observation of practice):

The market ratio of supply and demand passed the transition from:

1. The random market—they produce for themselves and sell only incidental surpluses, to
2. The producers market in which the supply is smaller than the demand ("queuing economy", buyers cannot find everything they want), to
3. The buyers market in which supplies exceed demand ("market economy", suppliers cannot find buyers for all what they offer, innovation is necessary), and further on to
4. The state supported buyers market in which supplies exceed demand a lot ("advanced market economy", innovation is unavoidable).

The continuous need for innovation encourages creativity of everybody, and therefore causes the evolution of democracy as a societal mechanism

---

9. Alan Mogensen was using the "Work Simplification" methodology he had created at Kodak with good success for six decades. But many managers and owners and commanding officers refused his services—in order to remain the boss. This is another clear case of a lack of requisite holism and ethics of interdependence. Lack of efficiency and effectiveness results all around the world, where bosses are persons who lack ethics of interdependence.

making room for creativity and holism, not only as democracy in politics (with many parties, civil society, and so on), but also in economics (with free entrepreneurship, choice of job, profession, and so on), on shop floor (with invention and innovation circles, quality circles, leadership instead of managership, i.e., cooperative instead of commanding management), in family, education, local community life—everywhere.

The innovative society differs from the (foregoing, historically) routine-based society:

- It applies all achievements of development of the worldwide civilization.
- It accepts and applies its own and foreign inventions and innovations rather quickly.
- It applies foreign knowledge to upgrade its own knowledge in order to effectively develop and use all the technologies of production, organization, education, etc.
- On this basis, it attains both, a high international competitiveness and quality of life.
- Its inventiveness and innovativeness, both as attributes and activities, reach the Western European level, so do their preconditions (at least).
- The creative co/workers, scientific and other inventors and innovators are well appreciated because they are the most useful co/citizens and co/workers.
- The uncreative individuals are in trouble, especially the ones under-using their natural and learned capabilities.

The dialectical system of attributes of an innovative society includes, therefore:

- A contemporary, creativity-based, and creativity-and-holism supporting democracy both, in the entire society and all organizations from families on.
- A contemporary, creativity enhancing market in which, as well as in the democracy, innovative persons and organizations prevail and reign.
- A contemporary perception of ownership, which tells clearly the responsibility and includes creative and innovative ambitions rather than seeking rent (as an income based on owning without creating) only.

- A contemporary perception of innovation, which says that innovation is every beneficial novelty accepted as such by customers and granting the suppliers a suitable profit/benefit, too.
- A contemporary way of running the business, the innovative business, which continuously strives for innovation of any kind discussed here earlier.
- A contemporary perception of entrepreneurship, i.e., innovative entrepreneurship, which means, that not every owner of an enterprise is an entrepreneur, but only the one who combines his or her business factors in an innovative way in order to produce innovation and live on it. Hence, private ownership is not enough, if owners are not entrepreneurial.
- Education and other societal subsystems, which are not economy and business, but rather create human resources, circumstances and preconditions for them to flourish, and therefore also support innovation rather than too much routine or even routine-loving.

Again, one sees that many types of knowledge, values, and KM are needed. This fact requires complying with the law of requisite holism and a well-developed ethics of interdependence of all specialists who must cooperate in the process.

14.10 SUGGESTION TO GOVERNMENTS OF COUNTRIES/REGIONS TRYING TO CATCH UP

14.10.0 The Problem and the Selected Viewpoint for Dealing with it in Chapter 14.10

Governments are rarely requisitely holistic, but working on a part of the real story only; they e.g., require businesses and higher education to do more for innovativeness, but fail to show their own best practices (Handbook, 2004). They can find instructive cases (Creech, 1994; Graham, Jr., 2006 and earlier for decades; Linden, 1990; Mogensen, 1981 and earlier since 1926; Jurše, 2003). They most probably did not detect and use them. The European Union (EU) has been seeing for quite some time its trouble with competitiveness, due to trouble with its innovation (Miege & Mathieux, 1987, 1989; EU, 1995). Newer research provides no basis for satisfaction either (EU, 2000; Lisbon, 2004; Sporočilo, 2006).

The problem depends neither on theoretical models only, nor on their inclusion in the macro-economic institutions/measures only (Ponovni, 2006; Prednostno, 2006; Predstavitev, 2006; Resolucija, 2006), but to an equal extent on the responses of people to them, which tend to be overseen. EU admits (Handbook, 2004; EU, 2004) e.g., serious differences in cultures of universities, on the one hand, and companies, on the other hand. But: EU requires universities only (!) to rethink their strategic orientation and to include working with businesses in it (Handbook, 2004: 38-39). The problem lies equally with a large majority of enterprises: among them, e.g., in Slovenia, about 93 percent are micro (under ten employees), about 6 (six) percent are small and medium sized, and 0.3 (three tenths) percent employ more than 250; it is similar in the EU (Mulej, 2006; Rebernik *et al.*, 2003, 2004). Thus, most enterprises lack absorption capacity due to their size.

This is not all the truth: EU does not tackle the role of government or other public sector organizations either; they demand innovation without being a role model and doing what they preach. This is what we will concentrate upon here.

14.10.1 Creativity, Insight, and Quality Enhancement as Factors in Innovation Processes in Government and Other Public Offices

Let us briefly summarize basic findings of authors quoted above when dealing with innovation in government and other public offices. What they all have in common has been admitted in the official international definition of innovation close to 40 years ago (Frascati Manual, 1971): every novelty applied with benefit by its users may be called innovation, not only a technological one. The users' benefit makes innovation different from invention (Mulej & Ženko, 2004a, b, and earlier). In other words, innovation is not a technological topic only, as one used to think, and many still need a warning and awareness building (Menih, 2006; IBM, 2006).

For a few decades it has been well documented that innovation of management comes before innovation management, and has done so always in history (Reich, 1984; Rosenberg & Birdzell, 1986): preconditions for innovation and innovation management must be created, and innovation of management towards more democracy as a humans' tool for requisite holism creates them (Mulej (ed.), 1984). We are not talking about political democracy only, here, about which we like the following statement: "Democracy is limitation of power" (Jež, 2006). We are talking about democracy as a creativity and holism enhancing human relation (Jurše, 1994; Lynn *et al.*, 2002; Mulej, 1981; Mulej & Jurše, 1994; Mulej & Mulej, 2006; Mulej *et al.*, 1994; Udovičič, 2004; Ženko, 1999; Ženko *et al.*, 2004).

Over four decades Creech (1994) converted many inventions to innovations, by and with his team and subordinates, as an army pilot and officer, on the basis of his democratic approach to quality: the pilot commands the airplane crew, but by listening to each other and making synergies, because they are all interdependent specialists and capable of their work and related creativity. Cooperation rather than one-way commanding helps coworkers be creative and make innovations. See Table 33 again.

Graham (2006 and earlier for decades) has specialized in innovation by helping administrative employees, including the ones in public offices, to rethink their own processes and procedures of work and cooperation: "Do we really do only the necessary and all the necessary jobs? Do we do them in the best following order and with sufficient consideration of direct and indirect influence over them and from them to other parts of our organization?" His methodology is called Paper Work Simplification. In 1998, Mulej worked with Graham on a case with local town municipality officers in Dayton, OH: their creativity stopped sleeping in a few minutes.

Linden (1990) demonstrated with a number of cases in different public offices, such as police, forestry, school, and government administration, that the market pressure can well be replaced, if the boss prefers innovation to routine-loving behavior, and enhances his or her subordinates' creativity and criticism in order to solve his/their own problems better. Such bosses need attributes, which can be learned:

- Strategic rather than short-term thinking and action;
- Keeping to the agreed goals, and letting coworkers have room to find their ways how to realize their parts of the shared objectives and tasks;
- Creating the felt need to innovate one-self into a more inventive, cooperative, creative and innovative team member (rather than complacency, routine-loving attitude, and blaming others);
- Starting with concrete steps (rather than "paralysis by analysis");
- Application of structural changes (in order to make more room for creativity to be freed);

- Consideration and mastering of risk (no avoiding of risk; it is very risky to take no risk);
- Application of political skills (persuasion, acquiring partners, detecting opponents in time, etc.).

These seven attributes are added to persistency and some entrepreneurial spirit. The latter used to be considered a natural gift for too long; now one knows it can be learned, too (Japti, 2005).

We added to these seven human attributes the 7F (Mulej et al., 1994) being:

- Focused (on what one does best);
- Fast (in changing promising suggestions to innovations);
- Flexible (to adapt to new circumstances creatively);
- Friendly (in relation with one's business partners, including coworkers);
- Fit (to physically and mentally stand all the efforts of the modern rapid changing conditions);
- Free (to work creatively), and;
- Fun (to enjoy one's work).

In 1926 Mogensen (1981) found, as a consultant in industrial engineering, that he had been wrongly educated. He was supposed to study motion in order to innovate/optimize a worker's work place organization and moves instead of that worker. Once, such a worker saw Mogensen's film about his way of working and said: "Do not use this film, I can do better." From this experience the "Work Simplification" methodology emerged and helped thousands in companies, army, and other civil services do a better job and be valued as creative coworkers—once bosses were able to admit that bosses are clever and creative, but not only them: subordinates can be so too, if trusted and supported.

Along these lines, in 1992, Mulej co-worked in a case with the Renselearville Institute in Albany, NY, with a group of local civil servants. After a few hours seminar and another few hours of practical application these persons developed very interesting suggestions. Then, their bosses joined them to listen to them and to decide how much office time the potential innovators may use to try to make innovations from their suggestions. The applied method was very close to the "Suggestion System"/"Submit Your Idea System", but involving a more active attitude on the bosses' part. They were asking for ideas and criticism as bases for inventing new solutions and turning them into innovations. In other cases, teamwork can be used based on e.g., either USOMID or the Six Thinking Hats methods, or their new synergy (Mulej & Mulej, 2006). See Ch. 13.5.

Jurše is a former researcher of TQM and leadership, now a municipality office head. She has been using such a cooperative approach in her office with a lot of good outcomes over several years. Knowing that the problem in the city administration could not be only the employees (maybe some of them) and that the main problem is the regulated environment in which they operate (Jurše, 1994), she stated three simple rules at the beginning of her work at the municipality, namely: lawfulness, transparency and a non-discriminatory approach to everything and everybody. Secondly, she started with her logo: "Show me how clever you are at work, not how hard you work."

But she has not been regarded as highly by other office heads as she has been by her team. This approach actually breaks routine, the old habits ("The way we have done things here for ages; why do we need change, if everything is OK?" …); it results in more responsible work of employees: one serves customers' needs more efficiently, effectively and equitably, making them satisfied (e.g., they reduced the time for administrative proceedings from 30 or more days to less than 10).

Jurše's employees could not be rewarded for their successes as in a modern company; they are viewed more like a noise in the system despite some rare praises. Therefore they could not spend much of their office time to work on their ideas; they had hard times getting their inventions accepted outside their own office, because they tended to be supposed to work rather than to think. In other words, public services tend to be managed much more in the commanding style dividing people into the thinking bosses and the obediently working subordinates. In such conditions, unhappily, space or courage are not provided for leadership or HRM (Jurše, 1995, 2003), which are now the two main topics in corporate management.

Results of such a managerial rather than leadership approach among employees are predictable. Jurše conducted her first pilot survey about employee satisfaction in her municipality in December 2002. Her analysis showed that the key problem lied in the bosses, not in the reward system as assumed. The worst findings:

- Lack of clear and concise setting of directions;
- No feedback about employee's work;
- No open and honest communication;
- Lack of participation in decision-making, and;
- Different rules for different groups of people.

At the end, the answer to the last question was that nobody believed that the result of the survey would improve the situation. Unfortunately that was true: three years later a new survey showed the same (or worse) results. Such a managerial relation gives bosses the right to be overburdened, and subordinates the right to be irresponsible (in terms of providing more than what is required explicitly). Bosses just do not perceive it, all too often. They lack requisite holism and ethics of interdependence.

Hence: in public services there are enough people able to be innovative, once their bosses want so, which especially the poorly capable ones do not want.

Problems may arise in the implementation processes of the invention-innovation chain (Krošlin, 2004). All of it must be mastered and accomplished in order for an idea/invention to succeed as an innovation. It is long, it demands persistence in fighting the obstacles, and it is poorly supported both, by the traditional culture and the given institutions. An appropriate innovation policy is thus necessary at state level; here, in e.g., the present situation in Slovenia, there is much scope for improvement (Bučar & Stare, 2003, 2006; Mlakar, 2000, 2007; Mlakar & Mulej, 2007). In relation to this an efficient requisitely holistic approach is needed (Gider, 2004; Mulej, 2003, 2003a, b, c). So are related measures that would foster innovation processes within the education system (Likar *et al.*, 2006; Pečjak, 2001; Trunk, 2000) as well as in businesses and other organizations. Knowledge of diffusion of novelties might be added with benefit, too (Germ Galič, 2003; Leder, 2004; Rogers, 1995). It can also be well completed up with heart-storming (Jensen, 2003, Table 35).

Preconditions for success include:

1. The bosses' capacity to admit that they are not the only clever people around, and democracy helps them better than one-way commanding in solving their unit's problems/jobs.

2. The subordinates' right to be coworkers and use office time to develop inventions to suggestions and further on to potential innovations and innovations.
3. The Suggestion System or the Submit Your Idea System are methodologies to start with, after a short seminar; later on team-based methods such as USOMID may have to be added.
4. Favoring innovating over exaggerated routine-loving work style belongs to basic values/culture/ethics/norms of both, bosses and subordinates as coworkers.
5. The bosses' asking for ideas and criticism is the basis for inventing new solutions and turning them into innovations (on a weekly basis).
6. Rethinking the processes and procedures aimed at rationalization of both, the routine and the creative parts of work process, promises much more than technological innovation; technology is normally bought rather then invented or even innovated in offices. Two basic questions include: (1) "Is this activity really needed? What (information) needs does it serve, if any?" (2) "Is this step in the procedure really unavoidable for (information) needs to be rationally met?"

Hence, much remains to be done in government offices and other public services, not only in research institutions and in businesses, especially in the new EU-member states such as Slovenia, or in other catching-up countries that make an important part of the forgotten four fifths of humankind. What do we suggest? Let us first take a look at the roots of the problem.

14.10.2 Roots of the Lack of Innovation Capacity in Government

We will not elaborate the theory of the two-generation cycles (see Mulej, 1994, 2004) here. Rather, we will point out the coincidence of three major processes:

Universities in Europe are proud of being very old institutions, having their roots in the middle Ages. Hence university culture was created when businesses did not need universities' scientific research (Fazarinc in Kobal, 2003), nor did universities depend on businesses, but rather on the government. The reforms of so far changed universities a lot, especially towards working on professions and specialization of their students, but they did not do enough concerning the tradition of separation from business. Innovation has remained a rather strange story, although a lot of invention has been taking place in universities.

Hence, few professionals in businesses develop enough innovation capacity. It is even worse with people with lower-level degrees training people for routine work rather than innovation.

The most entrepreneurial Europeans started leaving Europe, for the USA especially, in the 16-18th century. On this basis, in the 1860s the Civil War caused re-orientation of the USA. Instead of agriculture and slavery as the dominant economic and social activities and relations, democracy, entrepreneurship, manufacturing industries and urbanization prevailed, requiring much more innovation due to creating much more market pressure. Many people from around the world were offered a chance to work and make money in the USA, while in Europe, e.g., in Austria-Hungary, there were hard times. One speaks of the Vienna financial crisis of 1874. After it, Slovenia e.g., lost 20 (twenty) percent of its total population in the four decades before the World War I (Encyclopedia, 1964). The most entrepreneurial, not the idle, persons left. This percentage is higher than the one of people in the role of entrepreneurs today (see: Rebernik *et al.*, 2004). Germany and UK lost about 3% (Bošković, 2006a). They kept some entrepreneurial spirit, while Slovenia hardly kept any.

The role of the government has always been to provide stable business and social environment rather than innovation. Meeting rules has always been much more crucial than breaking rules to find new ways, while the latter is precondition of invention and innovation (see Thorpe, 2004). Even CAF, the concept of application of the Total Quality Management to the public services, has mostly not included innovation, e.g., in Slovenia (Kos, 2004; Perušič, 2002). In principle, competition helps (Competition, 2004), but the quoted cases prove that other incentives are available. The common denominator of the success reads: bosses consider their subordinates their co-working specialists rather than incapable ones. They can use points 1-7 in Section 14.10.1—once they decide to do so.

They will decide so more easily, if they use systems thinking. This will help them attain requisite holism by feeling and using ethics of interdependence.

14.10.3 A Dialectical System of Suggestions how to Make Public Officers Act as Role Models of Innovation

What could be done, if one uses systems thinking and hence strives for the requisite holism of suggestions? See Tables 1, 2, 8, and 38 for a summary of systems thinking and requisite holism. We have no room for details, here. Recall Ch. 12 if necessary. Thus, in principle, the way out of this problem is a team's or an individual's double capacity:

- To be a specialist in a single profession, in order to know enough within a selected viewpoint;
- To be capable of systems thinking as a methodology of creative interdisciplinary cooperation, rather than as a methodology of a sophisticated description of findings within a single selected viewpoint.

We suggest a new dialectical system of viewpoints (chosen on the basis of our research as a network of essential viewpoints) to make the approach requisitely holistic in order to make people in offices more responsive.

1. The feeling or ethics of interdependence, i.e., humans' need for each other as coworkers, is a precondition, which requires the practice of systems approach. Studying shared processes and procedures supports it, because the practice of interdependence is reflected.
2. The government, including public agencies and similar organizations, i.e., all civil services, acting as a big buyer on the modern buyers market, might be one way towards their acting as role models in relation to their suppliers and suppliers of their suppliers, making them all more innovative than so far.
3. But it is not a simple way: it requires the government people to change themselves innovatively in order to act as role models, e.g., by managerial, organizational and methodical innovations rather than by technological ones.
4. A further suggested solution's component is the application of innovation management, which helps diminish the risk level of innovating and avoids imposing of novelty by involvement of innovation users in its making from their own or else's inventions (see Huston & Sakkab, 2006).
5. One considers the theory of novelty diffusion (Rogers, 1995; see Table 34) completed with heart-storming methodology (Jensen, 2003; Table 35) to persuade and attract.

To (1): The feeling or ethics of interdependence

Governments are supposed to serve by governing. Democratic governments differ from e.g., feudal governments in addressing interests more holistically. The total population rather than single classes/partial groups and their own, rather narrow, interests must be represented in governmental decisions. This seems to be an ideal that is very difficult to attain, as long as single political parties compete in number of votes to get the upper hand over the others. It is so as well, if single parties (or their coalitions) feel that they have all the power because they have all the support, although they have received e.g., 51% of votes handed in, which often means only about 20–30 % of the really existing votes. Once they feel so, they also feel entitled to impose on the population what they find to be the right ways of governing and objectives. Imposing does not produce the feeling of interdependence, but the feeling of suppression, and produces opposition and refusing, even concerning very good suggestions, even innovations. Transition from a one-sided to a requisitely holistic approach is needed. But power-holders tend to poorly perceive it. Unfortunately, this practice implies work within public services, too.

To (2): The government, including public agencies/similar organizations, acting as a big buyer on the modern buyers market, might be one way towards a solution, although not a simple one.

As long as governments work by imposing, they need a lot of energy and time to make their population listen, and hear, and accept the governmental suggestion. Economic reforms, e.g., are inventions supposed to become innovations. It does not depend on their authors/promoters, whether or not they will become innovations, but on the customers, who should buy these suggestions. Once one uses the term customer, one comes to think of the market relations. If the customers of a governmental suggestion include, first of all, businesses, then one can see, that these business are also suppliers to the government, directly or indirectly, not only its customers. One can also see that currently governments, including all public organizations such as army, education, medical care, public agencies and similar organizations, form a big group of buyers, which might be able to become a consolidated buyer by public rules; this group purchases about 50% of GDP. The governments' purchasing rules should consider that the modern market situation is one of a buyers' market (Table 3). This provides buyers a bargaining power, which may include the precondition for all suppliers to meet. It may require them to be the most innovative ones and require the highest level of innovation from their own suppliers as well; it is expressed in quality, price, range, uniqueness, and care for natural environment (Table 4). This includes relations inside public offices: employees are both, suppliers (working on contractual basis) and insiders (more or less influential coworkers).

To (3): Government people should change themselves innovatively in order to act as role models, e.g., by managerial, organizational and methodical, not technological innovations only.

It is easier to require others to innovate than to innovate in one's own organization. If one requires innovation, one must be innovative to be trusted as a person/organization living to one's words. Governmental organizations can hardly produce technological innovation. But they may choose: every novelty that is found beneficial by its customers is innovation. The most frequent type of innovation tackles organizational processes and procedures, methods of working and innovating, as well as of management style going from "I think, you work" attitude to "We all think, we all work" attitude of all bosses as well as of their coworkers.

To (4): Application of innovation management diminishes the risk level of innovating and avoids imposing of novelty by involvement of novelty users in its making and (5) consideration of the novelty diffusion, completed with the heart-storming methodology to persuade and attract.

Traditional literature on innovation management hardly covers governments and hardly tackles innovation as a phase following invention making and—perhaps—transforming it into a potential innovation. This problem can be overcome, if governments stop using the economic (i.e., macro-aggregating) literature only, which simplifies and does not foresee any direct cooperation with rules' users; it presupposes that all people are entrepreneurial and rational profit-seeking individuals and organizations preferring the long-term to the short-term horizons in their decision-making. A way of overcoming this rarely realistic supposition is the application of the theory of diffusion of novelties: innovation is a communication process, too. Persuasion of to-be customers is a complex process reaching far beyond advertisement. In addition, decisions of customers do not tackle brain—rationality—only, but heart-storming is also an overseen and unavoidable topic. This applies to internal relations in public offices as well.

Application of Rosi's Dialectical Network Thinking (Rosi, 2004; Rosi & Mulej, 2006) may make these tasks easier to perform. So may Udovičič's approach to leading by activating the coworkers' inner motivation (Udovičič, 2004; Udovičič & Mulej, 2006) and Potočan's approach to organizing (Potočan, 2004). Mutual trust may result and help cooperation essentially (Rozman & Kovač, (eds.), 2004). But the quoted two sets of seven personal attributes are the most essential preconditions for innovation in public offices (Mulej, 2006a, b, 2007).

14.10.4 Some Concluding Remarks

The current obsolete model of societal management is hardly capable of converting the catching-up areas and societies from routine-based into innovative societies and economies, based on requisite holism and ethics of interdependence.

It is a lack of requisite holism, if only universities, but no other public services nor the government, are required to change in order for businesses to receive, accept, and use much more of the theoretical knowledge from the research institutions in solving their own practical problems. This requirement could be sufficient, if most enterprises were not SMEs and lacking ambition, the human, organizational, and material resources for the researchers' attempts to transfer knowledge for others to innovate. Public officers, from the members of parliament on, need knowledge and practice of innovation, to act as role models in order to help others make their country an innovative society rather than a neo-colony.

For centuries, the "academic world" and the "real world" have been developing essentially different values and turning them into cultures, ethics and norms, which have made them two worlds. A long time may be needed for an innovative and requisitely holistic change to come with no intervention, but on an evolutionary basis. Thus, our suggestion alone can hardly become an innovation, although it makes sense, if other conditions are not requisitely supportive. The feeling of interdependence is a real precondition, which requires the practice of a systems approach and leads to creative action. It must include government—politicians and officials/officers—all the time.

The government acting as a big buyer in the modern buyer's market might offer a solution, although no simple one. Competition among its various parts (ministries, offices, departments.), i.e., a fair struggle for each job, e.g., preparing of law, or release of tender, could be involved to induce the feeling of competition as well as cooperation between clerks (bosses and employees) to establish preconditions for innovation processes. It requires a lot of change in the culture of the government bodies. But the alternative is even more complex—the lack of requisite holism and the lack of an innovative change causing Europe to keep being and becoming even more an old lady rather than the most dynamic and innovation-based economy of the world. This is why we are stressing in this chapter the need for public services to become role models of innovating, and then to act as big buyers in the buyers' market rather than as a one-way commanding power over society at large.

How should one govern and run government, social services, and business organizations as business systems to attain prosperity by requisite holism?

14.11 Four times Ten Guidelines Enabling Innovative Business and Innovative Society

Experience of four decades demonstrated that we may suggest consideration of four times ten guidelines to humans, who are trying to attain requisite holism by ethics of interdependence and succeed in enabling innovative business of an organization or innovative society in an area. They have to do with Tables 1, 2, 4, 8, 9, 19, 28, 30, 31, 34, 35 and 36, first of all, and read as follows (Mulej et al., 1994: 59-87, updated here) to cover the basic, managerial, informational processes, and their synergies:

The ten guidelines for the managerial process to be requisitely innovative:

1. Innovativeness of the invention-innovation-diffusion policy and culture in society and organization;
2. Innovativeness of the invention-innovation-diffusion related objectives of managers and influential professionals;
3. Training of many coworkers for creativity, invention-innovation-diffusion processes and creative cooperation, along with refreshment of their professional skills;
4. Innovativeness of organizing of the invention-innovation-diffusion processes and routine work;
5. Availability of equipment, professional and financial support to inventors and innovators, who are testing, developing and marketing ideas looking like promising suggestions;

6. Availability of information sources;
7. Implementation of the potential innovations in the daily practice;
8. Marketing preparation of the potential innovation and its potential customers for accepting it, including a close teamwork of all marketing, research, development and daily-operation offices;
9. Motivation of coworkers by clear evaluation of suggestions, potential innovations, and innovations;
10. Legal and organizational order favoring the creation of inventions and innovations over a blind subordination of coworkers to bosses.

The ten guidelines for the informational processes to be innovative:

1. The purpose of (the innovative) business is the requisite effectiveness, including the requisite efficiency, while both of them receive the requisitely holistic support from the informing process;
2. The objective of the informing process is solving the real problems and avoiding the fictitious ones by attaining requisite holism by ethics of interdependence;
3. The informing process supports creativity, creation and invention-innovation-diffusion processes;
4. The essence of informing is the requisite holism of the enhancement of and the support to creation, invention, suggestion, potential innovation, and innovation diffusion processes;
5. Requisitely holistic informing and creation are based on cooperative management;
6. Requisitely holistic informing receives support from a requisitely holistic definition of both information needs and innovation needs;
7. Requisitely holistic informing receives support from a requisitely holistic insight in the daily routine of processes and other attributes of the daily work as well as related expectations;
8. Computer is a tool supporting rather than replacing creativity, invention and innovation processes;
9. It is better to view informing as the information viewpoint of the (innovative) business than as the information process of its own; the latter might make it self-sufficient;

10. In the establishment and the innovation of the information process obstacles against novelties, including innovations, must be expected, but not accepted.

The ten guidelines for the basic processes to be innovative:

1. The basic process (from purchasing of all business factors via production to selling) is a synergetic consequence of all partial processes in all subsystems and partial systems of the organization;
2. Beware: the basic process tends to be equally independent, dependent, and interdependent as the managerial and informational processes;
3. Beware: the basic process is the closest of all three processes to routine, standardization, prescriptions, rules, hence to a routine-loving behavior;
4. Beware: the routine-loving humans suit best the basic process of all three processes;
5. Making rules for the routines of the basic process to be requisitely holistically run and smooth is no routine work, but a creative and often an innovative one;
6. Beware: the natural need for determinism and hence for routine provides the basis for a Tayloristic, Fordistic, Weberian, and Fayolan management style, in which bosses think, and subordinates are supposed to listen only rather than to be coworkers;
7. Methods of creative work and cooperation can reduce the danger of too much love for routine;
8. Beware: the "Submit You Idea System", the "Suggestion System", or "Reinventing" can either support or ruin creativity of many; the practice of work depends on management practices considering the above guidelines or avoiding them;
9. Cooperative management enables more innovativeness of the basic process than a one-way commanding one;
10. Methods of creative cooperation enable more innovation than the individual methods (see guideline 8), due to better enabling requisite holism and ethics of interdependence.

The ten guidelines for the coordination of all three processes, which tend to be run in separated offices and other organizational units, for the entire business to be innovative:

1. It all depends on innovativeness of humans, especially of the most influential ones, such as opinion leaders, first of all of governors and managers;
2. Entire business follows the principle of synergy of marketing, research, development, and daily operations, and is aimed at innovation;
3. Innovativeness of all policies, strategies, tactics, and daily practices of all organizational units;
4. Organization governance follows the ethics of interdependence by the slogan "We all think, work and innovate all the time deliberately" on all levels of the organizational hierarchy;
5. Coordination of the managerial, informational, and basic processes with an innovative orientation;
6. Organized network of interdisciplinary creative cooperation aimed at requisite holism of the invention-innovation-diffusion and routine processes based on ethics of interdependence;
7. Requisitely usual systems thinking rather than a one-sided and routine-loving one of requisitely many organization members and partners;
8. Requisitely holistic and usual linking of the natural creativity and the logical daily work processes aimed at requisite holism in mastering both, the short-term and long-term tasks;
9. Requisitely holistic and quick transition from quality checking and similar out-dated methods of quality assurance, supposed to be resulting from innovation and routine, to total quality management aimed at systemic quality;
10. Requisitely usual and holistic creative self-management of organizational members and their teams concerning running and innovating of their own daily work.

In our experience, there is no reason for limiting the dialectical system of these forty guidelines to production on shop floor or to manufacturing companies. It all happens everywhere, including the government and social services, other services, and manufacturing.

14.12 Some Conclusions about Innovative Business and Innovative Society

The summarized dialectical system of suggestions means (Mulej, 2007):

For an accelerated transition to an innovative society, one starts best with ethics of interdependence and the non-technological innovation in the government.

Most of the government members in catching-up areas of the world had so far a poor chance to learn about and to practice innovation and its organization and managerial conditions; but they are the most influential societal group, once people find them credible. Then the government office people follow, and then all other public services, such as education, medicine, research. Now, businesses will follow government's advices more openly than so far, when routine-lovers were telling businessmen to be innovative. Everybody must be innovative in one's values/culture/ethics/norms for the given knowledge to be applied with new benefit, i.e., innovation, rather than to exist only (EU 1995, 1996, 2000, 2004).

And the last suggestion for which we still can have room in this chapter: let us consider the empirical finding that the most advanced and prosperous regions in the USA are the regions, in which most has happened for the "creative class" to prevail (Florida, 2005). In these regions the 3T model works best: there they are the most tolerant of differences between their inhabitants, thus they attract most talents, which attracts most technology investment, and vice versa. The three Ts are interdependent. All briefed attributes of KM and IIP management are backing them; for success, none may be missed. Cooperative leadership can be a natural gift or learned (Lester, 2005; Linden, 1990; McIntyre, 2003, 2005; Mesjasz, 2005), so can entrepreneurship (Japti, 2005); thus the basic managerial preconditions for IIP and related KM can be available.

We will add a supplement to critically show that the European Union has been trying to enhance innovation, but with

- Too much limitation to the technological innovation;
- Too much forgetting about the need to start with the innovation of management like always in history (although the book by Rosenberg and Birzzell had been published in 1986 already);

- Too much oversight that a government living in routine can hardly be credible with businesses, when it requires them to be innovative and provides for no own role model.

Thus, EU has neither met the law of requisite holism nor enhanced enough the ethics of interdependence in its population. Let us take a short look at a part of one of its documents.

Chapter 15

SUPPLEMENT: INTRODUCTORY THOUGHTS FROM THE EUROPEAN COMMISSION'S FIRST DOCUMENT AIMED AT ENHANCING INNOVATION AND SYSTEMS THINKING: GREEN PAPER ON INNOVATION (1995)

The European Union has been unhappy with the level of innovation in Europe for quite a long time, and still is so, including its newer statements (Competition, 2004; EU, 2000, 2004; Handbook, 2004; Lisbon, 2004; Resolucija, 2006; Sporočilo, 2006). Let us take a brief look at its suggestions in its first official document.

15.0 INTRODUCTORY REMARK

The objective of this Green Paper is to identify the factors—positive or negative—on which innovation in Europe depends, and to formulate proposals for measures which will allow the innovation capacity of the Union to be increased.

In the context of this document, innovation is taken as being a synonym for the successful production, assimilation and exploitation of novelty in the economic and social spheres. It offers new solutions to problems and thus makes it possible to meet the needs of both, the individual and the society. There is a wealth of examples, including the development of vaccines and medicines, improved safety in transport (ABS, airbags), easier communications (mobile phones, video-conferencing), more open access to know-how (CD-ROM, multimedia), new marketing methods (home banking), better working conditions, more environment-friendly techniques, more efficient public services.

According to the dictionary, the opposite of innovation is "archaism and routine". That is why innovation comes up against so many obstacles and encounters such fierce resistance. It is also why developing and sharing an innovation culture is becoming a decisive challenge for European societies.

15.1 INNOVATION, THE FIRM AND THE SOCIETY

Innovation has a variety of roles. As a driving force, it points firms towards ambitious long-term objectives. It also leads to the renewal of industrial structures and is behind the emergence of new sectors of economic activity. In brief, innovation is:

- The renewal and enlargement of the range of products and services and the associated markets;
- The establishment of new methods of production, supply and distribution;
- The introduction of changes in management, work organization, and the working conditions and skills of the workforce.

The innovative firm thus has a number of characteristic features, which can be grouped into two major categories of skills:

- Strategic skills: long-term view; ability to identify and even anticipate market trends; willingness and ability to collect, process and assimilate technological and economic information;
- Organizational skills: taste for and mastery of risk; internal cooperation between the various operational departments, and external cooperation with public research, consultancies, customers and suppliers; involvement of the whole firm in the process of change, and investment in human resources.

It is this global approach, which lies behind, for example, the success of Swatch watches. In practice, this amounts to four simultaneous innovations in:

- Conception (reduction in the number of parts);
- Production (assembly of the housing in a single part);
- Design (new concept for the presentation of the watches);
- Distribution (non-specialized sales outlets).

Research, development and the use of new technologies—in a word, the technological factor—are key elements in innovation, but they are not the only ones. Incorporating these means that the firm must make an organizational effort by adapting its methods of production, management and distribution.

Human resources are thus the essential factor. In this respect, initial and ongoing training plays a fundamental role in providing the basic skills required and in constantly adapting them. Many studies and analyses show that a better-educated, better-trained and better-informed workforce helps to strengthen innovation. The ability to involve the workforce to an increased extent, and from the outset, in the technological changes and their implications for the organization of production and work must be considered a deciding factor.

There is no hermetic seal between the innovative firm and its environment, by which it is influenced and which it helps to transform. It is the total sum of firms in an industry, the fabric of economic and social activities in a region, or even in society as a whole, which makes up the "innovation systems", whose dynamics are a complex matter. The quality of the educational system, the regulatory, legislative and fiscal framework, the competitive environment and the firm's partners, the legislation on patents and intellectual property, and the public infrastructure for research and innovation support services, are all examples of factors impeding or promoting innovation.

15.2 INNOVATION AND PUBLIC ACTION

The Commission has clearly identified—first in the White Paper on Growth, Competitiveness and Employment, and then in its 1994 Communication on An Industrial Competitiveness Policy for the European Union—that firms' capacity for innovation, and support for it from the authorities, were essential for maintaining and strengthening this competitiveness and employment. This Green Paper makes use of, adds to and extends that work with a view to arriving at a genuine European strategy for the promotion of innovation. While respecting the principle of subsidiarity, it will propose the measures to be taken at both national and Community levels.

> *In exercising their responsibilities, the authorities must promote the development of future-oriented markets and anticipate changes rather than react to them (...). The European Union must place its science and technology base at the service of industrial competitiveness and the needs of the market more effectively. Greater attention must be paid to dissemination, transfer and industrial application of research results and to bringing up to date the traditional distinction between basic research, pre-competitive research and applied research which, in the past, has not always allowed European industry to benefit from all the research efforts made.*

The Commission has paid attention to this aspect of updating in the new arrangements on research aid adopted in December 1995.

This responsibility of the authorities is particularly important with regard to technological innovation and the creation of businesses—fields in which the situation in Europe remains worrying compared with its competitors

In the Commission's opinion, Europe's research and industrial base suffers from a series of weaknesses. The first of these weaknesses is a financial one. The Community invests proportionately less than its competitors in research and technological development (...). A second weakness is the lack of coordination at various levels of the research and technological development activities, programmes and strategies in Europe. (...) The greatest weakness, however, is the comparatively limited capacity to convert scientific breakthroughs and technological achievements into industrial and commercial successes. (White Paper "Growth, Competitiveness, Employment. The Challenges and Ways Forward into the 21st Century", Chapter 4, European Commission, 1994).

Strengthening the capacity for innovation involves various policies: industrial policy, RTD policy, education and training, tax policy, competition policy, regional policy and policy on support for SMEs, environment policy, and other policies. Ways must therefore be found of identifying, preparing and implementing—in a coordinated fashion—the necessary measures covered by these various policies.

Thus as regard with SMEs, the Commission has outlined a new policy strategy in its report, *"Small and Medium-sized Enterprises, a Dynamic Source of Employment, Growth and Competitiveness in the European Union"*, which has been presented to the Madrid European Council in December 1995. These priority policies and measures to be undertaken, both by the European Union and the Member States, will form the basis of the next Multi-Annual Programme in Favour of SMEs and the Craft Sector for the period 1997 to 2000.

First and foremost, the authorities must establish a common strategy. This is a matter of ongoing monitoring and consciousness-raising. The Green Paper is contributing to these two objectives through the wide-ranging debate, which it aims to encourage amongst the economic and social, public and private players.

It touches upon the following:

- The challenges of innovation for Europe, its citizens, workers and firms, against a background of globalization and rapid technological changes;
- A review of the situation of innovation policies and the many obstacles to innovation.

Proposals or lines of action, while respecting the principle of subsidiarity, for government, regions and the European Union, aimed at removing these obstacles and contributing to the campaign for a more dynamic European society, which is a source of employment and progress for its citizens.

On this basis and on the basis of a few other references, which we have quoted earlier, see Table 37:

INNOVATION, as a process, is producing innovation as an outcome. Innovation, as an outcome, is every novelty that is accepted as useful by buyers and customers. Innovation management is a process aimed at efficient and effective control of innovation, both as a process and an outcome.

Table 37 *The internationally accepted modern definition of innovation*

If this definition is accepted, why is there no action to be undertaken by the government neither in this document nor in the one of 2000 (EU, 2000, 2004) nor in any other (Mulej, 2006a)?

Chapter 16

CAN AND MAY ONE AVOID INNOVATION, ETHICS OF INTERDEPENDENCE AND THE LAW OF REQUISITE HOLISM? NO LONGER!

World Bank's data let us see that in the century before 1970 the span of extremes of richness (in terms of the National Income per capita—NYPC) per countries has grown from 3:1 to +150:1, and after 1970 it grew in a quarter of a century to +400:1. In the recent decade it grew to +500:1 (Luxemburg and Mozambique making the extremes). It has grown to the benefit of the most innovative ones, of course (their list includes some oil owning countries, too, but their rich people also buy in the most innovative countries, mostly). The essential source of the difference has not been science alone, neither the technological development alone, neither commerce alone, but they all together along with free entrepreneurship using innovation to make profits. Innovation = invention + commercialization, i.e., a successful use of new findings for new benefits of both, consumers and suppliers (Afuah, 1998).[1]

---

1. Measurement of how advanced a country is in terms of the number of *patents* does not make much sense any more. The economy is global; only one (1) percent of the patents become innovations; this does not necessarily happen in the country where invention has been created and, perhaps, patented, too. The modern exploitation depends less on legal differences among persons and countries and more on economic ones with their common

In economic terms, the world is no longer divided into rich and poor parts with no clear knowledge of the origin of differences—the origin is the difference in the innovation culture:

- The innovative 20% of the population control 95% of the knowledge and investment in knowledge as well as hold the creative ambitions in their values and at the same time make the vast majority of the purchasing power and interest of the entire world, its main market;
- The routinized 80% of humankind are different in their cultures: 50% of humankind live on less than two US dollars a day, 25% even on less than one US dollar a day, in very difficult conditions in terms of drinking water; hence, they do no have comparable ambitions neither make much of an appealing market.
- Therefore the worldwide economic situation is tough for everybody, the least so for the most innovative ones. The crisis of 2008 proves this.

The difference in richness between the most innovative (the richest) ones and the least advanced (the most conservative) ones started growing seriously when in the West and in Japan, in the second half of the 19th century, they freed entrepreneurship. This has changed the equilibrium of supply and demand to produce a surplus of supply growing continuously more and more. Hence, crises became a normal situation, their pressures caused new and new solutions to surface, a small part of them (but an essential one) ended in new innovations. In their majority, they have, obviously, not been prepared and worked out holistically enough to outgrow a limited usefulness and use. Instead, innovations keep being encouraged and killed by further inventions and innovations permanently.[2]

This practical fact leads to both, progress and destruction at the same time (the Schumpeterian "creative destruction"). It puts the question which conditions are to be met for innovation to be able to flourish, as a continuous

---

denominator in innovations (where ever inventions may come from). Systems thinking, requisite holism and ethics of interdependence support them in the background.

2. Our research let us se that e.g., Slovenia has freed entrepreneurship and, for the first time in its history, attained a surplus of supply over demand causing the need for invention and innovation, for the first time in its history, only in the recent decades, after 1988. This is a delay of four generations or two two-generation cycles which have empirically proven to be a timescale *of changing of the noninnovative culture of a community/nation into an innovative one*. Now, Slovenia has many legal attributes of an innovative society, but keeps lacking the human behavior of an innovative type. (Bučar, 2001, 2004, 2006; Bučar & Stare, 2003).

process. This finding might be a basis for measures to be taken to modernize the prevailing VCEN (= most influential habits) and make them supportive of innovation.

Some people may think that such lags in development cannot be overcome. The experiences of Japan, Singapore, South Korea, as well as Spain, and USA—from the times of their catching-up efforts and overcoming their lags of those times, in different time periods, of course—demonstrate that such lags can be overcome. But waiting for the consciousness and sub-consciousness to change on their own may cause the process to be too slow for the forgotten four-fifths to ever catch up: it seems to take a two-generation period, lasting about 70 years, for the culture to change on its own (Mulej, 1994). The traditional legal support which merely permits innovation and entrepreneurship without accelerating their becoming a normal culture is not enough for the laggards, although it is good enough for the most advanced ones.

Since 1988-91, when the Soviet Bloc and Yugoslavia had fallen apart, one has spoken of the new countries emerging there (including Slovenia) as transitional economies. The same would be justified concerning China and other catching-up areas. Why? Superficially, one speaks of their transition from a centrally planned to a market economy, and this is correct concerning regulatory mechanism of most of them, but it does not say much about the reality of development economics. The countries under consideration are latecomers to the industrialization process. They kept the principles of guilds[3], of a central body's care for balance of supply and demand (S = D), and freed entrepreneurship for a good century later than the West. This may have saved some investment funds, but it wasted lots of creativity, which is worse. Namely, innovation results from the combination of creativity and entrepreneurship, which causes the contemporary differences between the parts of the World. The crisis of 2008 results from another type of one-sidedness—the exaggerated leaving aside of requisite holism.

---

3. In medieval Europe, guilds were associations of producers per branches in towns. Among other authorities, they took care of the supply - demand ration which was supposed to be balanced. The established competitors, members of the guild (e.g., of shoemakers, of bakers, etc.), were entitled to decide whether a new competitor may show up. Thus, the stability of supply and demand was assured at the detriment of productivity because competition was poor and innovation was therefore not necessary. A comparable role belonged to, e.g., the Chambers of Economy inside within their General Associations of Industries, in Yugoslavia, until her its end in 1991.

Hopefully, the model offered in this book can help laggards of the socio-economic development to innovate their cultures and catch up better than so far by using more of the law of requisite holism and of ethics of interdependence supportive of it.

Chapter 17

CONCLUDING REMARKS

17.0 THE SELECTED PROBLEM AND VIEWPOINT OF DEALING WITH IT IN CHAPTER 17

It is time to finish. We hope to have demonstrated, that innovation is unavoidable in catching-up areas because their future depends of high quality. Thus, requisite holism backed by ethics of interdependence is unavoidable.

17.1 GROWING DEPENDENCE OF HUMANKIND ON HIGH QUALITY AND HENCE ON INNOVATION

In human history, there have always been innovations, which solved the current human problems. Mostly, they have not been holistic enough, and caused new problems along with solving the earlier ones. Thus, over millennia, humankind came to the 20th century as the century of living on innovation. The 21st century is not promising to be any easier to handle. Both, the speed and style of life, are dictated by the innovative societies and their economies under the name of free market, competition and democracy. Crises are permanent, they demonstrate that entropy is a permanent tendency, which keeps requiring both, innovation and requisite holism. Systemic thinking helps better than one-sided thinking, cooperation on the basis of ethics of interdependence helps better than isolation, creativity helps better than routine-loving behavior.

Innovative business is supported in innovative societies, USOMID and similar method support effectiveness and efficiency of efforts aimed at excellent quality in order for the suppliers to satisfy their customers (better than their competitors do[1]). Innovation management must take care of all these preconditions in all details. They are not discussed here, but in other books which on the other hand do not include these background material. They are many now.

17.2 GROWING COMPLEXITY OF INNOVATION AND THEREFORE GROWING DEPENDENCE OF SUCCESS ON SYSTEMS THINKING

Let us answer the question from this chapter's title with quotes from distinguished authors:

It is in the nature of systemic thinking to yield many different views of the same thing and the same view of many different things. (Ackoff, 1999; a comment in: Flood, 1999)

In the present global market, systemic thinking has risen to a position of prominence never before attained. Globalization, networking and coalition building are examples of systemic behavior that stress the need to view and deal with issues holistically, rather than in isolation. (Sulaiman, 1999; a comment in: Flood, 1999)

"[In this book] you will see and hear, again and again, that:

We will not struggle to manage over things—we will manage within the unmanageable.

We will not battle to organize the totality—we will organize within the unorganizable.

We will not aim to know things—but we will know of the unknowable. (Flood, 1999: 192)

Systemic awareness begins with a spiritual appreciation of wholeness. Wholeness may be appreciated in terms of interrelatedness of events, and

---

1. Let us remind you of the following anecdote. Two friends were out there in the nature. They were taking some rest. All of a sudden, they were attacked by a wild animal. When they saw it, one person of them started running barefoot, the other took his time to put her his shoes on. "Do you intend to run more quickly than the animal, if you put your shoes on?"—"No, but more quickly than you, this is enough for me to survive in this competition."

spontaneous self-organization leading to emergence and new order. (Flood, 1999: 192)

Thus, we are living between mystery and mastery (Flood, 1999). Requisite holism is also somewhere there between these two extremes of knowledge about reality. Innovation management is a very good case of the permanent need for systems thinking, these quotes tell us.

17.3 NEED FOR CAREFUL CHOICE OF SYSTEMS THEORY

But, we must be careful, when talking about systems theory. There are five streams, no unified definition any longer:

1. The mathematical stream defining system as an ordered set, meaning that there is a set of components and a set of their relations; this is the most general one and applied by many specialists. It allows them to talk about whatever contents they like, even without any definition of it. It therefore allows for any level of holism, because every selected viewpoint allows its author to select the subsystem of the really existing attributes as he or she pleases. Under the name of holism one can attain a total holism or a requisite holism or a fictitious holism, the latter one being limited to a single selected viewpoint. The authors in this stream do not seem to care for clarity of their contents and the dictionaries quoting about 15 groups of the contents of the word system (Veliki, 1997; Webster, 1987, 1992).

2. The specialists of single disciplines of science, who do not work on mathematics, but use it within their own fields and their own selected viewpoints, without acknowledging any need for any interdisciplinarity. In terms of Table 1, they work on the traditional basis, attaining a fictitious holism. It may allow them deep insights, but no linking of them with insights from other viewpoints, used by specialists different from them. They do not seem to care for the said many groups of the contents of the word system either. They may attain many important outcomes, too, but in terms of complicatedness rather than complexity, systemic rather than systemic approach (Mayer, 2001; Potočan, 2003; Rosi, 2004; Schiemenz in Mulej and coauthors, 2000). See Table 1.

3. The specialists of single disciplines of science, who do not work on mathematics, but use it within their own fields and their own selected viewpoints, but acknowledge their need for interdisciplinarity. In terms of Table 1, they do not work on the traditional basis and do attain more holism rather than a fictitious one. It may allow them for deeper and broader insights than the ones from groups (1) and (2), due to linking them with insights from other, different specialists. These authors are many less than the members of groups (1) and (2). They seem to care for the many contents of the word system.

4. The specialists of systems theories, who explicitly work on systems thinking as a theory of thinking, but do not define the contents of the word system explicitly. They may have the impact over the influential worldwide organizations and countries that use the word or idea of systems thinking in their documents without defining it. But this is merely our impression, which is not explicitly visible, e.g., in references in this book.

5. The specialists of systems theories, who explicitly work on systems thinking as a tool aimed to make bridges between specialists of whatever disciplines and practices. We belong in this group. To us, the central point of the Bertalanffian systems theory is not the similarity—the isomorphism. It does allow for fruitful transfer of notions from one area to another, but it poorly cares for further cooperation between mutually different disciplines. To us, the basic notion, although isomorphism matters, is rather the interdependence of the disciplines different from each other, because they are different and hence needing each other for complementarity (Jackson in Mulej and coauthors, 2000) aimed at the requisite holism of insight, thus avoiding crucial oversights.

All these five streams make sense and contribute in the real world and in science, but do not equally fully use what has made systems thinking and theory. Its origin, which also makes a lot of sense today, not only many decades ago, requires systems thinking to reach beyond streams (1), (2), and (4) and join the streams (3) and/or (5). And: Bertalanffy originally used the word "teaching", which implies influence over people, rather than the word "theory", which implies generalization of findings only. Thus, cybernetics is equally crucial as systems theory.

Keeping to one single viewpoint rather than to a dialectical system of all essential viewpoints very rarely allows for requisite holism; a single viewpoint

No.	Systems/systemic/holistic thinking	Un-systemic/traditional thinking
1	Interdependences, relations, openness, interconnectedness, dialectical System	Independence, dependence, closeness, a single viewpoint/system
2	Complexity (& complicatedness)	Simplicity or complicatedness alone
3	Attractors	No influential force/s, but isolation
4	Emergence	No process of making new attributes
5	Synergy, system, synthesis	No new attributes resulting from relations
6	Whole, holism, big picture, holon	Parts and partial attributes only
7	Networking, interaction, interplay	No mutual influences

Table 38 *The seven interdependent basic sets of terms of systems/systemic/holistic vs. un-systemic thinking (as a dialectical system)*

uncovers only a subsystem or even a subset of the really existing attributes of any object under consideration. Ethics of interdependence supports requisite holism.

17.4 THE SEVEN PRINCIPLE ATTRIBUTES OF SYSTEMS THINKING APPLIED IN THIS BOOK

Most schools teach and educate for a profession. Thus, narrow specialists have never had much education/feeling/appreciation for systems thinking.[2] Very few around the globe had and/or have courses in systems theory.[3] And very few versions of the General Systems Theory/other systems theories cover the seven groups of attributes of systems thinking (see Mulej *et al.*, 2003). There, we suggested a definition of systems thinking, which we find requisitely holistic, and completed it later on, on the basis of responses received (Mulej *et al.*, 2006). We will only summarize it here in Table 38 (see Mulej *et al.*, 2003a; 2003b; 2003c; 2003d, 2004, 2005).

The following synergy of systems theories might be applicable on this basis (Table 39):

---

2. If they had enough education to practice systems thinking, they would most probably be able and willing to listen to and hear and accept views of their opponents as complementary rather than strange or dangerous. Rare specialists behave this way. Conference contributions show this: they rarely have coauthors from different professions.

3. At the economists' conference (CJE, 2003), e.g., a few speakers only used the word system and all of them found it a formality aimed at describing what ever one wants to, with no real importance and impact (oral discussions). Courses in systems theory, which we have come across over the years, often tend to forget about interdisciplinarity.

Life in the contemporary knowledge and innovation driven society, or suffering from neo-colonizing (of the less innovative ones) resulting from globalization and (huge) differences in innovation level		
Cybernetics of Conceptual Systems interfacing society and its individual members, hopefully supporting holism and innovation, including ethics of interdependence in both, their routine work and inquiry →	Humans' objective conditions (needs & possibilities) in interdependence with humans' subjective starting points (values/emotions & knowledge/talents & skills, in interdependence), hopefully aimed at holism and innovation, incl. ethics of interdependence in both, their routine work and inquiry	Dialectical Systems Theory providing for education and guidelines for managers and their co-workers to aim at holism and innovation, incl. ethics of interdependence in both, their routine work and inquiry ←
Management based on Critical Systems Thinking & Organization based on the Viable Systems Model	Mastering work processes by Soft Systems Methodology, Dialectical Network Thinking, and by routine and framework standardization	
Corporate social responsibility, adaptation to innovative society, and intrinsic motivation for holism and innovation, including methods of creative cooperation and excellent quality etc.		

Table 39 *Interdependence, complementarity, and synergy of several soft-systems theories in a general model of cooperation in inquiry, invention and innovation (as well as routine parts of the work and life processes)*

And there is at least one more conclusion resulting from Tables 34 and 39: systems thinking may also be implicit, informal. One may be able and willing to use the seven principles and never use the theoretical language of a systems theory. After all, systems thinking surfaced millennia earlier than systems theory! It faces changes in Table 38 demonstrating how much more complex life has become after World War II (UNESCO, 2006).

These data say clearly why the industrial paradigm must be replaced by the systemic behavior for humankind to survive. This is a crucial non-technological innovation.

How problematic a one-sided innovation may be due to oversight of another viewpoint and resulting attributes, the following case shows well (Tihec, 2006):

Attributes of the world	1945	2005
Total population (in billions)	2,2	6,5
Urban population (in % of the total population)	Under 29 %	49%
Literacy of adult population (%)	Under 50%	81,7%
Life expectancy at birth (in years)	Under 46	65
Countries with parliament	26	186
Women in parliament (in % of all members)	3	16
Fertility (number of children per woman)	Above 5	Under 3
Mortality of children (number of children per 1,000 not reaching 5 years of age)	224	86
Surface under woods (in million square kilometers)	50	39
Yearly consumption of water (in cubic kilometers)	797	2425
Yearly consumption of oil (millions of tons)	Under 470	Above 4000
Tourists (in millions)	Under 25	808

Table 40 *1945-2005: big differences*

Research detects that a flat screen TV set with a screen beyond 24 inches consumes three times the electric energy as a usual TV set. Such TV sets sell very well. Consequences include that in the United Kingdom alone the modern large screen TV sets will be responsible for a 70% increase of carbon in the air, by 2010. In the UK, they have 63 million TV sets consuming 9.6 KWh of electricity per year causing a million tons of additional carbon in the atmosphere. The increase of TV sets by 2010 is expected to reach 67 million TV sets, which may cause the electricity consumption to reach 15.7 million KWh per year and result in 1.7 million tons of carbon more in the atmosphere. Politicians must stop the population's desire to have larger and larger screens on their TV sets and prevent people from buying appliances consuming much energy, using all available measures.

Application of the law of requisite holism and consideration of ethics of interdependence may be measures of a systems approach in this kind of decisions of humans as politicians, businesspersons, and consumers. Then, one may hope for a greater capacity of humankind to prevent or diminish dangers resulting from wrong decisions due to humans' lack of requisite holism and ethics of interdependence; it is humans who are facing these dangers, and so it is their/our natural environment.

Hence: let us do our best to attain requisite holism and let us therefore develop and apply ethics of interdependence—by explicit or tacit/informal use of the (Dialectical) Systems thinking and Theory. For millennia, this has been the practice of the successful humans, and much less so of the failing ones.

17.5 SUGGESTION CONCERNING SYSTEMS EDUCATION[4, 5]

17.5.1 Introduction

Human behavior consists of monitoring, reflection, thinking, emotional and spiritual life, creation, routine, decision-making, communication, and action. It is based on human knowledge, values and circumstances (outer needs and possibilities). The contemporary unavoidable narrow specialization causes humans to capture only a part of the actually existing attributes, while other attributes may be subject to oversight that may lead to failures—from relatively small ones all way to world wars and to the end of human and other life on the planet Earth. This is why the (General) Systems Theory was created—to prevent oversights by fostering the worldview of the holism of approach and related methodology aimed at wholeness of outcomes of behavior.

L. von Bertalanffy explicitly wrote (1986 (ed.), 1979: VII) that his systems theory had been *originally intended to overcome current over-specialization*. He found this intention abused by making systems theory one of many specialized and technical disciplines. He might feel the same way much more today, when systems theory is fragmented in many theories and the original intention is taken care of in only very few of them.

Over-specialization differs from the unavoidable specialization: over-specialists are closed within their own single viewpoint; other specialists are open to interdisciplinary, creative cooperation in order to overcome one-sidedness and to come close to a holism of their approach, and the wholeness of outcomes of their behavior. Such a behavior was typical also for the practice of Norbert Wiener, the most famous among the pioneers of cybernetics.

---

4. In Brisbane, Australia, an international network of systems educators started to be created as a shared effort of IFSR (International Federation for Systems Research) and ISSS (International Society for Systems Sciences), during the 53rd meeting of ISSS. This chapter summarizes our suggestion to about this necessary new international body.

5. Document title: Criteria for programs of education in systems behavior validation; the 1st draft, 16 July 16, 2009, Brisbane, Aus, and 26 July 26, 2009, Maribor, Slovenia.

This fact about Systems Theory itself addresses the "uncommon sense" Bertalanffy has been defending (Davidson, 1983): he was fighting the common current practices of one-sidedness, because they were dangerous and still are so with a growing trend. If over-specialization is common, interdisciplinary cooperation leads the way to holism rather than individual disciplines and related values/culture/ethics/norms do alone. Though, more holism and less one-sidedness in the traditional and newer disciplines of science and practice are also helpful.

Equal conclusions can be made about cybernetics: it was created in interdisciplinary creative cooperation. It is now also fragmented in many disciplines, partly by specialization, and partly by over-specialization.

In practice, systems behavior, especially its part called systems thinking or systemic thinking, can exist naturally as a human attribute, which very few persons attain; others must and can learn it. Its basis is systems theory as a science, profession, and practice of integration (Hammond, 2003) of partially considered attributes and of mutually different and therefore complementary specialists of various disciplines, professions, and practices. Thus, researchers and/or practitioners can attain the requisite holism of their behavior and the requisite wholeness of its outcome (for more details about "requisite" see Mulej, 2007).

17.5.2 Basic Contents to be Covered in Teaching of Systemic Behavior

17.5.2.1 Capacity to be Attained

On the above basis and due to the fact that the currently surfacing long-term and increasingly threatening socio-economic and environmental crisis of humankind cannot be solved without application and integration of mutually different specialized capabilities of humans, it is extremely necessary for humans, especially the influential and decisive ones, to make their transition from a biased/over-specialized to a requisitely holistic behavior of specialists.

"Cooperative persons" are much better suited to lead this transition than "free-riders", both of these groups including about 15–20% of humans each, while the rest are passive with a "wait-and-see-attitude" (Lester, 2005).

The crisis tackles the humankind's running out of drinkable water, healthy air and soil, un-renewable natural resources, the dangerous climate change, the danger of higher sea-level chasing hundreds of millions of people to other locations, the disappearance of very necessary bio-diversity, and so forth. One-sided approaches cannot solve these problems, while the requisitely holistic ones can (Božičnik et al., 2008; Brown, 2009; Dyck & Mulej, 1998; Ečimovič et al., 2002; Harris, 2008; Korten, 2008; Martin, 2006; Taylor, 2008), but they can even do it better when in synergy of insights of several authors (see Chapter 15).

The educational programs enabling/supporting systemic behavior are necessary on all levels of primary, high school, and higher education to promote collaborative behavior aimed at requisite wholeness. This does not mean that every education for a specialized capability may be abandoned or even banned; it rather means that education must enable both, a specialization and the capability of integrating several specialized capabilities by the worldview of holism and related methods, especially of interdisciplinary creative cooperation.

17.5.2.2 Holism—Fictitious, Requisite, and Total

Holism—in the full sense of the word—includes totally all existing attributes with no narrow specialization within any profession or science. Thus, holism reaches beyond human natural capabilities, which have caused human specialization in many professions and sciences. The good consequence of it is the human capability of detailed insight within a single viewpoint, profession and/or science; the dangerous consequence is the oversight of all other existing and impacting attributes. Some of them are focused and covered by other viewpoints, professions and sciences, but effects resulting from their relations normally are not. E.g., chemicals are reported to be tested on safety with no regard for their synergies (Brown, 2008; Taylor, 2008). This is where systemic behavior enters the scene to cover the blank. Right now, this blank endangers humankind's existence: humans either accept the practice of systemic behavior or leave to our children and grandchildren (if any at all) a dying planet Earth (Taylor, 2008).

This means that holism/wholeness, if limited within a single viewpoint, profession and/or science, faces the danger of being a fictitious holism whose dangerous consequence is the oversight of other crucial attributes. Therefore this is helpful, but not good enough, often.

Therefore one needs to consider the "Mulej/Kajzer law of the requisite holism of behavior" (Mulej & Kajzer, 1998; Mulej, 2007) providing for a middle way between the impossible real holism and the dangerous fictitious holism. Practically this means that one must first collect insight into all viewpoints, professions and/or sciences that are found or deemed crucial, their interdependencies and resulting synergies. The system, as a synergetic network, of all crucial viewpoints is called the dialectical system (Mulej, 1974; Mulej et al., 2008).

In other words, Bertalanffy's "worldview of holism" should better be transformed into the "worldview of requisite holism". Author, researchers, managers and other practitioners must take responsibility for their selection of viewpoints, professions and/or sciences that are found or deemed crucial, their interdependencies and resulting synergies. Therefore they must first pay a requisitely holistic attention to their own starting points of this process, leading later on to the definition and realization of the objectives of their actions.

17.5.2.3 Topics to be Covered

Thus, the very first topic to be covered in systemic education is awareness of oversight resulting from the usual limitation of humans to a single viewpoint in terms of the usual over-specialization that has been the common sense so far. This is done with a case by (1) choosing any topic, (2) collecting all possible various viewpoints to consider, (3) collecting the relations between them, especially the interdependences rather than or along with one-way dependences, (4) collecting all visible synergies resulting from them, and (5) comparing the level of wholeness of insight and outcome attainable with this procedure versus the one limited to a single viewpoint. The latter is much simpler to do, but much more complex consequences result from oversights caused by the observer's/researcher's/manager's/user's approach.

On this basis the basic vocabulary about systems' and models' typologies and related boundaries, processes and similar necessary formalities should be covered in order to provide a shared language for a clear description of attributes and dynamics/processes and related structures should be added.

Methodology or several methodologies of requisitely holistic behavior, aimed at fighting the crucial oversights are the next crucial ingredient. Their selection may differ according to their suitability for students of different professions/specializations, especially in terms of details of study. Never should a single systems theory or cybernetic alone be covered at an elementary university level, and never should formalities of description of objects under discussion with systems and models be found sufficient. Everybody should receive insight and training in methodology of creative cooperation supported by teamwork and insight in 5–6 systems theories and cybernetics theories.

17.5.2.4 Depth of Elaboration

Depth of elaboration might well be divided into several levels suggested by Jones, Bosch, Drack, Horiuchi and Ramage (2009), which are 1. Sense-making; 2.1. Practical Mastery; and 2.2. Theoretical Mastery, and divided into A. Discipline-integrated; and B. Generic.

Program for K-12 exists and works well (discussion in the network of systems educators on July 15, 2009, ISSS, Brisbane, AUS). Thus it should be diffused. Its basis is the capacity of humans to listen to each other when and because they are not of the same opinion on the topic under consideration, thus developing their capability and experience of creative cooperation. Their teachers of various courses who work in parallel with the same students should develop their practice of linking their courses to each other for students to learn about interrelations, especially interdependences of various and different kinds of knowledge. The latter—all of them—transfer to students some knowledge about life, but specialize in different viewpoints and therefore different and often separately considered parts of the really existing and interdependent attributes.

Programs for higher education differ, as authors of systems theories and cybernetics differ, and there is rather fragmentation than a common shared basis. The latter may not be applied for anybody to impose a single approach. Insight into Encyclopedia (François, 2004) from such a viewpoint shows that the existing systems theories are actually complementary to each other rather than competing, because they have been authored from different selected aspects or viewpoints.

In general, systems theories could be divided into three groups, although they are rather interrelated than separated in real life:

1. The *"hard systems theories"* are meant to support natural and engineering sciences by precise, often mathematically supported descriptions of

natural phenomena and their application on very reliable products and other artefacts. In these cases requisite holism is focused on attributes of elements, processes and structures of these natural phenomena, products and other artefacts as well as processes of their making, both in nature and under human intervention. Their relations with their environment, both natural and social, are unavoidably added.

2. The *"soft systems theories"* are meant to support non-engineering sciences by framework, rarely mathematically but though realistically captured, descriptions of social and humanities' phenomena and their application on social life apart of making products and other artefacts. In these cases requisite holism is focused on attributes of elements, processes and structures of these social and humanities' phenomena, as well as processes of their making both, in humankind's social nature and under human intervention. Their relations with their environment, both natural and social, are unavoidably added

3. While both, hard and soft systems sciences, may often be confined behind the boundaries of traditional disciplines—with full right and a lot of benefit, these boundaries rarely enable them and their users to fully meet the objectives due to which systems theories and cybernetics have come into being. This is—let it be stressed again—prevention of oversights resulting from one-sidedness, which can neither cover attributes outside the selected territories of the traditional disciplines nor their relations, interrelations, interdependences and synergies. The *"integrating systems theories"* play the latter role. Their relations with their environment, both natural and social, are unavoidably added.

On this basis, a selection of systems theories suiting the selected specialization/profession should be added, and a link to that profession should be provided with applied cases. Basic attributes of all three kinds of systems theories should be presented, and the suitable ones elaborated in requisite details.

17.5.3 Some Conclusions

In practice, for ever, the systemic/requisitely holistic behavior has been a crucial attribute of successful persons, while the one-sided/biased/over-specialized one has resulted in failures, all way to world wars and the current danger of extinction of the contemporary civilization on the planet Earth.

Therefore, it is high time for system education to stop being exceptional and to become normal, because the informal systemic behavior is too rare to help humankind in time to create its own way out from the current blind alley. The latter was created by the recent period of the industrial paradigm in which over-specialization has been flourishing for too long; the same approach cannot solve the crucial humankind's problems of today (Taylor, 2008). Thus, the system and cybernetics movement needs support from the world-top bodies; actually, humans need it, unless our children should be condemned to live on a dying planet Earth.

The alternative to systemic behavior is visible from the following message provided by Bernard Scott, president of the Socio-cybernetics on August 23, 2009 to Matjaz Mulej: extremely much more money is spent on military purposes than for well-being of people (see: http://www.informationisbeautiful.net/visualizations/the-billion-dollar-gram/). One-sidedness and short-term views prevail; hence humankind is in danger of extinction, if systemic behavior does not become normal.

REFERENCES[1]

Ackoff, R.(1999). *Ackoff's Best: His Classic Writings on Management*, Wiley, New York, ISBN 0-471-31634-2.

Ackoff, R. (2001). Interviewed by Diane Staffors: "Interaction among departments is crucial," Kansas City Star, July 30, 2001. Article received by M. Mulej by e-mail from John Donges, jdonges@seas.upenn.edu. ISBN/ISSN not available.

Ackoff, L. R. (2003). Iconoclastic management authority advocates a "systemic" approach to innovation. *Interview by Robert J. Allio* (sent to M. Mulej by e-mail, July 11, 2003, from Ackoff Center: acasa@seas.upenn.edu), ISBN/ISSN not available.

Ackoff, R. L. and Rovin, S. (2003). *Redesigning Society*, Stanford Business Books, Stanford, CA, ISBN 0-80-474794-6.

Afuah, A. (1998 and 2003). *Innovation Management – Strategies, Implementation, and Profits*, Oxford University Press, New York etc., ISBN 0-19-511346-2 and 0-19-514230-6.

Aguilera, R. V., Rupp, D., Williams, C. and Ganapathi, J. (2007). "Putting the S back in CSR: a multi-level theory of social change in organizations," *Academy of Management Review*, Vol. 32, No. 3, ISSN 03637425, pp. 836–863.

Albach, H. (ed.) (1990). *Innovationsmanagement. Theorie und Praxis im Kulturvergleich*, Gabler, Wiesbaden, ISBN 3-409-13369-0.

Anderson, D. R., Sweeney, D. J. and Williams, T. A. (2003). *An Introduction to Management Science. Quantitative Approaches to Decision Making*, Thomson South-Western, Mason, ISBN 0-324-14563-2.

Anderson, V. and Johnson L. (1997). *Systems Thinking Basics: From Concepts to Causal Loops*, Pegasus communication, Williston, ISBN 1-883823-12-9.

Bahm, A. (1988). *Polarity, Dialectic and Organicity.* World Books, Albuquerque, NM, USA, ISBN 0-911714-18-9.

Bai, G., and Lindberg, L.-A. (1998). "Dialectical Approach to Systems Development," *Systems Research and Behavioral Science,* 15, ISSN 1092–7026, pp. 47–54.

---

1. In references EPF stands for: University of Maribor, Faculty of Economics and Business, Maribor.

Bailey, K. D. (2005). "Fifty Years of Systems Science: Further Reflections," *Systems Research and Behavioral Science*, 22, ISSN 1092-7026, pp. 355–361.

Balle, M. (1994). *Managing with Systems Thinking. Making Dynamics Work for You in Business Decision Making*, McGraw-Hill Book Company, New York etc., ISBN 0-07-707951-5.

Banathy, B. (2000). *Guided Evolution of Society: A Systems View* (Contemporary Systems Thinking), Kluwer Academic and Plenum Publishers, New York, ISBN 0 306-46362-2.

Barabba, P. (2004). *Surviving Transformation*, Oxford University Press, Oxford, ISBN 0-19-517141-1.

Basadur, M. and Gelade, G. A. (2006). "The Role of Knowledge Management in the Innovation Process," *Creativity and Innovation Management*, 15, ISSN 0963-1690, pp. 45–62.

Bausch, K. C. (2001). "The Emerging Consensus," in *Social Systems Theory*, Kluwer Academic/Plenum Publishers, New York–Boston–Dordrecht–London–Moscow, ISBN 0-306-46539-6.

Bausch, K. and Christakis, A. (2002 or 2003). "Chapter Six: Technology to Liberate rather then Imprison Consciousness," distributed with no reference to the entire book at the 47th ISSS conference, 2003, referenced here.

Beauchamp, T. and Bowie, N. (2004). *Ethical Theory and Business*, Prentice Hall, New York, ISBN 0-13-111632-0.

Beck, U. (2003). "Kaj je globalizacija," *Delo*, October and November, ISSN 0350-7521.

Beer, S. (1975). *Platform for Change*, Wiley, Chichester, ISBN 0-47-106189-1.

Beer, S. (1979). *The Heart of Enterprise*, Wiley, Chichester, ISBN 0-471-27599-9.

Belton, V. and Stewart, T. J. (2002). *Multiple Criteria Decision Analysis: an Integrated Approach*, Kluwer Academic Publishers, Boston–Dordrecht–London, ISBN 0-79-237505-X.

Bertalanffy, L. v. (1950). "An outline of General System Theory," *The British Journal for the Philosophy of Science*, 1, ISSN 0007-0882, pp. 134–165.

Bertalanffy, L. V. (1976). *General System Theory*, George Braziller, New York, ISBN 0-8076-0453-4.

Bertalanffy, L. v. (1968, edition 1979). *General Systems Theory. Foundations, Development, Applications. Revised Edition. Sixth Printing*, George Braziller, New York, ISBN (1968) 0-8076-0453-4.

Bickel, P. and Friedrich, R. (2001). *Environmental External Costs of transport*, Springer Verlag, Berlin– Heidelberg–New York, ISBN 978-3-540-42223-5.

Bolwijn, P. T. and Kumpe, T. (1990). "Manufacturing in 1990's – Productivity, Flexibility and Innovation," *Long Range Planning*, 23, No. 4, ISSN 0024-6301.

Bošković, D. (2006). "Drugačen ekonomski zemljevid," *Delo, Sobotna priloga*, September 23, ISSN 0350-7521, p. 11.

Bošković, D. (2006a). "Zapiranje obljubljene dežele," *Delo, Sobotna priloga*, April 01, 2006, 9, ISSN 0350-7521.

Božičnik, S. (2007). *Dialektično sistemski model inoviranja krmiljenja sonaravnega razvoja cestnega prometa*. Univerza v Mariboru, Ekonomsko-poslovna fakulteta, Maribor, ISBN not available.

Božičnik, S., Ečimovič, T. and Mulej, M., with co-authors (2008). *Sustainable Future, Requisite Holism, and Social Responsibility (Against the current abuse of free market society)*, ANSTED University, School of Environmental Sciences, Penang, ISBN 978-961-91826-3-5.

Brandon, P. and Lombardi, P. (2005). *Evaluating Sustainable Development*, Blackwell, Oxford, ISBN 0-63-206486-2.

Breu, K. and Hemingway, C. (2005). "Research Practitioner Partnering in Industry-Funded Participatory Action Research," *Systemic Practice and Action Research*, 18, ISSN (Print) 1094-429X, pp. 437–455.

Britovšek, M., et al. (1960). *Vodnik skozi čas in družbo*, Mladinska knjiga, Ljubljana, ISBN not available.

Brown L. R. (2008). *Plan B 3.0; Mobilizing to Save Civilization*, Earth Policy Institute, W. W. Norton and Co., New York–London, ISBN 978-0-393-06589-3 (cloth).

Bučar, M. (2001). *Razvojno dohitevanje z informacijsko tehnologijo*, University of Ljubljana, Faculty of Social Sciences, Ljubljana, ISBN 961-235-061-2.

Bučar, M. and Stare, M. (2003). *Inovacijska politika male tranzicijske države*, University of Ljubljana, Faculty of Social Sciences, Ljubljana, ISBN 961-235-119-8.

Bučar, M. (2004). "Efficacy of National R&D and Innovation Policy: the Case of Slovenia," in Rebernik, M., Mulej, M. and Krošlin, T. (eds.), *7 STIQE*, EPF, Institut za podjetništvo in management malih podjetij,ISBN 961-6354-41-8, pp. 21–26.

Bučar, M. and Stare, M. (2006). *Innovation and Innovation Policy in Slovenia*. http://trendchart.cordis.lu/tc_country_list.cfm?ID=19

Bukovec, B. (2006). "Management človeških virov in obvladovanje organizacijskih sprememb," *Organizacija*, 39, ISSN 1318-5454, pp. 117–123.

Burch, T. K. (1999). *Computer Modeling of Theory: Explanation for the 21st Century*, University of Western Ontario, London, Ontario, Canada, ISBN 07714 -2180-X.

Calleman, C. J. (2000). *Solving the Greatest Mystery of Our time: The Mayan Calendar*, Garev Publishing, Basingstoke, UK, ISBN 978-0970755803.

Calleman, C. J. (2004). *The Mayan Calendar and the Transformation of Consciousness*, Bear and Company, Rochester, Vermont, ISBN 1-59143-028-3.

Canuto, M. (2007). *Does the Sun Always Rise in the Suneast? The dawn of civilization – Maya Weekend*, University of Pennsylvania Museum, ISBN/ISSN not available.

Capra, F. (1997). *The Web of Life A New Scientific Understanding of Living Systems*, Anchor, New York, ISBN 978-0385476768.

Capra, F. (2002). *The Hidden Connections. Integrating the Biological, Cognitive, and Social Dimensions of Life into a Science of Sustainability*, Doubleday, New York etc., ISBN 0-385-49471-8.

Checkland, P. (1981). *Systems Thinking, Systems Practice*, Wiley, Chichester, etc., UK, ISBN 0-47-127911-0.

Chen, X. and Lai, B. (2005). "An Object-oriented Knowledge Link Model for General Knowledge Management," in: Nakamori (ed.), referenced here, pp. 388–393.

Chesbrough, H. (2006). *Open innovation: the new imperative for creating and profiting from technology*, Harvard Business School Press, Boston, ISBN 1-4221-0283-1.

Chomsky, N. (1997). *Somrak demokracije*, Studia humanitatis, Ljubljana, ISBN 961-6262-00-9.

Christakis, A. and Bausch, K. (eds.) (2003). *47th Annual Conference of the International Society for the Systems Science: Agoras of the Global Village, Conscious Evolution of Humanity: Using Systems Thinking To Construct Agoras of the Global Village. Conference proceedings and abstracts*, ISSS, at Hersonissos, Crete, Greece, ISBN 0-9740735-1-2.

Christakis, A. and Bausch, K. C. (2006). *How People Harness their Collective Wisdom and Power to Construct the Future in Co-Laboratories of Democracy*, IAP Information Age Publications, Greenwich, CO, ISBN 1-59311-481-8.

CJE (2003). "Economics for the Future. Proceedings," *Cambridge Journal of Economics*, Cambridge, UK, ISSN (Print) 0309-166X.

Clusella, M. M., Ortiz, M. E. and Luna, P. A. (2005a). "Systemic Epistemology: A Synthetic View for the Systems Science Foundation," in Gu and Chroust (eds.), referenced here.

Clusella, M. M., Herrera, S. I., Tkachuk, G. N. and Luna, P. A. (2005b). "How Systems Model Contribute to design of Information Systems," in Gu and Chroust (eds.), referenced here.

Collins, J. (2001). *Why Some Companies Make the Leap ... and others don't. Good to Great*, Random House Business Books, Sidney etc., ISBN 0-7126-7609-0.

Collins, J. and Porras, J. (1994). *Built to Last. Successful Habits of Visionary Comnpanies*, HarperBusiness, New York, ISBN 0-88730-671-3.

Competition (2004, author not signed). "Lisbon strategy. Competition needed for more innovation," *Innovation & Technology Transfer*, 4/04, July 2004, ISSN 1013-6452, pp. 3–4.

Cook, P. (1998). *Best Practice Creativity*, Gower Publishing Limited, Hampshire, ISBN 0-566-08027-3.

Cooper, P. and Vargas, C. (2004): *Implementing Sustainable Development: From Global Policy to Local Action*. Rowman and Littlefield, Lanham. ISBN 0-74-252361-6.

Cottam, R., Ranson, W. and Vounckx, R. (2005). "Life and Simple Systems," *Systems Research and Behavioral Science*, September–October, ISSN 1092-7026, pp. 413–430.

Cornell, A. (2005). "Short-termism checks growth," *The Press, Christchurch* (New Zealand), P. B10, Executive, December 2, ISSN not available.

Crane, A. and Matten, D. (2004). *Business Ethics*, Oxford University Press, Oxford, ISBN 0-19-925515-6.

Creech, B. (1994). *The Five Pillars of TQM. How to Make Total Quality Work for You*, Truman Talley Books, Dutton, NY, ISBN 0-452-27102-9.

Crowther, D. and Caliyut, K. T. (eds.) (2004a). *Stakeholders and Social Responsibility*, ANSTED University, Penang, ISBN 983-41879-1-2.

Crowther, D., Barry, D. M., Sankar, A., Goh, K. G. and Ortiz Martinez, E. (2004b). *S. R. W. Social Responsibility World of RecordPedia 2004*, Ansted Service Center (Ansted University Asia Regional Service Center), Penang, ISBN 983-41879-0-4.

Crowther, D. and Ortiz Martinez, E. (2004). "Corporate Social Responsibility: History and Principles," in Crowther, *et al.* (2004b), pp. 102–107, referenced here.

Čančer, V. (2000). "Environmental Management of Business Processes," *Management*, 5, ISSN 1331-0194, pp. 83–97.

Čančer, V. (2003). *Analiza odločanja*, EPF, ISBN 961-6354-29-9.

Čančer, V. (2004). "The Multicriteria Method for Environmentally Oriented Business Decision-Making," *Yugoslav Journal of Operations Research*, 14, ISSN 0354-0243, pp. 65–82.

Čančer, V. and Knez-Riedl, J. (2005). "Why and How to Evaluate the Creditworthiness of SMEs' Business Partners," *International Small Business Journal*, 23, ISSN (print) 0266-2426, pp. 141–158.

Čančer, V. and Mulej, M. (2005). "The roles of creative thinking and decision-making tools for building knowledge societies," in Gu and Chroust, referenced here.

Čančer, V. and Mulej, M. (2005a). "Systemic decision analysis approaches as requisite tools for developing creative ideas into innovations," in: Mulej, *et al.* (eds.), referenced here.

Čančer, V. and Mulej, M. (2006). "Systemic decision analysis approaches: requisite tools for developing creative ideas into innovations," *Kybernetes*, 35, ISSN 0368-492X, pp. 1059–1070.

Čelofiga, S. (2008). *Dejavniki inoviranja v podjetju (Factors of innovating in company)*, University of Maribor, Faculty of Economics and Business, Maribor, ISBN not available.

Daft, R. (2001). *Organization Theory and Design*, South-Western College, Cincinnati, ISBN 0-324-02100-3.

Daft, R. (2003). *Management*, Tomson, Mason, ISBN 0-03-035138-3.

David, F. A. and Richardson, G. P. (1997). "Scripts for group model building," *System Dynamics Review*, Vol. 13, No. 2, ISSN 0883-7066, pp. 107–129.

Davidson, M. (1983). *Uncommon Sense. The Life and Thought of Ludwig von Bertalanffy, Father of General Systems Theory*, J. P. Tarcher, Inc., Los Angeles, CA, USA, ISBN 0-87477-165-X.

De Bono, E. (2003). *Creative Thinking*, Talk to the 3rd New Moment Ideas Campus, Piran, August 2003, New Moment Ideas Company, Ljubljana (taped by N. Mulej and her team), ISBN/ISSN not available.

De Bono, E. (2005). "Šest klobukov razmišljanja," *New Moment, 28* (entire journal), ISSN 1580-1322.

De Bono, E. (2006). "Lateralno razmišljanje (Lateral Thinking. Slovenian edition)," *New Moment*, 30 (entire journal), ISSN 1580-1322.

Dees, G. and Emerson, J. (2002): *Strategic Tools for Social Entrepreneurs.* John Wiley and Sons, New York. ISBN 0471150681.

Delavnica (2006). See: Fatur, P. (2006), refenced here.

Delgado, R. R. and Banathy, B. H. (1993): *International Systems Science Handbook*, Systemic Publications, Madrid, Spain, ISBN 84-604-6236-6.

Dent, C. B. (1999). "Complexity, the New World View," *Emergence*, 1, ISSN 1521-3250, pp. 5–20.

De Zeeuw, G. (1998). "Improving on differences among viewpoints," in Boog, B., Coenen, H, Keune, L. and Lammers, R. (eds.), *The Complexity of Relationships in Action Research*, Tilburg University Press, Tilburg, ISBN 9036198682, pp. 153–175.

Dodič Fikfak, M. (2009). "Prestrukturiranje goispodarstva in kazalci zdravja," in Hrast, A. and Mulej, M. (eds.): 4. mednarodna konferenca: *Družbena odgovornost in izivi časa 2009 z naslovom: Delo – most za sodelovanje: odnosi do zaposlenih in različnih stsarostnih generacij. Zbornik prispevkov. 4th international conference Social responsibility and current challenges 2009 "Work – abridge to cooperationb: Relations with co-workers and different age generations. Conference proceedings*, IRDO Inštitut za razvoj družbene odgovornosti, Maribor.

Drack, M. and Apfalter, W. (2007). "Is Paul Weiss' und Ludwig von Bertalanffy's Systems Thinking Still Valid Today?," *Systems Research and Behavioral Science*, 24, ISSN 1092-7026, pp. 537–546.

Drack, M. (2008). "Bertalanffy's Early Systems Approach," in Metcalf (2008), referenced here.

Dremelj, M. and Ogorelc Wagner, V. (2006). "Pravična trgovina – zgodovina, načela in trendi," in Hrast, A., Mulej, M. and Knez-Riedl, J. (eds.) (2006), referenced here.

Drucker, P. (1985). *Innovation and Entrepreneurship*, Harper Trade, New York, ISBN 0-06-015428-4.

Drucker, P. (1987). *The Frontiers of Management. Where Tomorrow's Decisions are being Shaped Today.* Heinemann, London, ISBN 0-434-90394-9.

Drucker, P. (1989). *The new realities: in government and politics, in economics and business, in society and world view*, Harper & Row Publishers, New York, ISBN 0-06-016129-9.

Dyck, R., Mulej, M. and co-authors (1998). *Self-Transformation of the Forgotten Four-Fifth*, Kendall/Hunt, Dubuque, Iowa, ISBN 0787244996.

Ečimovič, T., Mulej, M., Mayur, M. and 30 co-authors (2002). *System Thinking and Climate Change System*, Korte, SEM Institute for Climate Change, www.institut-climatechange.si, ISBN 961-236-380-3.

Ečimovič, T., Esposito, M., Mulej, M. and Haw, R. (2008). "The individual and corporate social responsibility," in Hrast, A. and Mulej, M. (eds.), referenced here.

Ečimovič, T. (ed.) (2008). *Proceedings of the 20th WACRA Conference*, www.institut-climatechange.si, ISBN not available.

Ečimovič, T., Esposito, M., Flint, W., Haw, R. B., Mulej, M., Shankaranarayana, M. A., Wilderer, P. A. and Williams, L. (2007). *Sustainable (Development) Future of Mankind*, Korte: SEM Institute for Climate Change, www.institut-climatechange.si, ISBN 978-961-91826-2-8.

Edwards, A. and Orr, D. (2005). *The Sustainability Revolution: Portrait of a Paradigm Shift*, New Society Publishers, Gabriola Island, ISBN 0865715319.

Edwards, J. S. (2005). Knowledge Management Systems and Business Processes. *ISKSS – International Journal of Knowledge and Systems Sciences*, 2, ISSN 1349-7030: 9–18.

Elohim, J. L. (1999). *A Message from professor Elohim*, Poster exposed during the 11th WOSC Conference, Uxbridge, ISBN/ISSN not available.

Elohim, J. L., Hofkirchner, W., *et al.* (2001). *Unity through Diversity. Conference on the Occasion of 100 years anniversary of Ludwig von Bertalanffy*, Technical University of Vienna, ISBN not available.

Embley, L. L. (1992). *Doing well while doing good*, Prentice Hall, Englewood Cliffs, New Jersey, NJ, ISBN 0-13-219874-6.

EMCSR (2002). Trappl, R. (ed.) (1972–, biannually): *European Meeting on Cybernetics and Systems Research. Proceedings*, different publishers, ISBN 3-85206-160-1.

Enciklopedija (1959). Etika. Data in *Enciklopedija leksikografskog zavoda*, Jugoslovenski leksikografski zavod, Zagreb, pp. 617–618. ISBN not available.

Enciklopedija (1964). Slovenci, *Enciklopedija leksikografskog zavoda,* Zagreb, pp. 58–74, Jugoslovenski leksikografski zavod, ISBN not available.

Engels, F. (1953). *Dialektika prirode*, Cankarjeva založba, Ljubljana, ISBN not available.

Eriksson, D. M. (2003). "Identification of Normative Sources for Systems Thinking: An Inquiry into Religious Ground-Motives for Systems Thinking Paradigms," *Systems Research & Behavioral Science, 20*, ISSN: 1092-7026, pp. 475–488.

Espejo, R. and Harnden, R. (eds.) (1989). *The Viable System Model: Interpretations and Applications of Stafford Beer's VSM*, Wiley, Chichester, ISBN 0-471-92288-9.

Esposito, M. (2008). *Corporate Social Responsibility – Overview and bridging*. Presentation at IRDO 2008, 3rd International Conference, "Social Responsibility and Current Challenges 2008 – Social responsibility as contribution to stakeholders' long term success in market", June 5–6, Maribor, Slovenia, ISBN 978-961-91826-4-2.

Esposito, M. and Henderson J. (2008). *CSR – Corporate social responsibility primer*, presentation at the Vienna Symposium 2008, URL: http://blog.swissmc.ch/smcsc/files/2008/09/csr-primer.pdf, May 5, 2009.

European Commission (2009a). "Overview of the links between corporate social responsibility and competitiveness," *European competitiveness report 2008*, Office for Official Publications of the European Communities, Luxembourg, ISSN 1682-0800, pp. 106–121.

European Commission (2009b). *Corporate social responsibility*, a definition by the European Commission's Directorate-General for Enterprise and Industry, URL: http://ec.europa.eu/enterprise/csr/index_en.htm, May 5, 2009.

Expert Choice, Inc., *Expert Choice*, available: www.expertchoice.com, consulted April 2005, ISBN/ISSN not available.

EU (1995). *Green Paper on Innovation*. European Commissions, European Union, http://europa.eu/documents/comm/green_papers/pdf/com95_688_en.pdf, ISBN/ISSN not available.

EU (1996). *Living in an Information Age. People First*, European Commissions, European Union, ISBN/ISSN not available.

EU (2000). *Communication from the Commission to the Council and the European Parliament. Innovation in a knowledge-driven economy*, Commission of the European Communities, Brussels, xxx, COM(2000) 567 final, http://www.euractiv.com/en/future-eu/lisbon-agenda/article-117510, ISBN/ISSN not available.

EU (2001). Green Paper on Promoting a European Framework for Corporate Social Responsibility, European Commissions, European Union, ISBN/ISSN not available.

EU (2004). *Innovation Management and the Knowledge-Driven Economy*, European Commission, Brussels, ftp://ftp.cordis.lu/pub/innovation-policy/studies/studies_innovation_management_final_report.pdf, ISBN/ISSN not available.

EU (2005, and earlier). *Sustainable Development*, EU, http://ec.europa.eu/sustainable/.

EU (2006a). *Commission of the European Communities: Implementing the partnership for growth and jobs: Making Europe a pole of excellence on corporate social responsibility*, Com (2006), European Commissions, European Union, ISBN/ISSN not available.

EU (2006b). CSR Europe (2006). *A European Roadmap for Business Towards sustainable and competitive enterprise*, European Commissions, European Union, ISBN/ISSN not available.

Fagerberg, J., Mowery, D. C. and Nelson, R. R. (eds.) (2005). *The Oxford Handbook of Innovation*, Oxford University Press, Oxford, ISBN 0-19-926455-4.

Fatur, P. (2006). "Analiza invencijsko-inovacijske dejavnosti v slovenskih podjetjih," in Delavnica (2006): *Zakaj in kako pospeševati inovacijsko dejavnost v podjetju*, Gornja Radgona, April 06, 2006 (Three topics: Fatur, P.: Analysis of invention-innovation management in Slovenian enterprises; Baebler, M.: Model of monitoring and rewarding innovations in enterprises; Bokan, T.: Innovation in ELTI Gornja Radgona. Handouts with power-point presentations.), Gospodarska zbornica Slovenije, Ljubljana, ISBN/ISSN not available.

Ferrell, O., Fraedrich, J. and Ferrell, L. (2004). *Business Ethics*, Houghton Mifflin Company, Boston, ISBN 0-618-39573-3.

Feucht, H. (1996). *Implementierung von Technologiestrategien*, Peter Lang, Frankfurt am Main, Germany, ISBN 3631497873.

Flis, V. (2006). "Informed Consent to Diagnostic Procedures and medical Treatment," in Hrast, A., Mulej, M. and Knez-Riedl, J. (eds.), referenced here.

Flood, R. L. (1999). *Rethinking the Fifth Discipline. Learning with the unknowable*, Routledge, London, UK, and New York, NY, USA, ISBN 0415185297.

Florida, R. (2005). *Vzpon ustvarjalnega razreda*, IPAK, Velenje (Translation into Slovenian), ISBN 961-91632-0-6.

Forrester, J. W. (1961). *Industrial Dynamics*, Pegasus Communication, Williston, ISBN 978-1883823368.

François, Ch. (1992). *Diccionario de Teoria General de Systems e Cibernetica. Conceptos y Terminos*, editado por GESI Association Argentina de Teoria General de Systemas y Cibernetica, Buenos Aires, Argentina, ISBN 987-9901.

François, Ch. (ed.) (2004). *International Encyclopedia of Systems and Cybernetics*, 2nd Ed. Saur, Munich, ISBN 3-598-11630-6.

Frandberg T. (2005). "Living Systems and its Philosophy Considered at the Level of the Earth," *Systems Research and Behavioral Science*, September–October, ISSN 1092-7026, pp. 373–383.

Frank, H. (1962). *Kybernetik. Brücke zwischen Wissenschaften*, Umschau Verlag, Frankfurt am Main, Germany (6 editions until 1966), ISBN not available.

Frascati Manual (1971, version 2002). OECD, Paris, ISBN 92-64-19903-9.

Freeman, R., Pierce, J. and Dodd, R. (2005). "Shades of Green: Business, Ethics and the Environment," in Gini, A., *Case Studies in Business Ethics*, Prentice Hall, Chicago, ISBN 0131127462.

Fromm, E. (1994). *Vom Haben zum Sein: Wege und Irrwege der Selbsterfahrung*, Wilhelm Heine Verlag, Munich, Germany, ISBN 9783886792405.

Garriga, E. and Mele, D. (2004). "Corporate social responsibility theories: mapping the territory," *Journal of Business Ethics*, Vol. 53, ISSN 1573-0697, pp. 51–71.

Gerber, M. E. (2004). *Mit o podjetniku. Zakaj večina podjetij ne uspe in kako to spremeniti*, Lisac & Lisac and Gea College, Ljubljana, ISBN 961-6312-51-0.

Germ Galič, B. (2003). *Dialektični sistem kazalnikov inoviranja in kakovosti poslovanja*, EPF, ISSN not available.

Geyer, F., Hornung, B., et al. (eds.) (2003). *The Fourth International Conference on Sociocybernetics: Sociocybernetics – the Future of the Social Sciences, Society from Ancient Greece to Cyberspace and Beyond. Abstracts and Program*, ISA, RC 51, held At Kerkyra, Corfu, June 30 – July 5, 2003, ISBN/ISSN not available.

Gider, F. (2004). "Sistem 20 ključev – izkušnje z uporabo celovitega sistema za povečanje konkurenčnosti podjetij v Sloveniji," *Organizacija* 37, ISSN 1318-5454, pp. 379–384.

Gigch, J. P. v. (2003). "The Paradigm and the Science of Management and of the Management Science Disciplines," *Systems Research & Behavioral Science*, 20, ISSN 1092-7026, pp. 499–506.

Glor, E. D. (1998). "What Do We Know About Enhancing Creativity and Innovation?," *The Innovation Journal – The Public Sector Innovation Journal*, 3, No. 1, ISSN 1715-3816.

Gomez, P. and Probst, G. (1987). *Die Orientierung (Nr. 89) – Vernetztes Denken im Management*, Schweizerische Volksbank, Bern, ISBN not available.

Gomez, P. and Probst, G. (1997). *Die Praxis des ganzheitlichen Problemlösens*, 2. überarb. Aufl., Verlag Paul Haupt, Bern–Stuttgart–Wien, ISBN 3-258-05575-0.

Goerner, J. (2004). „An Introduction," *World Future*, 60, ISSN 1946-7567 pp. 273–286.

Goerner, S., Dyck, R. G. and Lagerroos, D. (2008). *The New Science of Sustainability. Building a Foundation for Great Change*, Triangle Center for Complex Systems, Chapel Hill, N.C., ISBN 978-0-9798683-1-3.

Gorenak, Š. (2008). "Popolnoma odgovorno upravljanje kot element dolgoročnih konkurenčnih prednosti," in Hrast, Mulej (eds.), referenced here.

Graham, B., Jr. (2006). *Paper Work Simplification*, Ben Graham Inc. Dayton, OH, ISBN not available.

Grayson, J. and O'Dell, C. (1988). *American Business: A Two Minute Warning. Ten Lessons American Managers Must Learn to Survive into the 21st Century*, Free Press, New York, ISBN 0029126800.

Greer, B. (2000). *Ethics and Uncertainty: The Economics of John M. Keynes and Frank H. Knight*, Elger, Cheltham, ISBN 1-84064-445-1.

Gregory, W. and Richardson, K. (eds.) (2005). *Systems Thinking and Complexity Science*, 11th Annual ANZSYS/Managing the Complex V Conference, Christchurch, NZ, ISBN 0976681447.

Gregory, W. (ed.) (2006). *Social Sustainability*, a Stream in the 12 ANZSYS Conference '06, Blue Mountains, AUS, ISBN/ISSN not available.

Gregory, W., Midgley, G. and guest editors (2003). "Systems Thinking for Social Responsibility," *Systems Research and Behavioral Science*, 20, ISSN 1092-7026, pp. 103–216.

Grun, E., *et al*. (eds.) (2006). *Contribuciones a la primera reunion regional de ALAS*, 7 al 9 de Agosto, 2006, Buenos Aires, Argentina, ISBN/ISSN not available.

Gu, J. and Chroust, G. (eds.) (2005). *IFSR 2005: The New Roles of Systems Sciences for a Knowledge-based Society*, The First World Congress of the International Federation for Systems Research, Kobe (on CD), ISBN 4-903092-02-X.

Gu, J. and Tang, X. (2005). "Metasynthesis and Knowledge Creation," in Nakamori (ed.), referenced here, pp. 118–122.

Gu, J.a, Nakamori, Y., Wang, Zh. and Tang, X. (eds.) (2006). *Towards knowledge synthesis and creation. Proceedings of the Seventh International Symposium on Knowledge and Systems Sciences, Beijing, China, September 22–25, 2006* (Lecture Notes in *Decision Sciences*, 8), Global-Link Publisher, Hong Kong etc., ISBN/ISSN not available.

Guernsey, J. (2006). *Ritual and Power in Stone: The Performance of Rulership in Mesoamerican Izapan Style Art*, University of Texas, ISBN 978-0292713239.

Hammond, D. (2003). *The Science of Synthesis. Exploring the Social Implications of General Systems Theory*, University Press of Colorado, Boulder, CO, ISBN 0-87081-722-1.

Handbook University Enterprise Cooperation. Tempus-Top Project, www.esmu.be, downloaded on August 23, 2004, European Centre for Strategic Management of Universities (ESMU), ISBN 92-9157-050-8.

Hanley, S.W., *et al*. (2006). *Environmental Economics in Theory and Practice*, Oxford University Press, Oxford, ISBN 033397137X.

Harris E. E. (2008). *Twenty-first Century: Democratic Renaissance, From Plato to Neoliberalism to Planetary Democracy*, The Institute for Economic Democracy Press, Sun City, AZ., and Fayetteville, PA, in cooperation with Institute on World Problems, and Earth Rights Institute, ISBN 978-1-933567-15-0.

Hartley, R. (2004). *Business Ethics*, John Wiley and Sons, New York, ISBN 978-0-471-66373-7.

Hays, P. and Wasilewski, J. (2005). "Global *Agoras* for Creating a Knowledge Society: The Example of the Center of Excellence (COE) Boundary-spanning Dialogue Approach (BDA) project at International Christian University (ICU)," in Nakamori (ed.), referenced here, pp. 73–80.

Hawkins, B. (2000). *How to Generate Great Ideas*, Kogan Page, London, ISBN 0749427612.

Helsinki University of Technology, *Web-HIPRE Help*, available: http://www.hipre.hut.fi, consulted April 2005, ISBN/ISSN not available.

Hilton, B. (2008). *The Global Way. The Integral Economics of the Post-Modern World*, Trafford Publ., Canada, ISBN 1-4251-6232-0.

Hindle, K. (2004). "Choosing Qualitative Methods for Entrepreneurial Cognition Research: A Canonical Development Approach," *Entrepreneurship Theory and Practice*, 29, ISSN 1042-2587, pp. 575–607.

Hofkirchner, W. (2001). See Elohim and Hofkirchner, referenced here.

Hofkirchner, W. (2005). "Ludwig von Bertalanffy Forerunner of Evolutionary Systems Theory," in Gu and Chroust (eds.), referenced here.

Horbny, A. S., Gatenby, E. V. and Wakefield, H. (1963). *The Advanced Learner's Dictionary of Current English, second edition*, Oxford University Press, London, ISBN not available.

Hornung, B. R. (2005). "Steering and Control of Innovations in Complex Societies," in: Mulej, *et al.* (eds.) (2005), referenced here.

Hrast, A. and Mulej, M. (2008). "Affluence – cause for dangers for sustainable future – to be solved by social responsibility," in Ečimovič, *et al.*, referenced here.

Hrast, A., Mulej, M. and Knez-Riedl, J. (eds.) (2006). *Družbena odgovornost in izzivi časa 2006*, IRDO Inštitut za razvoj družbene odgovornosti, Maribor, ISBN 961-91828-0-4.

Hrast, A., Mulej, M. and Knez-Riedl, J. (eds.) (2007). *Družbena odgovornost 2007*, Maribor, IRDO Inštitut za razvoj družbene odgovornosti, ISBN 978-961-91828-1-9.

Hrast, A. and Zavašnik, A. (eds.) (2007). *Uvajanje družbene odgovornosti v poslovno prakso malih in srednje velikih podjetij v Sloveniji: Priročnik s primeri dobre prakse*, GZS – Območna zbornica Maribor, Maribor, ISBN 978-961-6666-07-7.

Hrast, A. and Mulej, M. (eds.) (2008). *Družbena odgovornost 2008. Zbornik 3. IRDO Konference o družbeni odgovornosti*, Maribor, IRDO Inštitut za razvoj družbene odgovornosti, ISBN 978-961-91828-2-6.

Huston, L. and Sakkab, N. (2006). "Connect and Develop. Inside Procter & Gamble New Model for Innovation," *Harvard Business Review*, March 2006, ISSN 00178012, pp. 1–9.

Hwang, S. W. (1997). "The Implication of the Nonlinear Paradigm for Integrated Environmental Design and Planning," *Journal of Planning Literature,* 11, ISSN 0885-4122, pp. 167–180.

IBM (2006). *Global Innovation Outlook 2.0*, IBM, Armonk, NY, ISBN/ISSN not available.

IDIMT: Chroust, G., Doucek, P. and Hofer, Ch. (eds) *IDIMT, Proceedings of the conference Interdisciplinary Information Management Talks*, Universitaetsverlag Trauner, Linz, (yearly since 1993), ISBN 978-3-85499-448-0.

Ilich, I. (ed.) (2004). *Pregovori in reki. Proverbs and Sayings*, DZS, Ljubljana, ISBN 86-341-3639-6.

ISSS45 (2001). *Systems Science in Service to Humanity. 45th International Conference*, International Society for the Systems Sciences, Harold G. Nelson, President, edited by: Kiara Sage Raven, Sophea Hieam. Asilomare, CA, ISBN/ISSN not available.

ISSS (2003). *47th Annual Conference of the International Society for the Systems Sciences: Agoras of the Global Village. Conscious Evolution of Humanity: Using Systems Thinking To Construct Agoras of the Global Village*, A. Christakis, president and co-chair, K. Bausch, co-chair, Hersonissos, Crete, ISBN 0974073504.

Jackson, M. (1991). *Systems Methodology for the Management Sciences*, Plenum Press, New York, NY, USA, ISBN 0-306-43877-1.

Jackson, M. (2003). *Systems Thinking. Creative Holism for Managers*, Wiley, Chichester, ISBN 0-470-84522-8.

James, O. (2007). *Affluenza – a contagious middle class virus causing depression, anxiety, addiction and ennui*, Vermillion, an imprint of Ebury Publishing, Random House Ltd., UK, etc., ISBN 9780091900113.

Jan, J., *et al.* (1990). *Inovacijska praksa* (Innovation Practice. In Slovene), Delavska enotnost, Ljubljana, ISBN/ISSN not available.

JAPTI (2005). *Korak višje! Program razvoja podjetnosti in ustvarjalnosti mladih*, JAPTI, Javna agencija za podjetništvo in tuje investicije, Ljubljana, ISBN/ISSN not available.

Jennings, M. (2005). *Business: Its Legal, Ethical and Global Environment*, South-Western Pub, Brentford, ISBN 0324204884.

Jensen, R. (2003). "Burjenje srca. Sanjska družbe II, " *New Moment*, 22 (all journal), ISSN 1580-1322.

Jere Lazanski T. (2008). "Systems thinking and Complex Systems Modeling," *Academica Turistica I*, no. 3–4, December 2008, ISSN 1855-3303, pp. 79–83.

Jež, B. (2006). "Telefonski nagobčniki," *Delo*, 26, ISSN 0350-7521, p. 5.

Jones, J., Bosch, O., Drack, M., Horiuchi, Y. and Ramage, M. (2009). "On the Design of System-Oriented University Curricula," *Research Report (Humanity Science)*, 43, 1, March, pp. 121–130, Shibaura University of Technology, ISSS not available

Journal IJKSS. *International Journal of Knowledge and Systems Science*, ISSN 1349-7030.

Jurše, K. (1994) *Popolna kakovost kot pogoj družbene in poslovne eksistence v razmerah inovativne družbe*. Ljubljana, University of Ljubljana, Faculty of Economics. ISBN not available.

Jurše, K. (1994). "Total quality government," in Rebernik, M. and Mulej, M. (eds.). *STIQE '94: proceedings of the 2nd International Conference on Linking Systems Thinking, Innovation, Quality and Entrepreneurship*, EPF. ISBN 86-80085-54-5.

Jurše, K. (1995). "Challenges of the systems approach to leadership," *Systemica*, 11, ISSN 1813-4769, pp. 1–6.

Jurše, K. (2003). *Is leadership the appropriate management concept to enhance corporate sustainability?*, Corporate sustainability: conference proceedings, Graz, IFF/IFZ, ISBN/ISSN not available.

Kajfež Bogataj, L. (2009). "Zgodovina je polna poznih lekcij iz zgodnjih svaril," *Le Monde diplomatique in Slovene*, 2009 October, p. 30.

Kajzer, Š. (1982). *Istraživanje strukture i kompozicije mikro-ekonomskih sistema unudruženom radu*, Sveučilište u Zagrebu, Zagreb, ISBN not available.

Kajzer, S. and Mulej, M. (1997). „Systemtheoretisch fundierte Ethik als ueberlebungskonzept in turbulenten Zeiten der innovativen Wirtschaft und Gesellschaft," in Schwaninger, M. (ed.) (1999): *Ueberlebungskonzepte fuer turbulente Zeiten auf der Grundlage von Systemtheorie und Kybernetik*, Jahrestagung 1997 der GSW e.v., Humblot, Berlin, Germany, ISBN not available.

Kameoka, A. and Wierzbicki, A. (2005). In Gu and Chroust (eds.), referenced here.

Kawalek, J. P. (2004). "Systems Thinking and Knowledge Management: Positional Assertions and preliminary Observations," *Systems Research and Behavioral Science*, 21, ISSN: 1092-7026, pp. 17–36.

Keane, John (2000). *Civilna družba: stare podobe, nova videnja*, Znanstveno in publicistično središče, Ljubljana, ISBN 961-6294-20-2.

Kejžar, I., et al. (1995). *Modra knjiga. Plače v Sloveniji*, Moderna organizacija, Kranj, ISBN 86-81049-86-0.

Kekes, J. (1988). "Self-Directions: The Core of Ethical Individualism," in K. Kolenda, *Organisations and Ethical Individualism*, Praeger, New York, ISBN 0275927601.

Kiel, L. D. and Elliot, E. (1997). *Chaos Theory in the Social Sciences. Foundations and Applications*, The University of Michigan Press, Ann Arbor, MI, USA, ISBN 0-472-10638-4.

Kieselbach, T. (2009). "Health in restructuring: Empirical evidence and policy recommendation," in Hrast, A. and Mulej, M. (eds.). 4. mednarodna konferenca: *Družbena odgovornost in izivi časa 2009 z naslovom: Delo – most za sodelovanje: odnosi do zaposlenih in različnih stsarostnih generacij. Zbornik prispevkov. 4th international conference Social responsibility and current challenges 2009 Work – abridge to cooperationb: Relations with co-workers and different age generations. Conference proceedings*, IRDO Inštitut za razvoj družbene odgovornosti, Maribor.

Kline, M. (2000). "Creative Thinking," in Sakan, D. and Mulej, N. (eds.), New Moment Ideas Campus, held in Piran, *New Moment*, 14 (entire journal edition), ISSN 1580-1322.

Kljajić, M. (2008). "Significance of simulation and systems approach methodology in development of complex systems," in: Symposium on Engineering and Management of IT-based organizational systems, Baden-Baden, Germany, 2008. *Engineering and management of IT-based organizational systems: a system approach*. Tecumseh, Ontario: The international institute for advances studies in system research and cybernetics, 2008, ISBN 978-1-897233-48-1, pp. 34–38.

Knez-Riedl, J. (1998). *Objektivizacija presoje bonitete podjetja s pomocjo panoznih dejavnikov*, EPF, ISBN not available.

Knez-Riedl, J. and Rebernik, M. (1998). "Linking Economic and Environmental Concerns by Systems Thinking," in Hofer, S. and Beneder, M. (eds.). *IDIMT '98: 6th Interdisciplinary Information Management Talks*, Universitaetsverlag Rudolf Trauner, Linz, Austria, ISBN 3-85320-955-6.

Knez-Riedl, J. (2000). *Pojmovanje in presojanje bonitete podjetja*, Zbirka Srebrna knjiga, 15. Zveza računovodij, finančnikov in revizorjev Slovenije, Ljubljana, ISBN 86-7591-069-X.

Knez-Riedl, J. (2000a). "Individualnost in sodelovanje," *Naše Gospodarstvo*, 46, ISSN 0547-3101, pp. 126–133.

Knez-Riedl, J., Mulej, M. and Ženko, Z. (2001). "Approaching sustainable enterprise," in Lasker, G. E. and Hiwaki, K. (eds.). *Sustainable development and global community*, International Institute for Advanced Studies in Systems Research and Cybernetics, ISBN 1-894613-11-2.

Knez-Riedl, J. (2003). "Kakovost, inovativnost in boniteta podjetja – prešibko upoštevanje kakovosti poslovanja in inovativnosti pri presojanju bonitete podjetja," *Organizacija*, 36, ISSN 1318-5454, pp. 620–627.

Knez-Riedl, J. (2003a): "Corporate social responsibility and communication with external community (Korporacijska društvena odgovornost i komuniciranje sa vanjskim okruženjem)," *Informatologia*, 36, ISSN 1330-0067, pp. 166–172.

Knez-Riedl, J. (2003b). "Social responsibility of a family business," *Revija za management in razvoj*, 5, ISSN 1408-9343, pp. 90–99.

Knez-Riedl, J. (2003c). "Corporate social responsibility and holistic analysis," in Chroust, G. and Hofer, Ch. (eds.). *IDIMT-2003: proceedings*, (Schriftenreihe Informatik, Bd 9), Universitaetsverlag R. Trauner, Linz, pp. 187–198. ISBN 3-85487-493-6.

Knez-Riedl, J. and Mulej, M. (2001). "Developing a Sustainable/Holistic Firm," in Ečimovič, T. (ed.): *18th International Conference of WACRA Europe, Vienna/Krems, Austria: Sustainable development through research and learning: the book of abstracts*, Komenda: SEM Institute for Climate Change, ISBN not available.

Knez-Riedl, J. (2002). "Družbena odgovornost malih in srednje velikih podjetij," in Rebernik, M. et al., *Slovenski podjetniški observatorij*, 2. part, ISSN 1854-8040, pp. 91–112.

Knez-Riedl, J. (2004). "Slovenian SMEs: from the environmental responsibility to corporate social responsibility," in Sharma, S. K. (ed.), *An enterprise odyssey: building competitive advantage* (Zagreb International Review of Economics & Business), pp. 127–139, ISBN 953-6025-10-8.

Knez-Riedl, J. and Hrast, A. (2005). "Innovation in the context of the corporate social responsibility (CSR)," in Bulz, N., Stoica, M, Mulej, M., Grigorescu, A., Dyck, R. G., Likar, B., Trček, D., Si-Feng, L, Medvedeva, T. A., Potočan, V., Vallée, R., Jiménez-

López, E., Lebe, S. S. and Schwaninger, M. (eds.) (2005). *Proceedings of The WOSC 13th International Congress of Cybernetics and Systems, July 6–10, 2005, Maribor, Slovenia*, Maribor: Faculty of Economics and Business, 6, pp. 45–54. ISBN 961-6354-57-4.

Knez-Riedl, J. (2006). "Družbena odgovornost in univerza," in Hrast, A., *et al.*, referenced here.

Knez-Riedl, J. and Hrast, A. (2006). "Managing corporate social responsibility (CSR): a case of multiple benefits of socially responsible behaviour of a firm," in Trappl, R. (ed.), *Cybernetics and systems 2006: proceedings of the Eighteenth European Meeting on Cybernetics and Systems Research*, Austrian Society for Cybernetic Studies, Vienna, pp. 405–409, ISBN 3-85206-172-5.

Knez-Riedl, J., Mulej, M. and Dyck, R. G. (2006). "Corporate Social Responsibility from the Viewpoint of Systems Thinking," *Kybernetes*, 35, ISSN 0368-492X, pp. 441–460.

Knez Riedl, J. (2007a). "Kako DOP povečuje konkurenčnost," in *Projekt CSR – Code to Smart Reality*, GZS-OZ Maribor, ISBN/ISSN not available.

Knez-Riedl, J. (2007b). "Družbena odgovornost podjetja in evropski strateški dokumenti," in *Projekt CSR – Code to Smart Reality*, Maribor, GZS OZ Maribor, ISBN/ISSN not available.

Knez-Riedl, J. (2007c). "Obvladovanje celovite (družbene) odgovornosti," *Razgledi MBA*, 12 [i.e. 13], 1/2, ISSN 1408-1660, pp. 37–43.

Kobal, E. (2003). *Strast po znanju in spoznavanju. Pogovori z velikimi slovenskimi znanstvenicami in znanstveniki*, Ustanova Slovenska znanstvena fundacija, Ljubljana, ISBN 961-6381-02-4.

Koch, R. (1998). *The 80/20 Principle*, Currency, New York, ISBN 1-85788-167-2.

Kohlberg, L. (1976). *Moral Stages and Moralization*, McGraw Hill, New York, ISBN not available.

Koletnik, F. (1998). *Poklicna etika ocenjevalcev vrednosti,* Slovenski inštitut za revizijo, Ljubljana, ISBN not available.

Korade Purg, Š. (2006). „Načelo delovanja javnih uslužbencev," in Hrast, A., Mulej, M. and Knez-Riedl, J. (eds.), referenced here.

Kordes, U. (2004). *Od resnice do zaupanja,* Studia Humanitatis, Ljubljana, ISBN: 961-6262-58-0.

Korn, J. (2003). "Letter to the Editor," *Systems Research and Behavioral Science,* 20, ISSN 1092-7026, pp. 533–536.

Korten D. S. (2009). *Agenda for a New Economy; From Phantom Wealth to Real Wealth*, Berrett-Koehler Publ., Inc., San Francisco, CA, ISBN 978-1-60509-289-8 (pbk), 978-1-60509-290-4 (PDF e-book).

Kos, A. (2004). *Možnost sodelovalnega vodenja v organih državne uprave*, EPF, ISBN not available

Krošlin, T. (2004). *Vpliv dejavnikov invencijsko-inovacijskega potenciala na uspešnost podjetja*, EPF, ISBN not available.

Kukoleča, S, (1969). „Sistemi i nauka o sistemima," *Revija Sistem, oddelek O,* ISSN 1318-9077.

Kurent, V. (2006). „Primeri lokalnih partnerstev (podjetja, občine, društva, šole)," in Hrast, A., Mulej, M. and Knez-Riedl, J. (eds.), referenced here.

Laurent, J. (2003). *Evolutionary Economics and Human Nature,* Edward Elgar Publishing, Cheltenham, ISBN 1-84064-923-2.

Leder, B. (2004). *Inoviranje trženja turizma na slovenskem podeželju,* Maribor, EPF, ISBN not available.

Lee, R. (2005). "The Contemporary Reordering of Social Knowledge: The Role of Systems Approaches and Complexity Studies," in Mulej, *et al.* (eds.) (2005), referenced here.

Lessem, R. (1991). *Total Quality Learning. Building a Learning Organization,* Basil Blackwell, Oxford and Cambridge, MA, ISBN 0631168281.

Lester, G. (2005): "Researchers Define Who we Are When We Work Together and Evolutionary Origins of the 'Wait and See' Approach," *Complexity Digest 2005-05,* ISSN not available. Available online: http://www.comdig.com/index.php?id_issue=2005.05.

Likar, B. (2001). *Inoviranje. (Innovating),* Koper, Visoka šola za management, ISBN 961-6268-51-1.

Likar, B., Antunovič, P., Berginc, J., Černjak, D. S., Demšar, J., Fatur, P., Križaj, D., Mulej, M., Pečjak, V., Sitar, S., Trček, D. and Trunk-Širca, N. (2002). *Uspeti z idejo,* Pospeševalni center za malo gospodarstvo, Ljubljana, ISBN 961-90592-2-0.

Likar, B., Macur, M. and Trunk-Širca, N (2006). "Systemic Approach for innovative education process," *Kybernetes,* 35, ISSN 0368-492X, pp. 1071–1086.

Likar, B., Križaj, D. and Fatur, P. (2006). *Management inoviranja,* Koper: Univerza na Primorskem, Fakulteta za management, ISBN 961-6573-28-4.

Lind, A. and Lind, B. (2005): "The Practice of Information System Development and Use: A Dialectical Approach," *Systems Research and Behavioral Science,* September–October, ISSN 1092-7026, pp. 453–464.

Linden, R. (1990). *From Vision to Reality,* LEL Enterprises, Charlottesville, VA, ISBN 0934961069.

"Lisbon Strategy – Not good enough" (2004). *Innovation and Technology Transfer,* 2, ISSN 0033-6807: 3–4.

Lynn, M. L., Mulej, M. and Jurše, K. (2002). "Democracy without empowerment: the grand vision and demise of Yugoslav self-management," *Management Decision,* 40, ISSN (printed) 0025-1747, pp. 797–806.

Loeckenhoff, H. (2005). "Innovation for Societal Evolution: A Transdisciplinary Approach to Guided Change," in Mulej, *et al.* (eds.) (2005), referenced here.

Logical Decisions, *Logical Decisions® for Windows,* available: www.logicaldecisions.com, consulted April 2005, ISBN/ISSN not available.

Lunati, T. (1997). *Ethical Issues in Economics from Altruism to Co-operation to Equity,* MacMillan Press, Houndsmills, ISBN 0333673662.

Malačič, J. and co-authors (2006). Študija o kazalcih ustvarjalnosti slovenskih regij, Služba za regionalni razvoj R. Slovenije, ISBN not available.

Martin G. (2006). *World Revolution through World Law,* IED Institute for Economic democracy, Sun City, AZ, in cooperation with Institute on World Problems, Radford, VA, ISBN 0-9753555-2-X (pbk.).

Mauss, Marcel (1996). *Esej o talentu in drugi eseji. Uvod v delo Marcela Mause/Claude Levi-Strauss,* SKUC, Znanstveni institut Filozofske fakultete – Studia humanitatis, Ljubljana, ISBN 961-6085-14-X.

Mayer, J. (2001). "Nastajanje celostnega pogleda – kljuc za ustvarjalnost tima," *Organizacija,* 34, ISSN 1318-5454, pp. 429–434.

McGregor, J. (2006). "The world's most innovative companies," *BusinessWeek,* April 24, ISSN (printed) 0007-7135, pp. 63–74.

McIntosh, M., Leipziger, D., Jones, K. and Coleman, G. (1998). *Corporate citizenship,* Financial Times Management, London, ISBN 0-273-63106-3.

McIntyre, J. J. (2003). "Participatory Democracy: Drawng on C. West Churchman's Thinking When Making Public Policy," *Systems Research and Behavioral Science,* 20, ISSN 1092-7026, pp. 489–498.

McIntryre, J. (2005). "Critical Praxis to Address Fixed and Fluid Identity and Politics at the Local, National and International Level," *Systemic Practice and Action Research,* 18, ISSN (print) 1094-429X, pp. 223–259.

McWilliams, A., Siegel, D. and Wright, P. (2006). "Corporate social responsibility: strategic implications," *Journal of Management Studies,* Vol. 43, ISSN 0022-2380, pp. 1–18.

Mendibil, K., Hernandez, J., Espinach, X., Garriga, E. and Macgregor, S. (2007). *How can CSR practices lead to successful innovation in SMEs?,* Publication from the RESPONSE project, Strathclyde, 141, ISBN/ISSN not available.

Menih, K. (2006). Slovenska podjetja se morajo primerjati tudi s tujo konkurenco, ne le z domačo. (A journalist's report on 10th IBM Forum held in the first days of April 2006), *Večer,* 9, ISSN 0350-4972.

Mesjasz, Cz. (2005). "Do We Know What We Do Not Know? Main Weaknesses of Discourse on Management of 'Complex Learning Organizations,'" in Mulej, *et al.* (eds.) (2005), referenced here.

Mesner Andolšek, D. (1995). *Organizacijska kultura,* Gospodarski vestnik, Ljubljana, ISBN 86-7061-098-1.

Metcalf, G. (2005). "Will Systems Work? A Search for Models for the 21st Century," in Gu and Chroust (eds.), referenced here.

Metcalf, G. (ed./president) (2008): *ISSS 2008. "Systems that make a Difference." University of Wisconsin – Madison, WI, July 13–18, 2008,* ISBN 978-1-906740-01-6.

Meyer, B. and Medeni, T. (2005). "Knowledge creation and systems research: implications from memory science," in Nakamori (ed.), referenced here, pp. 115–117.

Midgley, G. (2004). "Systems Thinking for the 21st Century," *International Journal of Knowledge and Systems Sciences,* 1, ISSN 1349-7030, pp. 63–69.

Miege, R. and Mahieux, F. (eds.) (1989): *Training in Innovation Management,* Commissions of European Communities, Directorate General for Telecommunications, Information Industries, and Innovation, ISBN not available.

Miller, J. G. (1978). *Living Systems,* McGrawHill, New York, ISBN 0-07-042015-7.

Minati, G. (2005). "Time, processes and cycles," in Nakamori (ed.), referenced here, pp. 376–380.

MIT (2009). "Lab for Corporate Social Innovation", situated at the Massachusetts Institute of Technology, Sloan School of Management, URL: http://sloanleadership.mit.edu/courses/15.975.php, May 5, 2009.

Mlakar, B .and Korošec Lajovic H. (2006). "Certificiranje družbene odgovornosti," in Hrast, A., Mulej, M. and Knez-Riedl, J. (eds.), referenced here.

Mlakar, P. (1998). "EQA as an Informal Way of Spreading Systems Thinking," in Hofer, S. and Doucek, P. (eds.), *IDIMT '98,* referenced here.

Mlakar, P. (1998). "EQA as an Informal Way of Spreading Systems Thinking," in Hofer, S. and Beneder, M. (eds.), *IDIMT '98: 6th Interdisciplinary Information Management Talks,* Universitaetsverlag Rudolf Trauner, Linz, Austria, ISBN 3-85320-955-6.

Mlakar T. (2000). *Zdravstveno varstvo kot organizacijski sistem v luči teorije živih sistemov in dialektične teorije sistemov,* EPF, ISBN not available. (Medical Care as an Organisational System in the Light of Living Systems Theory and Dialectical Systems Theory, M. A. Thesis.)

Mlakar, T. (2007). Kontrolna teorija sistemov – nov model sistemskega razmišljanja (s preveritveno aplikacijo na zdravstveni system), EPF, ISBN not available. (Control systems theory – a new model of system thinking [with testing application in the medical care system].)

Mlakar, T. and Mulej, M. (2007). "Complementarity of the Living Systems and the Dialectical Systems theories: The case of public medical care in Slovenia," *Cybernetics and Systems,* 38, ISSN (printed) 0196-9722, pp. 381–400.

Mlakar, T. and Mulej, M. (2008). "On the concept of the 'Control systems theory' as a new model of systemic consideration," *Kybernetes,* 37, ISSN 0368-492X, pp. 215–225.

Mogensen, A. (ed.) (1981 and earlier for decades). *Work Simplification. Executive Conference Material,* Lake Placid, NY, Work Simplification Inc., ISBN not available.

Mogensen, A. and Rausa, R. (1989). *Mogy. An Autobiography. Father of Work*

Simplification, Idea Associates, Chesapeake, VA, ISBN 0-9623050-0-6.

Molander, E. A. and Sisavic, M. (1994). "Contrasting Paradigms and Movements: Systems Theory and Total Quality Management," *Systems Research*, 11, ISSN (printed) 0953-5314, pp. 47–58.

Mulej, M. (1967). *Entropija ekonomskog sistema na razini radne organizacije*, University of Zagreb, Zagreb, ISBN not available.

Mulej, M. (1971). *Teorija sistemov*, Visoka ekonomsko-komercialna šola Maribor, ISBN not available.

Mulej, M. (1975). *Dialektična teorija sistemov*, Lecture notes (soon a few articles in Slovene journal followed), ISBN/ISSN not available.

Mulej, M. (1976). "Toward the Dialectical Systems Theory," in Trappl, R., Hanika, P. and Pichler, F. (eds.), *Progress in Cybernetics and Systems Research*, Vol. 5, OeSGK, Vienna (Published 1978), ISSN 0275-8717.

Mulej, M. (1977). "A note on dialectical systems thinking," *International Cybernetics Newsletter*, ISSN 0146-1591, p. 63.

Mulej, M. (1979). *Ustvarjalno delo in dialektična teorija sistemov* (Creative Work and the Dialectical Systems Theory. In Slovenian), Razvojni center, Celje, ISBN not available.

Mulej, M. (1981). *O novem jugoslovanskem modelu družbene integracije*, Maribor, Založba Obzorja, ISBN not available.

Mulej, M. (1981b). *Dialektično sistemsko programiranje delovnih procesov*, Maribor, Univerza v Mariboru, Visoka ekonomsko-komercialna šola, ISBN not available.

Mulej, M. (1982). "Dialektično sistemsko programiranje delovnih procesov – metodologija USOMID," *Naše Gospodarstvo*, 28, ISSN 0547-3101, pp. 206–209.

Mulej, M. (1994). "Three Years of Support for a Theory: Two-Generation Cycles in the Transition from a Preindustrial to a Modern Society," *Cybernetics and Systems*, 5, ISSN 0196-9722, pp. 861–877.

Mulej, M. (1996). "Different International Environments – Differents Initiatives for Systems Thinking," in Pfeiffer, R. (ed.) (1997): *Systemdenken und Globalisierung. Folgerungen fuer lernende Organisation im internationalen Umfeld*, Duncker & Humblot, Berlin, ISBN 3-428-09116-7.

Mulej, M. (2000). *Basics of Systems Theory. Applied to Innovation*, EPF, International Program FEBA, EPF, ISBN not available.

Mulej, M. (2003). "Promotion of innovation in small and medium-sized enterprises," in Schwarz, E. J. (ed.), *Technologieorientiertes Innovationsmanagement. Strategien fuer kleine und mittelstaendische Unternehmen*, Gabler, Wiesbaden, pp. 107–137. ISBN 3-409-12399-7.

Mulej, M. (2003 b). "Evropska unija in globalizacija zahtevata od Slovenije, ljudi in organizacij v njej sistemsko / celovito razmišljanje, odločanje in delovanje," in Artač, V., Alič, M., Komic, I., Zavrl, S. and Švarc, J. (eds.), *Zbornik Letne konference* kakovosti Gorenjske, GZS, OZ za Gorenjsko, Kranj, ISBN/ISSN not available.

Mulej, M. (2003c). "Pravilnik ali Poslovnik?," in Rebernik, M., Mulej, M. and Rus, M. (eds.) (2003), *Zbornik posvetovanja 24. PODIM – Podjetništvo, inovacije, management: Dostop do virov za prenos invencij / Access to resources for Invention Transfer*, EPF, Inštitut za podjetništvo in management malih podjetij, pp. 281–295, ISBN 961-6354-33-7.

Mulej, M. (2004). "Entrepreneurship Experience in Slovenia/Yugoslavia before the Break of Communism in Europe", in Schwarz, E. and Wdowiak, M. (eds.), *International Entrepreneurship – Poland, Slovenia and Austria*, E-book, Klagenfurt, Austria, University of Klagenfurt, ISBN not available.

Mulej, M. (2005). "Lack of systemic thinking and innovation: case of Slovenia – a transitional country," in Gu and Chroust (eds.), referenced here.

Mulej, M. (2005a). "Workshop: new roles of systems science in a knowledge society: introductory provocation," in Gu and Chroust (eds.), referenced here.

Mulej, M. (2006). "Systems theory – a worldview and/or a methodology. Ashby Memorial Lecture," in Trappl, R. (ed.), *European Meeting on Cybernetics and Systems Research EMCSR '06*, Austrian Society for Cybernetics Research, Vienna, pp. XXV–XXVII, ISBN 3 85206 172 5.

Mulej, M. (2006a). *Absorbcijska sposobnost tranzicijskih manjših podjetij za prenos invencij, vednosti in znanja iz univerz in inštitutov*, University of Primorska, Faculty of Management, Koper, ISBN not available.

Mulej, M. (2006b). "Has Slovenia sufficient absorption capacity to be innovative (knowledge) society?," in Haček, M. and Zajc, D. (eds.), Slovenski politološki dnevi 2006, Portorož, 29.–31. maj 2006. *Slovenija v evropski družbi znanja in razvoja: zbornik povzetkov*, Ljubljana: Slovensko politološko društvo, pp. 54–55, ISBN not available.

Mulej, M. (2006c). "Introductory Address," in Hrast, A., Mulej, M. and Knez-Riedl, J. (eds.), referenced here.

Mulej, M. (2007). "Systems theory – a worldview and/or a methodology aimed at requisite holism/realism of humans' thinking, decisions and action," *Systems Research and Behavioral Science*, 24, ISSN 1099-1743, pp. 347–357.

Mulej, M. (ed.) (1984). "Vgrajevanje inventive v politiko in prakso OZD. 5. PODIM. Proceedings," *Naše Gospodarstvo*, Vol. 30, No. 1–2, ISSN 0547-3101.

Mulej, M. (ed.) (1995). *Proizvodno inovacijski management* (Production Innovation Management), University of Maribor, Slovenia, ISBN not available.

Mulej, M. (ed.) (1997). "Inoviranje in ekonomija (Innovating and Economy. In Slovene)," articles based on the 17th PODIM Conference on Innovation, *Naše Gospodarstvo*, Vol. 43, No. 1–6, ISSN 0547-3101.

Mulej, M. (ed.) (2002). *Management inovacijskih procesov*, EPF, ISBN not available.

Mulej, M., *et al.* (eds) (2005). *Proceedings of The WOSC 13th International Congress of*

Cybernetics and Systems (11 Symposia), July 6–10, 2005, and *The 6th International Conference of Sociocybernetics, Theme: Sociocybernetics and Innovation*, July 5–10, 2005, held in Maribor, Slovenia. WOSC, and ISA, RC 51 on Sociocybernetics, in cooperation with SDSR, and EPF, Institute for Entrepreneurship and Small Business Management, on CD, ISBN 961-6354-58-2.

Mulej, M., *et al.* – partly new co-authors in every edition (1982, 1983, 1984, 1985, 1986). *Usposabljanje za ustvarjalnost. Metodologija USOMID,* Ekonomski center, Maribor (NB. 3rd edition was in Serbian, Radnicki univerzitet, Subotica, and Ekonomski center, Maribor); all editions reworked, ISBN not available.

Mulej, M., Devetak, G., Drozg, F., Ferš, M., Hudnik, M., Kajzer, Š., Kavčič, B., Kejžar, I., Kralj, J., Milfelner, R., Možina, S., Paluc, C., Pirc, V., Pretnar, B., Repovž, L., Rus, V., Senčar, P. and Tratnik, G. (1987). *Inovativno poslovanje,* Gospodarski vestnik, Ljubljana, ISBN not available.

Mulej, M., de Zeeuw, G., Espejo, R., Flood, R., Jackson, M., Kajzer, Š., Mingers, J., Rafolt, B., Rebernik, M., Suojanen, W., Thornton, P. and Uršič, D. (1992). *Teorije sistemov,* Maribor, EPF, ISBN 86-80085-27-8.

Mulej, M., Rebernik, M. and Kuster, T. (1992b). *(Dialektična) teorija sistemov. Gradivo za vaje,* EPF, ISBN not available.

Mulej, M., Hyvarinnen, L., Jurše, K., Rafolt, B., Rebernik, M., Sedevčič, M. and Uršič, D. (1994a). *Inovacijski management, 1. del, Inoviranja managementa.* EPF. ISBN 86-80085-47-2.

Mulej, M. and Jurše, K. (1994). "Popolna kakovost in potrošniška družba," in Pavlin, N. (ed.), *Organizacija, informatika, kadri pri vodenju in upravljanju družb,* Moderna organizacija, Kranj, ISBN 86-81049-64-X, pp. 393–399.

Mulej, M., Kajzer, S., Treven, S. and Jurše, K. (1997). "Sodobno gospodarstvo med odpori do inovacij in zivljenjem od njih," *Naše Gospodarstvo,* Vol. 43, No. 3–4, ISSN 0547-3101.

Mulej, M. and Kajzer, S. (1998a): "Tehnološki razvoj in etika soodvisnosti," *Raziskovalec,* 28, ISSN 0351-0727, pp. 26–35.

Mulej, M. and Kajzer, S. (1998b). "Ethic of interdependence and the law of requisite holism," in Rebernik, M. and Mulej, M. (eds.) (1998), *STIQE '98*. ISRUM, *et al.,* Maribor, pp. 56–67, ISBN 3-85320-955-6.

Mulej, M., Kajzer, S., Vezjak, M. and Mlakar, P. (1998): "Teaching on/for Systems Thinking," in Hofer, S. and Beneder, M. (eds.), *IDIMT '98: 6th Interdisciplinary Information Management Talks,* Universitaetsverlag Rudolf Trauner, Linz. ISBN 3-85320-955-6.

Mulej, M. and Ženko, Z. (1998). "Sodobnost izobraževanja – za ozko specializacijo brez medstrokovnega sodelovanja ali z njim?," in Lipičnik, M., *et al.* (eds.), *Zbornik 1. Kongresa Transport, promet, logistika,* Drustvo za poslovno logistiko Slovenije et al., Maribor, ISBN 86-435-0247-2.

Mulej, M. and Kajzer, S. (1998). "Ethics of Interdependence and The Law of Requisite

Holism," in Rebernik, M. and Mulej, M. (eds.) (1998), *STIQE '98. Proceedings of the 4th International Conference on Linking Systems Thinking, Innovation, Quality, Entrepreneurship and Environment,* Institute of Systems Research Maribor et al., Maribor, Slovenia, pp. 129–140, ISBN 86-80085-88-X.

Mulej, M., Kajzer, S., Vezjak, M. and Mlakar, P. (1999). "Applied systems thinking and the law of requisite holism. Invited Keynote paper," in *IDIMT '99*, ed. by Ch. Hofer, Universitaetsverlag Rudolf Trauner, Linz, ISBN 3-85487-046-9.

Mulej, M., Espejo, R., Jackson, M., Kajzer, S., Mingers, J., Mlakar, P., Mulej, N., Potočan, V., Rebernik, M., Rosicky, A., Schiemenz, B., Umpleby, S., Uršič, D., and Vallee, R. (2000): *Dialektična in druge mehkosistemske teorije (podlaga za uspešen management),* EPF, ISBN 961-6354-01-9.

Mulej, M. and Potočan, V. (2000). "Economic Reasons Opposing Bertalanffian Thinking in Practice: The Law of Requisite Holism in Decision Making," in Rebernik, M. and Mulej, M. (eds.), *STIQE '00. Proceedings of the 4th International Conference on Linking Systems Thinking, Innovation, Quality, Entrepreneurship and Environment,* Institute for Systems Research et al., Maribor, ISBN/ISSN not available.

Mulej, M., Potočan, V., Kajzer, Š. and Ženko, Z. (2001). "A Complementing Criticism of Ludwig von Bertalanffy's general systems theory," in Elohim, J. L., Hofkirchner, W. (eds.), referenced here.

Mulej, M., Bastič, M., Belak, J., Knez-Riedl, J., Pivka, M., Potočan, V., Rebernik, M., Uršič, D., Ženko, Z. and Mulej, N. (2003). "Informal Systems Thinking or Systems Theory," *Cybernetics and Systems,* 34, ISSN 0196-9722, pp. 71–92.

Mulej, M., Knez-Riedl, J., Potočan, V. and Ženko, Z. (2003a). "Upravljanje – kdaj je razmišljanje o njem in v njem sistemsko/celovito?," in Rozman, R. and Kovač, J. (eds.), *4. znanstveno posvetovanje o organizaciji: Upravljanje. Zbornik referatov,* Univerza v Mariboru, Fakulteta za organizacijske vede Kranj, Univerza v Ljubljani, Ekonomska fakulteta, Zveza organizatorjev Slovenije, Kranj, ISBN 961-6430-60-2.

Mulej, M. and Ženko, Z. (2003). "Inovativno podjetništvo kot osebna lastnost in vpliv vlade nanj," *Organizacija,* 36, ISSN 1318-5454, pp. 273–281.

Mulej, M., Knez-Riedl, J., Potočan, V. and Ženko, Z. (2003a). "ržavni ukrepi za pospešeno uveljavljanje inovativne družbe in inovativnega poslovanja," *Organizacija,* 36, ISSN 1318-5454, pp. 358–367.

Mulej, M., Knez-Riedl, J., Potočan, V. and Ženko, Z. (2003b). "Upravljanje – kdaj je razmišljanje o njem in v njem sistemsko/celovito?," *Organizacija,* 36, ISSN 1318-5454, pp. 443–445.

Mulej, M., Kajzer, S., Ženko, Z. and Potočan, V. (2003c). "Bertalanffijska teorija sistemov kot podlaga za etiko soodvisnosti, nujno v inovativni druzbi in globalnem gospodarstvu," *Naše Gospodarstvo,* 49, ISSN 0547-310, pp. 196–215.

Mulej, M., Ženko, Z., Potočan, V., Kajzer, S., Umpleby, S. and Ečimovič, T. (2003b). "The system of seven basic groups of systems thinking principles and eight basic assumptions of a general systems theory," in Chroust, G. and Hofer, Ch. (eds). *IDIMT 2003: proceeding*, Schriftenreihe Informatik, Bd. 9, Universitaetsverlag Rudolf Trauner, Linz, pp. 137–152, ISBN 3-85487-493-6.

Mulej, M. and Potočan, V. (2004). "What Do EU, United Nations, International, Standards Organization, OECD, etc., Mean By Systems Thinking?," in Trappl, R. (ed.), *Cybernetics and Systems 2004*, referenced here.

Mulej, M., Potočan, V., Ženko, Z., Kajzer, S., Uršič, D., Knez-Riedl, J., Lynn, M. and Ovsenik, J. (2004). "How To Restore Bertalanffian Systems Thinking," *Kybernetes*, 33, ISSN 0368-492X, pp. 48–61.

Mulej, M., Likar, B. and Potočan, V. (2005). "Increasing the Capacity of Companies to Absorb Inventions from Research Organizations and Encouraging People to Innovate," *Cybernetics and Systems*, 36, ISSN 0196-9722, pp. 491–512.

Mulej, M., Rebernik, M., Knez-Riedl, J., Ženko, Z., Uršič, D., Potočan, V., Rosi, B. and Krošlin, T. (2005a): "From Knowledge to Innovation Society," in Gu and Chroust (eds.), referenced here. Also in Nakamori (ed.) (2005), referenced here, pp. 467–474.

Mulej, M., Kajzer, Š. and Potočan, V. (2005b). "Common sense of the uncommon sense called systems thinking and systems theory: holism," in Gu and Chroust (eds.), referenced here.

Mulej, M. and Ženko, Z. (2004a). *Introduction to Systems Thinking with Application to Invention and Innovation Management*, Management Forum, Maribor, ISBN 961-90717-2-7.

Mulej, M. and Ženko, Z. (2004b). *Dialektična teorija sistemov in invencijsko-inovacijski management (Kratek prikaz)*, Management Forum, Maribor, ISBN 961-90717-1-9.

Mulej, M., Kajzer, S., Potočan, V. and Rosi, B. (2005). "Requisite Holism by Co-Operation of Systems Theories: An Invention or Innovation in Inquiry of Innovation," in Mulej, M., *et al.* (eds.) (2005), referenced here.

Mulej, M., Čančer, V., Hrast, A., Jurše, K., Kajzer, S., Knez-Riedl, J., Mulej, N., Potočan, V., Rosi, B., Uršič, D. and Ženko, Z. (2007): *The Law of Requisite Holism and Ethics of Interdependence: Basics of The Dialectical Systems Thinking (Applied To Innovation In Catching-Up Countries)*, on GESI Website, Buenos Aires (in process of translation into Spanish, ISBN not available)

Mulej, M., Potočan, V., Ženko, Z., Prosenak, D. and Hrast, A. (2007). "What will come after the investment, innovation and affluence phases of social development?," in Sheffield, J. and Fielden, K. (eds.) (2007), *Systemic developmen : local solutions in a global environment: proceedings of the 13th Annual Australia and New Zeland Systems (ANZSYS) Conference*, December 2–5, 2007, Auckland, New Zeland, Goodyear: ISCE, ISBN 978-0-9791688-9-5.

Mulej, M., *et al.* (eds.) (2005). *Proceedings of The WOSC 13th International Congress of Cybernetics and Systems (11 Symposia), July 6–10, 2005, and The 6th International Conference of Sociocybernetics, Theme: Sociocybernetics and Innovation*, July

5–10, 2005, Maribor, Slovenia. WOSC – World Organsation of Systems and Cybernetics, and ISA – International Sociological Association, Research Committee on Sociocybernetics, in cooperation with SDSR – Slovenian Systems Research Society and EPF, Institute for Entrepreneurship and Small Business Management (EPF-IP), ISBN 961-6354-57-4.

Mulej, M. and Mulej, N. (2006). "Innovation and/by Systems Thinking by Synergy of Methodologies 'Six Thinking Hats' and 'USOMID', in Rebernik, M., *et al*. (eds.), *PODIM 26, Cooperation between the economic, academic and governmental spheres: Mechanisms and levers, 30–31 March 2006,* Maribor, EPF, Institute for Entrepreneurship and Small Business Management, ISBN 961-6354-33-7.

Mulej, M., Kajzer, S., Potočan, V., Rosi, B. and Knez-Riedl, J. (2006). "Interdependence of systems theories – potential innovation supporting innovation," *Kybernetes*, 35, ISSN 0368-492X, pp. 942–954.

Mulej, M., Potočan, V., Ženko, Z. and Kajzer, V. (2006). "Etika soodvisnosti kot ozadje družbene odgovornosti," in Hrast, A., Mulej, M. and Knez-Riedl, J. (eds.), referenced here (The Ethics of Co-dependence as a Background of Social Responsibility).

Mulej, M., and co-authors Fatur, P. Knez-Riedl, J., Kokol, A., Mulej, N., Potočan, V., Prosenak, D., Škafar, B. and Ženko, Z. (2008). *Invencijsko-inovacijski management z uporabo dialektične teorije sistemov (podlaga za uresničitev ciljev Evropske unije glede inoviranja),* Inštitut za inovacije in tehnologijo Korona plus, d.o.o., Ljubljana, ISBN 978-961-90592-8-9.

Mulej, M., Potocan, V., Zenko, Z. and Knez-Riedl, J. (2008). "Social Responsibility – an Innovation Toward Requisite Holism as a Basis for Humans to Make a Difference in Affluence," in Metcalf, G. (ed.), *ISSS 2008: Systems that make a Difference. Proceedings,* International Society for Systems Sciences, Madison, WI, ISBN 978-1-906740-01-6.

Mulej, M. and Prosenak, D. (2007). "Society and Economy of Social Responsibility – The Fifth Phase of Socio-economic Development?," in Hrast, A., Mulej, M. and Knez-Riedl, J. (eds.), referenced here.

Mulej, N. (1998). "Marketing as a Formal or Informal Systems Thinking," in Hofer, S. and Beneder, M. (eds.), *IDIMT '98: 6th Interdisciplinary Information Management Talks,* Universitaetsverlag Rudolf Trauner, Linz, Austria, ISBN 3-85320-955-6.

Mulej, N. (ed.) (2004). "New Moment Ideas Campus: Think Like Leonardo," *New Moment, 22* (entire journal), ISSN 1580-1322.

Mustajoki, J., Hämäläinen, R. P. and Salo, A. (2005). "Decision Support by Interval SMART/SWING – Incorporating Imprecision in the SMART and SWING Methods," *Decision Sciences*, 36, ISSN 1540-4595, pp. 317–339.

Mueller-Merbach, H. (1992). "Vier Arten von Systemansaetzen, dargestellt in Lehrgespraechen," *Zeitschrift fur Betriebswirtschaft*, 62, ISSN 0044-2372, pp. 853–876.

Myers, N., (ed.) (1991). *Gaia – modri planet*. Mladinska knjiga, Ljubljana (Translation into Slovenian), ISBN 86-11-07195-6.

Nakamori, Y. (ed.) (2005). *The Second International Symposium on Knowledge Management for Strategic Creation of Technology; Proceedings,* JAIST Press (N.B.: The editors' name does not appear in the book. Nakamori wrote the preface, which made us name him in this role here.), ISSN 1477-8238.

Nakamori, Y. and Wierzbicki, A. (2005). "Systems for Integrating and Creating Knowledge," in Nakamori (ed.), referenced here, pp. 296–305.

Nelson, H. G. (2003). "The Legacy of West Churchman: A Framework for Social Systems Assessment," *Systems Research and Behavioural Science,* 20, ISSN 1092-7026, pp. 463–474.

Nesci (1999). *Managing the Complex – Mastering Corporate Complexity: Doing It Not Just Talking About It. The Role of Coherence,* Conference held in Boston, Mass, USA, www.Proceedings.net/MTC/, ISBN/ISSN not available.

Nesci (2000). *NESCI conference on complexity in organizations; track on knowledge management,* 31st May to 3rd June, Boston, USA, ISBN/ISSN not available.

Nixon, B. (2004). "Speaking Plainly – A New Agenda for the 21st Century," in Crowther D. and Caliyurt, K. T. (eds.) (2004), referenced here.

Nussbaum, B., Berner, R. and Brady, D. (2005). "Special Report. Get Creative! How to Build Innovative Companies. And: A Creative Corporation Toolbox," *Business Week,* 8/15, ISSN (print) 0007-7135, pp. 51–68.

Oblak, H. and Mulej, M. (1998). *Organiziranje in poslovanje prometnih podjetij,* Univerza v Mariboru et al., ISBN 961-90303-5-4.

OECD (2000). *Framework to measure sustainable development,* OECD, Paris, ISBN 9789264180635.

Osawa, Y. and Miyazaki, K. (2005). "An Analysis of R&D Project Performance from Research to Commercialization – Evidence from a Japanese Electric Company," in Nakamori (ed.), referenced here, pp. 160–167.

Oshry, B. (1996). *Seeing Systems. Unlocking the Mystery of Organizational Life,* Berrett-Koehler Publishers, San Francisco, CA, USA, ISBN 1881052990.

Palacios-Marqués, D. and Garrigós-Simón (2005). "A measurement Scale for knowledge management in the biotechnology and telecommunications industries," *International Journal of Technology Management,* 31, 3–4, ISSN 14742748, pp. 358–374.

Palčič, R. and Mulej, M. (1991). *Doubling the Monthly Output in Six Months by O.D.* – invited paper to the 11th Organizational Development World Congress, Berlin, Germany (unpubl.), ISBN/ISBN not available.

Palčič, R. and Mulej, M. (1994). "A Success Story: Modern Management Producing Productivity in a Plant in a less-eveloped Area of Slovenia," *Public Enterprise,* 14, ISSN 0351-3564, pp. 121–131.

Parhankangas, A., Ing, D., Hawk, D. L., Dane, G. and Kosits, M. (2005). "Negotiated Order and Network Form Organizations," *Systems Research and Behavioral Science*, September–October, ISSN 1092-7026, pp. 431–452.

Parsloe, E. (1995). *The Manager as Coach and Mentor,* Institute of Personnel and Development, London, ISBN 0852925867.

Parsons, T. (1965). "An outline of the social system," in Parsons, T., Shils, E. A., Naegle, K. D. and Pitts, J. R. (eds.), *Theories of society. Foundations of modern sociological theory*, pp. 30–79, Free Press, New York–London, ISBN 0029244501.

Pavlin, S. (2005). "Upravljanje znanja kot posebno raziskovalno področje," *Organizacija*, 38, ISSN 1318-5454, pp. 361–367.

Pečjak, V. (2001). "Poti do novih idej – Tehnike kreativnega mišljenja," *New Moment*, 16 (entire journal), ISSN 1580-1322.

Petelinšek, A. (2006). "Sodobne dileme novinarske etike," in Hrast, A., Mulej, M. and Knez-Riedl, J. (eds.), referenced here.

Perušič, S. (2002). *Reforma državne uprave s poudarkom na kakovosti,* Maribor, EPF, ISBN not available.

Peters, T. (1995). *The Pursuit of WOW,* Macmillan, New York, NY, USA, ISBN 0-333-65084-0.

Peters, T. (1997). *The Circle of Innovation,* Knopf, New York, NY, USA, ISBN 0-340-71720-3.

Petzinger, T. (2000). *The New Pioneers. The Men and Women Who Are Transforming the Workplyce and the Marketplace,* A Touchstone Book, Simon & Schusters, New York, ISBN 0-684-86310-3.

Pichler, F. (1997). "On the Concept of Holarchy by Arthur Koestler," in Hofer, S. and Doucek, P., *IDIMT '97,* R. Oldenbourg, Wien, ISBN 3-486-24526-0.

Pivka, M. and Mulej, M. (2004). "Requisitely Holistic ISO 9000 Audit Leads to Continous Innovation/Improvement," *Cybernetics and Systems*, 35, ISSN 0196-9722, pp. 363–378.

Pivka, M. and Uršič, D. (1998). "Quality Systems between Tzheory and Practice – Slovenian Practice," in Rebernik and Mulej (eds.), referenced here.

Pivka, M. and Uršič, D. (2001). *ISO 9000 v slovenskih podjetjih,* Management Forum, Maribor, ISBN not available.

Pless, N. (1998). *Corporate Caretaking. Neue Wege der Gestaltung organsationeller Mitweltbeziehungen,* Metropolis Verlag, Marburg, ISBN 3-89518-217-6.

Podnar, K. and Golob, U. (2006). "Pričakovanja o družbeni odgovornosti podjetij in njihov vpliv na vedenjsko intencijo posamerznikov," in Hrast, A., Mulej, M. and Knez-Riedl, J. (eds.), referenced here.

Ponovni (2006). *Ponovni zagon lizbonske strategije na Evropski in nacionalni ravni ter izvajanje programa reform v RS,* Ljubljana, Urad RS za makroekonomske analize in razvoj (Paper for public discussion in Parliament of Slovenia "Restarting the

Lisbon strategy in the European and national levels and implementation of the national program of reforms in Republic of Slovenia, on 07 April), ISBN/ISSN not available.

Potočan, V. (1998). "Holistic Business Decision Making as an Practical Example of Training for Systems Thinking," in Hofer, S. and Beneder, M. (eds.), *IDIMT '98: 6th Interdisciplinary Information Management Talks,* Universitaetsverlag Rudolf Trauner, Linz, Austria, ISBN 3-85320-955-6.

Potočan, V. (2000). "New Perspectives on BDM," *Management,* 5, ISSN Y502-7055, pp. 13–28.

Potočan, V. (2002). "Sustainable development," *Management,* 7, ISSN Y502-7055, pp. 67–77.

Potočan, V. (2003). *Organizacija poslovanja,* Doba, Maribor, ISBN 961-6084-13-5.

Potočan, V. (2004). *Izvedbeni management,* EPF, ISBN 961-6354-49-3.

Potočan, V. (2005). "Efficiency or Effectiveness?," *Organization,* 38, ISSN 1318-5454, pp. 570–576.

Potočan, V., Mulej, M. and Kajzer, S. (2002). "Standardisierung der Entscheidungsprozesse in komplexen und komplizierten Geschaeftssystemen: Zwischen der echten und der scheinbaren Ganzheitlichkeit," in Milling, P. (ed.), *Entscheiden in komplexen Systemen,* Dunkler & Humblot, Berlin (Wirtschaftskybernetik und Systemanalyse, Band 20), pp. 221–234, ISBN 3-428-10683-0.

Potočan, V. and Mulej, M. (2003). "On requisitely holistic understanding of sustainable development," *Systemic Practice and Action Research,* 16, ISSN 1573-9295, pp. 421–436.

Potočan, V. and Mulej, M. (2005). "Ethics of Sustainable Development in Corporate Governance," *Global Business & Economics Anthology 2005,* ISSN Y505-3765, pp. 323–334.

Potočan, V. and Mulej, M. (2003). "On Requisitely Holistic Understanding of Sustainable Development from Business Viewpoints," *Systemic Practice and Action Research,* 16, ISSN 1573-9295, pp. 421–436.

Potočan, V. and Mulej, M. (2006a). "Social responsibility of a sustainable enterprise," in R. Kovač, referenced here, pp. 41–44.

Potočan, V. and Mulej, M. (2006b). "Enterprise competitiveness and ethics of sustainable enterprise," in *Kickstarting small island enterprise competitiveness: Conference Proceedings CD.* [Compact disc ed.]. [S.l.], The University of the West Indies at Cave Hill, Department of Management Studies, pp. 299–314, ISBN/ISSN not available.

Potočan, V. and Mulej, M. (2006c). "Systemic understanding of trust and ethics of interdependence in innovative business," in XVI World Congress of Sociology, Durban, South Africa, July 23–29, *ISA 2006 Congress The quality of social existence in a globalising world* [Compact disc ed.], Malvern, Naren Bhimsan, ISSN 0038-0202.

Potočan, V. and Mulej, M. (eds.) (2007a). *Transition into an Innovative Enterprise*, University of Maribor, Faculty of Economics and Business, Maribor, ISBN 978-961-6354-64-6.

Potočan, V. and Mulej, M. (2007b). "Ethics of a Sustainable Enterprise and the Need for it," *Systemic practice and action research*, 20, ISSN (print) 1094-429X, pp. 127–140.

Potočan, V., Mulej, K. and Kajzer, S. (2005). "Business Cybernetics," *Kybernetes*, 34, ISSN 0368-492X, pp. 1496–1516.

Potočan, V. and Mulej, M. (2009). "Business Cybernetics – provocation number two," *Kybernetes*, 38, ISSN 0368-492X, pp. 93–112.

Prednostno (2006). *Prednostno področje lizbonske strategije – vlaganje v izobraževanje, raziskave in inovacije*, Ljubljana, RS, Ministrstvo za visoko šolstvo, znanost in tehnologijo, see reference Ponovni.

Predstavitev (2006). *Predstavitev prednostnega področja lizbonske strategije – sprostitev malih in srednjih podjeti*, Ljubljana, RS, Ministrstvo za gospodarstvo, see reference Ponovni.

Problems of… (1979–2001, biannualy): De Zeeuw, G., *et al.* (eds.). *Conferences on several selected topics*, Dutch Systems Group, University of Amsterdam et al., ISBN/ISSN not available.

Prosenak, D. and Mulej, M. (2008). "O celovitosti in uporabnosti obstoječega koncepta družbene odgovornosti poslovanja (About holism and applicability of the existing concept of corporate social responsibility)," CSR, *Naše Gospodarstvo*, 54, ISSN 0547-3101, pp. 10–21.

Prosenak, D., Mulej, M. and Snoj, B. (2008). "A requisitely holistic approach to marketing in terms of social well-being," *Kybernetes*, 37, 9/10, ISSN 0368-492X.

Pucelj, M. and Likar, B. (2006). "Vodenje kot ključni dejavnik za ustvarjanje inovacijske kulture," *Organizacija*, 39, ISSN 1318-5454, pp. 132–140.

Rant, M. (2006). "Negativni vplivi strukture upravljanja na poslovno odločanje in družbeno odgovorno delovanje: sedanji trendi v Evropi," in Rozman and Kovač, referenced here, pp. 45–50.

Rebernik, M. and Mulej, M. (2000). "Requisite holism, isolating mechanisms and entrepreneurship," *Kybernetes*, 29, ISSN 0368-492X, pp. 1126–1140.

Rebernik, M. and Repovž, L. (2000). *Od ideje do denarja*. Gospodarski vestnik, Ljubljana, ISSN 86-7061-216-X.

Rebernik, M., Mulej, M. (co-chairs and eds.) (2006, and earlier, since 1992, biannually). *International Conference on Linking Systems Thinking, Innovation, Quality, Entrepreneurship and Environment,* Maribor, Institute for Entrepreneurship and Small Business Management, at Faculty of Economics and Business, University of Maribor, and Slovenian Society for Systems Research, ISBN 961-6354-47-7.

Rebernik, M., Knez-Riedl, J., Močnik, D., Tominc, P., Širec Rantaša, K., Rus, M., Krošlin, T. and Dajčman, S. (2004). *Slovenian Entrepreneurship Observatory 2003,* EPF, Institute for Entrepreneurship and Small Business Management, ISBN 961-6354-46-9.

Rebernik, M., Tominc, P., Glas, M. and Širec Rantaša, K. (2004). *Global Entrepreneurship Monitor. Slovenija 2003. Spodbujati in ohranjati razvojne ambicije,* EPF, Inštitut za podjetništvo in management malih podjetij, ISBN not available.

Reich, R. (1984). *The Next American Frontier,* Penguin Books, New York, ISBN 0812910672.

Resolucija. (2006). *Resolucija evropskega parlamenta o prispevku k spomladanskemu zasedanju Evropskega sveta 2006 v zvezi z lizbonsko strategijo. Preliminary edition P6_TA-PROV(2006)0092 B6-0162/2006,* Preparations for the Spring Session of the European Council: Lisbon Strategy, Strasbourg, see reference Ponovni.

Risopoulos, F. (2005). "How Can Communication Support Complex Problem Solving for Innovation?," in Mulej, *et al.*, (eds.) (2005), referenced here.

Rodríguez, P., Siegel, D. S., Hillman, A. and Eden, L. (2006). "Three lenses on the multinational enterprise: politics, corruption, and corporate social responsibility," *Journal of International Business Studies,* Vol. 37, ISSN 0047-2506, pp.733–746.

Rogers, E. M. (1995). *Diffusion of Innovation. Fourth Edition,* The Free Press, New York, ISBN 0029266718.

Rooke, D. and Torbert, W. R. (2005). "7 Transformations of Leadership," *Harvard Business Review*, 83, ISSN 00178012, pp. 67–76.

Rosen, J. and Kineman, J. J. (2005). "Anticipatory Systems and Time: A New Look at Rosennean Complexity," *Systems Research and Behavioral Science*, September–October, ISSN 1092-7026, pp. 399–412.

Rosenberg, N. and Birdzell, L. E. (1986). *The Past: How the West Grew Rich,* Basic Books, New York, ISBN 0465031099.

Rosi, B. (2004). *Prenova omrežnega razmišljanja z aplikacijo na procesih v železniški dejavnosti,* EPF, ISBN not available.

Rosi, B., Kramberger, T., Lisec and Lipičnik, M. (2006). "Short Term Road Ice Forecast – a Case study," *Proceedings of IDIMT- 2006,* September 13–15, 2006, held in Budweis, Czech Republic, The Prague University of Economics and The University of Linz, ISBN 3-85499-049-9.

Rosi, B. and Mulej, M. (2005). "Z več dialektično omrežnega razmišljanja lahko postane slovenski železniški primetni system evropsko konkurenčnejši," *Organizacija*, 38, ISSN 1318-5454, pp. 169–175.

Rosi, B. and Mulej, M. (2006). "The dialectical network thinking – a new systems theory concerned with management," *Kybernetes*, 35, ISSN 0368-492X, pp. 1165–1178.

Rosicky, A. (2000). "Informacija in sistem: bistvo, forma in pomen," in Mulej, *et al.* (2000), referenced here.

Rozman, R. and Kovač, J. (eds.) (2004), "Zaupanje v in med organizacijami," *Proceedings of the 5th scientific conference on organisation (In Slovenian)*, University of Maribor, Fakulteta za organizacijske vede, Kranj, Zveza organizatorjev Slovenije, Kranj; University of Ljubljana, Ekonomska fakulteta, Ljubljana, ISBN 9612400032.

Rozman, R. and Kovač, J. (eds.) (2005). "6. znanstveno posvetovanje o organizaciji, Brdo pri Kranju, 3. junij 2005," *Konflikti v in med organizacijami: zbornik referatov*, UM, Fakulteta za organizacijske vede, Kranj, Zveza organizatorjev Slovenije, Kranj; UL, Ekonomska fakulteta, Ljubljana, ISBN 961-240-047-4.

Rozman, R. and Kovač, J. (eds.) (2006). *Družbena odgovornost in etika v organizacijah. Proceedings of the 7th scientific conference on organisation. (In Slovenian),* Univerza v Mariboru, Fakulteta za organizacijske vede, Kranj, Zveza organizatorjev Slovenije, Kranj; Univerza v Ljubljani, Ekonomska fakulteta, Ljubljana. ISBN 961-240-091-1.

Saaty, T. L. (2001). *Decision Making with Dependence and Feedback – The Analytic Network Process,* RWS Publications, Pittsburgh, ISBN 0962031798.

Salo, A. and Gustafsson, T. (2004). "A Group Support System for Foresight Processes," *International Journal of Foresight and Innovation Policy*, 1, ISSN 1740-2816, pp. 249–269.

Senge, P. (2006). *The Fifth Discipline: The Art and Practice of the Learning Organization* (revised and updated), Currency Doubleday, New York, ISBN 0-385-51725-4.

Singer, P. (1999). *Practical Ethics,* Cambridge University Press, Cambridge, ISBN 052143971X.

Sirkin, H. L., Keenan, P. and Jackson, A. (2005). "The Hard Side of Change Management," *Harvard Business Review*, 83, ISSN 0017-8012, pp. 109–118.

Smith, A. (1759, ed. 2000). *The Theory of Moral Sentiments,* Prometheus Books, New York, ISBN 1573928003.

Smith, A. (1776, ed. 1998). *An Inquiry into the Nature and Causes of the Wealth of Nations,* Modern Library, New York, ISBN 0199535922.

Social Investment Forum (SIC) (2001). *Report of Socially Responsible Investing Trends in the US,* SIC, Washington, ISBN/ISSN not available.

Sporočilo (2006). *Sporočilo komisije spomladanskemu evropskemu svetu: Čas za višjo prestavo. Novo partnerstvo za rast in delovna mesta. KOM(2006) 30 končno, Del I, II,* Bruselj, Komisija evropskih skupnosti 25 January, see reference Ponovni, ISSN 1725-6976.

Sruk, V. (1986). *Etika in morala,* Cankarjeva založba, Ljubljana, ISBN not available

Sruk, V. (1999). *Leksikon morale in etike,* Univerza v Mariboru, EPF, ISBN 86-80085-92-8.

Steiner, G. (2006). "The Planetary Model as a Framework for Organizing Innovation," *Naše Gospodarstvo,* 52, ISSN 0547-3101, pp. 18–23.

Steiner, G. (2006a). "Indigenous Innovation and Social Responsibility: Stakeholder-management as Tool to Cope with Change," in Hrast, A., Mulej, M. and Knez-Riedl, J. (eds.), referenced here.

Sterman, J. D. (2000). *Business Dynamics – Systems Thinking and Modeling for a Complex World,* Irwin McGraw-Hill, Boston et al., ISBN 007238915X.

Stern, N. (2006). *The Stern Review. The economics of climate change,* available on: http://www.hm-treasury.gov.uk/stern_review_report.htm [27.3.2009], ISBN 0-521-70080-9.

Stern, N. (2007), interviewed by M. Stein: "Special Report. The Climate Change, The Economic Argument," Research*eu, 52, ISSN 1830-7981, pp. 14–15.

STIQE (1992–, biannually): Rebernik, M. and Mulej, M. (eds.). *International Conferences Linking Systems Thinking, Innovation, Quality, Entrepreneurship, Environment,* Slovenian Society for Systems Research, Maribor, and EPF, ISBN not available.

Sutherland, J. and Canwell, D. (2004). *Key Concepts in Management,* Palgrave, Hampshire, ISBN 1403915334.

Svetlik, I. and Ilič, B. (eds.) (2004). *Razpoke v zgodbi o uspehu,* Založba Sophia, Ljubljana, ISBN 961-6294-55-5.

Swanson G. A. (2005). "The Study of Pathology and Living Systems Theory," *System Research and Behavioural Science, September–October 2005,* ISSN 1092-7026.

Schermerhorn, J. and Chappell, D. (2000). *Introducing Management,* John Wiley and Sons, New York, ISBN 047113581X.

Shea, G. (1998). *Practical Ethics,* American Management Association, New York, ISBN 9780814423394.

Schiemenz, B. (ed.) (1994). *Interaktion, Modellierung, Kommunikation und Lenkung in komplexen Organisationen,* Duncker und Humblot, Berlin, Germany, ISBN 3-428-08200-1.

Schmidt, J. (1993). *Die sanfte Revolution. Von der Hierarchie zu selbststeuernden Systemen,* Campus, Frankfurt am Main, Germany, ISBN 3593350084.

Schnaber, P. (1998). "Moralische Urteile sind (k)eine Geschmacksache. Weshalb handeln wir so, wie wir handeln?," *Neue Zuercher Zeitung, Zeitfragen,* 211, ISSN 0376-6829, p. 86.

Schwaninger, M. (2006). *Intelligent Organizations. Powerful Models for Systemic Management,* Springer Berlin–Heidelberg, ISBN 3-540-29876-2.

Šek, K. (2007). Kako ustvarjalne lastnosti zaposlenih vplivajo na inovativnost v organizacijah, Univerza v Mariboru, Ekonomsko-poslovna fakulteta, Maribor, ISBN not available.

Škafar, B. (2006). *Inovativnost kot pogoj za poslovno odličnost v komunalnem podjetju,* Univerza v Mariboru, Ekonomsko-poslovna fakulteta, Maribor, ISBN not available.

Stanford (2009). "Corporate Social Innovation Program," situated at Stanford University, Graduate School of Business, URL: http://www.gsb.stanford.edu/corprel/corp_programs/corp_innov_pgm.html, May 5, 2009.

Steiner, G. (2008). "Supporting sustainable innovation through stakeholder management: a systems view," *International Journal of Innovation and Learning,* Vol. 5, No. 6, ISSN 1471-8197, pp. 595–616.

Steiner, G. (2009). "The Concept of Open Creativity: Collaborative Creative Problem Solving for Innovation Generation – a Systems Approach," *Journal of Business and Management,* Vol. 15, No. 1, ISSN 1535-668X.

Štoka Debevc, M. (2006). "Družbena odgovornost v Sloveniji – izsledki vladnih ugotovitev," in Hrast, A., Mulej, M. and Knez-Riedl, J. (eds.), referenced here.

Strike, V. M., Gao, J. and Bansal, P. (2006). "Being good while being bad: social responsibility and the international diversification of US firms," *Journal of International Business Studies,* Vol. 37, ISSN 0047-2506, pp. 850–862.

Tavčar, M. (2006). "Družbena odgovornost ali podjetniška etika," in Rozman, K., referenced here, pp. 65–72.

Tajlor, F. W. (1967). *Naučno upravljanje,* Izd. pod. Rad, Beograd (Translation into Serbian), ISBN not available.

Taylor, G. (2008). *Evolution's Edge: The Coming Collapse and Transformation of our World,* New Society Publishers, Gabriola Island, Canada, ISBN 978-0-86571-608-7.

Thorpe, S. (2003). *Vsak je lahko Einstein. Kršite pravila in odkrijte svojo skrito genialnost!,* Mladinska knjiga, Ljubljana (Translation into Slovenian), ISBN 86-11-16537-3.

Tihec, S. (2006). "Ploščati televizor porabi več," *Večer, priloga Kvadrati* 159, ISSN 0350-4972, p. 23.

Toth, G. (2008). *Resnično odgovorno podjetje,* GV Založba, Ljubljana, in Društvo komunikatorjev Slovenije, Ljubljana.

Trappl, R. (ed. and co-eds.) (2006, and earlier, since 1972 – biannually). EMCSR (European Meeting on Cybernetics and Systems Research), different publishers of proceedings, sponsor: Austrian Society for Cybernetics, ISSN 0196-9722.

Treven, S. (2005). *Premagovanje stresa,* Ljubljana, GV Založba, ISBN 86-7061-403-0.

Treven, S. and Mulej, M. (2005). "A Requisitely Holistic View of Human Resources Management in Innovative Enterprises," *Cybernetics and Systems*, 36, ISSN 0196-9722, pp. 1–19.

Treven, S. and Mulej, M. (2005b). "Stress Management in Work Setting of an Innovative Society by Systems Theory/Thinking," in: Mulej, *et al.* (eds.) (2005), referenced here.

Trevino, L., Hartman, L. and Brown, M. (2000). "Moral Person and Moral Managers," *California Management Review*, 42, ISSN 0008-1256, pp. 128–142.

Trevino, L. and Nelson, K. (2004). *Managing Business Ethics,* Wiley and Sons, New York, ISBN 0 471 23054 5.

Troncale, L. (2002). "Integrated Science General Education (ISGE): 'Stealth' systems science for every university," in Trappl, R. (ed.), *Cybernetics and Systems 2002*, Austrian Society for Cybernetic Studies, Vienna, ISSN (print) 0196-9722, pp. 43–48.

Trstenjak, A. (1981). *Psihologija ustvarjalnosti,* Slovenska matica, Ljubljana, ISBN not available.

Trunk-Širca, N. (2000). "Načrtovanje lastnega razvoja – vsak je lahko uspešen ali sistematično uvajanje učenja 'Učiti se biti': prispevek o izbirnem predmetu," in Erčulj, J. and Trunk Širca, N., *S sodelovanjem do kakovosti: mreže učečih se šol*, Ljubljana, Šola za ravnatelje, 2000, pp. 245–255, ISBN/ISSN not available.

Udovičič, K. (2004). *Metode nematerialne motivacije za inoviranje managerjev v tranzicijskem podjetju (Udejanjanje intrinzičnosti v inovativnem poslovodenju človeških sposobnosti),* EPF, ISBN not available.

Udovičič, K. and Mulej, M. (2006). "Managers' requisite holism between personal and organizational values," *Kybernetes*, 35, ISSN 0368-492X, pp. 993–1004.

Ulrich, P. (1997). *Integrative Wirtschaftsethik,* Verlag Paul Haupt, Stuttgart, ISBN 3258062765.

Umpleby, S. (2002). "Organization of Regulation of the Global Economy," *Naše Gospodarstvo*, 45, ISSN 0547-3101, pp. 22–47.

Umpleby, S. (2005). "Some Influences on the Cybernetics Movement in the United States," in Mulej, *et al.* (eds.) (2005), referenced here.

Umpleby, S. (2005). "A History of the Cybernetics Movement in the United States," in Mulej, *et al.* (eds.) (2005), referenced here.

Umpleby, S. and Vallée, R. (2000): "Kibernetika ničtega, prvega, drugega in tretjega reda. (Cybernetics of the zero, first, second, and third order cybernetics)," in Mulej, *et al.* (2000), referenced here.

UN (1992). *Rio Declaration,* UN, Rio de Janeiro, ISBN not available.

UNESCO (1999, 2000). *UNESCO Courier,* UNESCO, Paris, ISSN 1993-8616.

UNESCO (2006). 1945–2005: velike razlike, from: www.unesco.org/en/courier/interalia, UNESCO Novi Glasnik, November (Slovenian edition), ISSN 1993-8616, p. 6.

Uršič, D. (1996). *Inoviranje podjetja,* Studio Linea, Maribor, ISBN 961-90303-3-8.

Uršič, D., Stare, M., Bučar, M., Mulej, M. and Pivka, M. (2000). *National uptake study: Slovenia, (Process re-engineering in Europe: choice, people and technology),* Directorate General XII, Science, Research and Development, Brussels, ISBN/ISSN not available.

Veliki anglesko-slovenski slovar (1994). Electronic, ISBN not available.

Vezjak, M., Stuhler, E. and Mulej, M. (eds.) (1997): *Environmental Problem Solving. From Cases and Experiments to Concepts, Knowledge, Tools, and Motivation. Proceedings of the 12th International Conference on Case Method Research and Case Method Application* (held 1995 in Maribor), Rainer Hamp Verlag, Munich, ISSN 0940-2829.

Vincke, Ph. (1992). *Multicriteria Decision-Aid,* John Wiley & Sons, Chichester, ISBN 978-0-471-93184-3.

von Hippel, E. (2007). "An emerging hotbed of user-centered innovation," *Harvard Business Review,* Vol. 85, ISSN 0017-8012, pp. 27–28.

Vrana, T. (2006). "Dileme in razvojne možnosti družbene odgovornosti v neprofitnem sektorju," in Hrast, A., Mulej, M. and Knez-Riedl, J. (eds.), referenced here.

Waddock, S. and C. Bodwell (2007). *Total Responsibility Management: The Manual,* Greenleaf Publishing Limited, Sheffield, ISBN 1874719985.

Waldman, D. A., Sully de Luque, M., Washburn, N. and House, R. J. (2006). "Cultural and leadership predictors of corporate social responsibility values of top management: a GLOBE Study of 15 countries," *Journal of International Business Studies,* Vol. 37, ISSN 0047-2506, pp. 823–837.

Waldrop, M. (1992). *Complexity. The Emerging Science at the Edge of Order and Chaos,* Penguin Books, London, UK, ISBN 0671872346.

Warfield, J. N. (2003). "A Proposal for Systems Science," *Systems Research and Behavioral Science,* 20, 6, ISSN 1092-7026, pp. 507–520.

WBCSD (1998). *CSR: Meeting changing expectations,* World Business Council for Sustainable Development – Corporate Social Responsibility, URL: http://www.wbcsd.ch/templates/TemplateWBCSD5/layout.asp?type=p&MenuId=MTE0OQ&doOpen=1&ClickMenu=LeftMenu, May 15, 2009.

WBCSD (2004). *A Business Guide to Development Actors,* WBCSD, Geneva, ISBN 2-940240-64-7.

WBCSD (2005). *Business for Development,* WBCSD, Geneva, ISBN 2-940240-81-7.

WCED (1987, ed. 1998). *Our common future,* Oxford University Press, Oxford, ISBN 019282080X.

Webster (1978). *Webster's School & Office Dictionary, new edition,* Banner Press, New York, ISBN not available.

Webster (1992). *Webster Comprehensive Dictionary. International Edition,* J. G. Ferguson Publ. Co., Chicago, ISBN 0894341367.

Wiener, N. (1972). *Kibernetika ili upravljanje i komunikacija kod živih bića i mašina* (Translation into Serbo-Croate; Original: 1948), Izdavacko-informativni centar studenata, Beograd, ISBN not available.

Wiener, N. (1964). *Kibernetika i drustvo. Ljudska upotreba ljudskih bića* (Translation into Serbo-Croate), Biblioteka sasveždža, Beograd, ISBN not available.

Wierzbicki, A. and Nakamori, Y. (2006). "Nanatsudaki: A New Model of Knowledge Creation process," in Gu, *et al.* (eds.), referenced here.

Wilby, J. and Allen, J. K. (eds.) (2005). *ISSS 2005. Proceedings of the 49th Annual Conference, The Potential Impacts of Systemics on Society,* The International Society for the Systems Sciences, Cancun, Mexico, ISBN/ISSN not available.

Wilby, J. (ed.) (2006). *Complexity, Democracy and Sustainability: proceedings of the 50th annual conference,* International Society for the Systems Sciences (at Sonoma University, CA), ISBN/ISSN not available.

Wilby, J. (2005). "Combining a Systems Framework with Epidemiology in the Study of Emerging Infectious Disease," *Systems Research and Behavioral Science,* 22, ISSN (print) 1092-7026, pp. 385–398.

Williams, C.A. and Aguilera, R.V. (2008). "Corporate social responsibility in a comparative perspective," in Crane, A., McWilliams, A., Matten, D., Moon, J. and Siegel, D. (eds.), *The Oxford handbook of corporate social responsibility,* pp. 452–472, Oxford University Press, Oxford, ISBN 0199211590.

Wilson, E. O. (1998). "Die zehn Gebote liegen in den Genen. Das Biologische Fundament der Moral," *Neue Zuercher Zeitung, Zeitfragen,* 211, ISSN 0376-6829, p. 85.

Wood, R. (2000). *Managing Complexity. How Businesses Can Adapt and prosper in the Connected World,* Economist Books, London, ISBN 1861971125.

Wu, J. (1996). *Systems Dialectics,* Foreign Languages Press, Bejing, ISBN 7119018809.

Zahn, E. (ed.) (1995). *Handbuch Technologiemanagement,* Schaeffer-Poeschel, Stuttgart, Germany, ISBN 3-7910-0758-0.

Ženko, Z. (1999). *Comparative Analysis of Management Models of Japan,* USA, and Europe, EPF, ISBN not available.

Ženko, Z., *et al.* (2002). "Informal Systems Thinking or Systems Theory – the Case of Ideas about Slovenia," in Trappl, R. (ed.), *Proceedings of EMCSR '02,* OeSGK, Vienna, referenced here.

Ženko, Z., Mulej, M. and Marn, J. (2004). "Innovation before entry into the EU; the case of Slovenia," *Postcommunist Economies*, 16, ISSN (print) 1463-1377, pp. 169–189.

http://www.informationisbeautiful.net/visualizations/the-billion-dollar-gram/

www.ingramcontent.com/pod-product-compliance
Lightning Source LLC
Chambersburg PA
CBHW080755300426
44114CB00020B/2734